The Operation of Contemporary Blast Furnaces

Yuncai Liu

The Operation of Contemporary Blast Furnaces

冶金工业出版社
Metallurgical Industry Press

Springer

Yuncai Liu
Shougang Group
Beijing, China

Translated by
Jianliang Zhang
School of Metallurgical and Ecological
Engineering
University of Science and Technology
Beijing Beijing, China

Kexin Jiao
School of Metallurgical and Ecological
Engineering
University of Science and Technology
Beijing Beijing, China

Wudi Huang
School of Metallurgical and Ecological
Engineering
University of Science and Technology
Beijing Beijing, China

ISBN 978-981-15-7073-5 ISBN 978-981-15-7074-2 (eBook)
https://doi.org/10.1007/978-981-15-7074-2

Jointly published with Metallurgical Industry Press
The print edition is not for sale in China (Mainland). Customers from China (Mainland) please order the print book from: Metallurgical Industry Press.
ISBN of China (Mainland) edition: 978-7-5024-8441-5

This Springer imprint is published by the registered company Springer Nature Singapore Pte Ltd.
The registered company address is: 152 Beach Road, #21-01/04 Gateway East, Singapore 189721, Singapore

Foreword

A special precious memento was presented for me on October 2018 from Yuncai Liu, my old friend and classmate in university 60 years ago. The memento was his new professional work (in Chinese) titled *The Operation of Contemporary Blast Furnaces*. I read it immediately and was impressed very much. Then, an idea emerged that this book should be translated into English as soon as possible to introduce Chinese iron-making technology and experience to foreign iron makers and experts. Fortunately, this issue was strongly supported by The Metallurgy School, University of Science and Technology Beijing. Then, a team, containing about 10 faculty members and doctoral graduated students, was established to accomplish such a program.

In this book, the distinguishing features are searching out:

First of all, it contains his plentiful blast furnace operation experience accumulated for more than 60 years. Since graduating from the university in 1956, Prof. Liu had served for the Capital Steel Company all his life from the foreman to the general engineer, in charge of iron-making, coke-making, sinter-making and pellet-making. So, colorful practical experience had been collected and condensed in his book perfectly, involving not only positive but also negative concrete examples and lessons. It shows an objective model of combining theory with practice and could reflect the advanced China's iron-making technology in miniature. It should be available for all of the blast furnace men in the world.

Secondly, besides serving for the Capital Steel Company, at the same time, he was also invited by many other iron-making plants frequently to consult and help to deal with technical problems and accidents. So, his experience had wide and far foundation with ground gas.

Thirdly, he was passionate about taking part in academic and technical events, attending relative domestic and international conferences and presenting lots of reports or papers as well as composing some other professional works, some of which are listed in the references. It should be pointed out that, as you can see in this book, he used the relevant basic knowledge of physical chemistry, mathematics, automation, artificial intelligence to study and explain the invisible complicated processes inside the blast furnace. Most of such research results had been identified by the peer experts.

In a word, he left us the last valuable heritage, but unfortunately, he passed away suddenly when the English version is going to be published, that must be his last wish. Anyway, he had already learned the draft after finishing and the possibility of composing that might partly satisfy his spirit of the sky!

I hope the English version would benefit international iron-making technology development and communicate with each other.

Wudi Huang
Emeritus Professor and Former Provost
University of Science and Technology Beijing
Beijing, China

Preface

I had been engaged in blast furnace iron-making over 60 years, served as a technician at the beginning and then successively held the positions of metallurgical engineer, deputy director, director. Finally, I retired from the position of the deputy chief engineer of the Capital Steel Corporation. I had worked as a blast furnace foreman for more than 2 years as well. I was deeply impressed by the complicated and comprehensive process of blast furnace iron-making.

In 2016, I finished the book, *The Operation of Contemporary Blast Furnaces*, which elaborated on the iron-making technical improvement of blast furnace operation in recent decades, as well as the joyful success and painful setbacks that my colleagues and I experienced. Such experiences, integrating with in-depth interactive discussions, as well as some inventions from my colleagues and numerous experienced workfellows in iron-making industry, had inspired and benefited me a lot!

The Operation of Contemporary Blast Furnaces systematically facilitated the operational principles and methods to the blast furnace operator. The blast furnace operator should pursue the best technical methods to deal with the actual situations of each blast furnace. What's more, the primary data have also been provided in this book to help IT engineer to understand the operational conditions of blast furnace.

The translation of this book from Chinese into English version would be published jointly by Metallurgical Industry Press and Springer. It was suggested by my classmate and good friend Prof. Wudi Huang and initiated by the doctoral advisors, Prof. Tianjun Yang and Prof. Jianliang Zhang of the University of Science and Technology Beijing. Chapters 1, 2, 10, 11 and 12 were translated by Prof. Wudi Huang, and the other chapters were translated by Yimin Huang, Jing Yu, Xinyu Liu, Tongxu Zhang and Lin Zhang who are doctoral students of Prof. Tianjun Yang and Prof. Jianliang Zhang. My old friend Arthur Cheng reviewed Chap. 3, and all other chapters were reviewed by Prof. Wudi Huang. The management of translation was organized by lecturer Ph.D. Kexin Jiao and his assistant Ph.D. Xiaoyue Fan, and facilitated by Prof. Tianjun Yang and Prof. Jianliang Zhang. I sincerely appreciate the contributions from all of the participants, and without their supports, it would be impossible to realize this publication.

Beijing, China Yuncai Liu

Contents

List of Figures

Chapter 1
Introduction

Abstract The process of contemporary integrated steelworks consists of raw materials, iron-making, steelmaking, steel rolling and so on as a continuous chain. Every stage of the chain must be operated in order; otherwise, the whole production stream would be out of gear. But iron-making production is different from others for the reason that its importance is not because it lies in front of the production processes, but because it plays a special role in iron and steel enterprises.

1.1 Position of Iron-Making in the Integrated Steelworks

The special role of iron-making in iron and steel enterprises:

(1) Especially, the iron-making is not only the first stage of the production processes, but also plays an important role to receive about 3–4 t (per ton steel) of raw materials, fuels and other auxiliary additives, in which the injurious ingredients could be dislodged by blast furnace (BF); thus, it can protect the following procedures (steelmaking, rolling, etc.) from injury. This is why qualified hot metal is very important for steelmaking.
(2) Because the hot metal could not be reserved for a long time, it must be produced continuously and charged into the converter instantly.
(3) Generating gas, as a fuel resource, is another function of the blast furnace. The energy utilization rate is just 65–75% for iron-making itself. The rest gas becomes an important fuel of different facilities. If the blast furnace process interrupts or reduces the gas supply, the users must be impacted.
(4) The cost of iron-making system.

The BF system cost possesses about 60–70% of iron and steel products. For the process of blast furnace–converter, the energy consumption of iron-making is about 70% of the total consumption in the whole iron and steel manufacture process, which means the stability and balance of BF determine better benefit of the iron and steel enterprises.

Smart entrepreneurs always try their best to keep the iron-making production stable and normal, create conditions to keep the blast furnace production smooth and obtain the maximum benefits of iron and steel enterprises.

Y. Liu, *The Operation of Contemporary Blast Furnaces*,
https://doi.org/10.1007/978-981-15-7074-2_1

1.2 The Conditions for BF Production Level

The main determinate factors for BF process are burden, equipment and operation, and among which the burden is the first. As early as the 1950s–1960s, some experts from Russia had already told that the effect of raw materials for BF process would be 70%. Since then, the practical experience of us confirmed that is almost true, especially the quality of coke, although it is hard to evaluate numerically.

During February–March 1995 in the Capital Steel, some BF accidents such as channeling and bridging happened frequently and are hard to retrieve. Because of the unqualified coke, the BFs sustainedly flooded tuyeres, and are hard to handle by checking. Figure 1.1 shows the frequent channeling, bridging and checking at No.3 BF.

It can be seen in Fig. 1.1 that whether before or after the checking, the channeling still continues and is hard to check, because of worrying about tuyere flooding with slag. Although the checking did thoroughly, the channel still could not be removed. Especially after the first and second checking, the stock surface is downward 3–4 m with channels in the dislocated stock. If qualified coke was used, the channels would be able to remove, but unfortunately, the channels existed all the way and checking did not play a part with unqualified coke. Because of the ceaseless channels, a lot of tuyeres breaking and hearth accumulation must happen (to be discussed in Chap. 2). Frequent channels cause health temperature diversified up and/or down (to be discussed in Chap. 4), and channel managing will be discussed in Chap. 7.

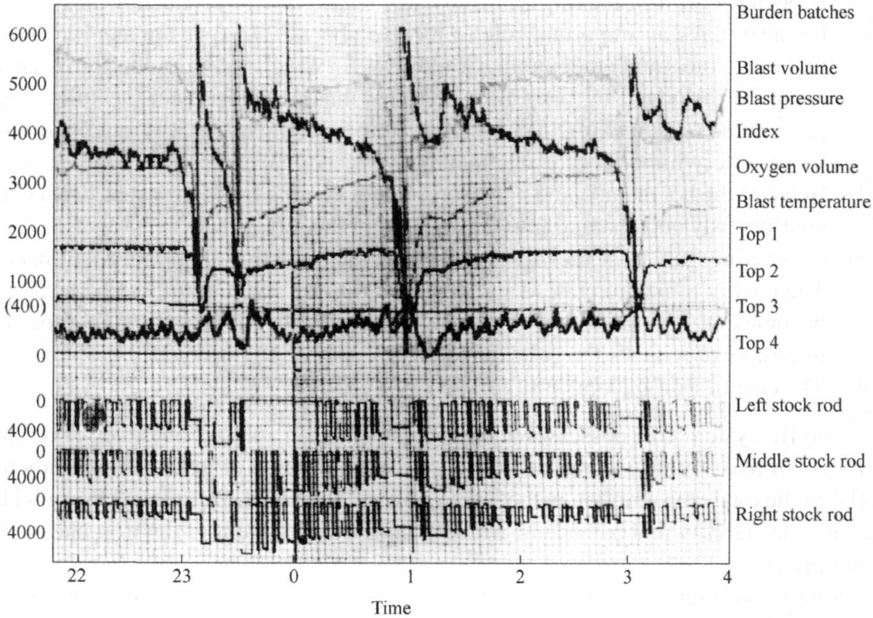

Fig. 1.1 Channeling, bridging and checking

The second factor is equipment, the importance of which is inseparably interconnected with operation. For example, the only hot blast stove with a silicon brick top and upper body could be able to supply hot blast at as high as 1300 °C for BF. But such high blast temperature could not be accepted without reasonable operation; even the hot blast stove system is excellent. Generally speaking, equipment is more important than operation. The contemporary equipment has been making BF modernizing. The progress of equipment is the material foundation for the iron-making process advancing.

The traditional charging facilities such as double to four bells were replaced rapidly by bell-less, which could distribute the burden to the scheduled location over stock line in the throat. This is impossible in the past hundred years! As a result of an excellent function of bell-less, the gas stream distribution through the stock could be controlled well, then to improve the gas utilization as well as BF performance. The special seal of bell-less hopper could permit higher top pressure operation with simplified facilities. Therefore, the technology of high top pressure developed sooner and better, leading to obvious results from the smelting process accelerating. To compare with traditional large bell and hopper systems, bell-less facilities are more superior at manufacturing, installation and transportation.

The application of copper stave cooler with soft water has been extending BF campaign, improving steady, safety as well as cutting maintenance down. Soft water and its closed-cycle recovery system could save water consumption, which is most important for less water resource areas, such as North China. That is why it has been developing all the time!

1.3 The Task of BF Operation

In the production process, the operation is dealing with many aspects to make use of raw materials and fuel reasonably for a steady and sustainable high level of productivity, low consumption and more benefit of BF. The regular performance is the precondition of stabilization, only which could yield qualified hot metal for next steelmaking to ensure the optimum benefit of the integrated steelworks.

1.3.1 BF Regular Performance and Stability

The BF operation influences raw material consumption very much. Regular performance could prevent furnace flushing without flushing additions, such as fluorites, manganese ores and so on, that not only could save a lot of coke, but also prolong the BF campaign. At the same time, the profile changing becomes reasonable that no crucible burning out might happen and the senseless relining titanium mineral consumption is also unnecessary.

To control the gas distribution and to increase its utilization rate are very important for reducing the coke ratio, protecting the furnace wall and decreasing heat loss.

The intensification of the smelting process is decided by many factors. It is impossible to reach optimum productivity without qualified operator's domination; even the burden and facilities are excellent. So, the intensification level is determined by operation under certain material conditions.

The outstanding BF experts always take regular performance first, to keep the stability and nip the trouble in the bud for the optimum smelting process. But some other kinds of experts, who are familiar with handling accidents, do not care about slipping and bridging and look the slipping and bridging without seeing. They boast of such experience, but could never be an outstanding BF expert, just a valiant fighter!

The third kind of BF experts, who usually pay attention to regular performance, rather than stability, always takes advantage of every opportunity to aspire higher productivity by increasing blast volume or ratio of ore-to-coke till tight furnace. They do not understand that only temporary regular performance is not enough and constant regulation, videlicet stability, must be kept.

The comprehension of stability not only is restricted in the scope of BF, but also covers the whole iron-making process without making ups and downs. Otherwise, higher productivity and better benefit could not be gained. Some BF men take the wrong way to force blast increasing for higher productivity, aggressing in one shift, defending in the next two shifts. As a matter of fact, regular performance would be destroyed more often than not. This is not a wise way.

1.3.2 To Make Use of Burden Reasonably

It is an important topic of BF operation that how to make use of a burden. Every work and plant have their situation, so to constitute a proper standard of blending and proposition is necessary for utilizing the burden optimistically and controlling the slag composition, which is the foundation of BF process. The slag analysis has a critical influence on the quality of hot metal, and the better the slag, the better the hot metal. The burden optimizing could improve the permeability of stock column and promote indirect reduction of iron ores.

1.4 The Premise of Regular Performance and Stability of BF

The premises of regular performance and stability include:

(1) The quality of burden

The level of BF intensification is determined by burden quality, without which the high level would be impossible. So, "beneficiated burden" is a matter of primary importance for BF. Smart experts always confront the diversifying raw materials to take the skill to adjust the furnace operation and keep the process regular and steady, even the highest productivity could not be obtained at the moment. This might be an optimum choice in such disadvantageous situation.

The "BF Iron-making Technology Design Specification" (GB 50,427) [1, 2], drafted by the senior designer Mr. Xiang, Zhongyong, et al., shows the standard of burden (Tables 1.1, 1.2, 1.3, 1.4 and 1.5), and then the required standard and regulation for burden had been established.

If the above requirements are satisfied, the regular performance and high productivity of BF could be realized on the material foundation. The role of burden will be discussed in detail in the following chapters.

(2) Keep the proper profile

All of the hearth accumulation, burning out or bottom rising would bring disastrous results for the BF process. The irregular profile, too thick or too thin, would destroy the normal smelting process to decrease the productivity substantially.

Table 1.1 Sinter quality requirement

Furnace volume, m^3	1000	2000	3000	4000	5000
Fluctuation of iron content, %	$\leq \pm 0.5$	$\leq \pm 0.5$	$\leq \pm 0.5$	$\leq \pm 0.5$	$\leq \pm 0.5$
Fluctuation of basicity	$\leq \pm 0.08$	$\leq \pm 0.08$	$\leq \pm 0.08$	$\leq \pm 0.08$	$\leq \pm 0.08$
Standard rate of iron content and basicity, %	≥80	≥85	≥90	≥95	≥98
FeO, %	≤9.0	≤8.8	≤8.5	≤8.0	≤8.0
Fluctuation of FeO, %	$\leq \pm 1$	$\leq \pm 1$	$\leq \pm 1$	$\leq \pm 1$	$\leq \pm 1$
Tumbler index(6.3 mm), %	≥68	≥72	≥76	≥78	≥78

Note The basicity is CaO/SiO2

Table 1.2 Pellet quality requirement

Furnace volume, m^3	1000	2000	3000	4000	5000
Iron content, %	≥63	≥63	≥64	≥64	≥64
Tumbler index(6.3 mm), %	≥86	≥89	≥92	≥92	≥92
Abrasion index(−0.5 mm), %	≤5	≤5	≤5	≤4	≤4
Cold crushing strength, N/pellet	≥2000	≥2000	≥2000	≥2500	≥2500
RDI (+3.15 mm), %	≥65	≥80	≥85	≥89	≥89
Expansion rate	≤15	≤15	≤15	≤15	≤15
Fluctuation of iron content, %	$\leq \pm 0.5$	$\leq \pm 0.5$	$\leq \pm 0.5$	$\leq \pm 0.5$	$\leq \pm 0.5$

Note Excluding special ores

Table 1.3 Sized ore quality requirement

Furnace volume, m^3	1000	2000	3000	4000	5000
Iron content, %	≥62	≥62	≥64	≥64	≥64
Decrepitation, %	–	–	≤1	≤1	≤1
Fluctuation of iron content, %	≤ ± 0.5	≤ ± 0.5	≤ ± 0.5	≤ ± 0.5	≤ ± 0.5

Table 1.4 Raw material grain size requirement

Sinters		Sized ores		Pellets	
Size, mm)	5–50	Size, mm	5–30	Size, mm	6–18
>50 mm, %	≤8	>30 mm, %	≤10	9–18 mm, %	≥85
<5 mm, %	≤5	<5 mm, %	≤5	<6 mm, %	≤5

Note The grain size of limestone, dolomite, fluorite, manganese ore and silica is the same as that of sized ore

Table 1.5 Coke quality requirement

Furnace volume, m^3	1000	2000	3000	4000	5000
M40, %	≥78	≥82	≥84	≥85	≥86
M10, %	≤8.0	≤7.5	≤7.0	≤6.5	≤6.0
Post-reaction strength CSR, %	≥58	≥60	≥62	≥65	≥66
Reactivity index CRI, %	≤28	≤26	≤25	≤25	≤25
Coke ash content, %	≤13	≤13	≤12.5	≤12	≤12
Sulfur content in coke, %	≤0.7	≤0.7	≤0.7	≤0.6	≤0.6
Coke grain size, mm	75–20	75–25	75–25	75–25	75–30
Above upper limit, %	≤10	≤10	≤10	≤10	≤10
Below lower limit, %	≤8	≤8	≤8	≤8	≤8

(3) Active hearth

The hot blast blows into hearth through tuyeres to burn the coke and create the space for stock descending. The heat and gas are produced from combustion, bringing about smelting reactions. It is reasonable to say that hearth is the engine of BF. Also, hearth is a reservoir of hot metal and slag. If the hearth could not work normally, not only the productivity would decrease, but also the hot metal quality would degrade. It is easy to yield a lot of irregular hot metal when the hearth works in a terrible situation. So, the active hearth is necessary for high yield, high grade and low consumption.

(4) Proper gas distribution

Gas, as the carrier of heat and chemical energy, plays an important role in heating and reducing burden. The proper gas distribution, especially in the hearth, must match the blasting speed and furnace volume to combine the "adjusting in upper and lower zones." The two zones could not be separated to adjust and must be combined harmoniously with each other, as a whole of the furnace.

(5) Stability of hearth temperature

The hearth temperature means enough thermo-energy in crucible and proper hot metal analysis which are usually expressed by silicon content or temperature of hot metal, sometimes just observing the flowability of hot metal and slag without measuring facility. If the fluctuation of hearth temperature is too big, not only the productivity would decrease and the coke rate would increase, but also injure the next steelmaking. Too low hearth temperature is an advantageous condition for the accident of BF.

(6) The proper composition of slag

The proper slag composition determines the hot metal quality and slag flowability, which depends on the burden blending and also on the hearth temperature fluctuation. If the burden composition fluctuates, the only way to keep slag analysis stable is adjusting the furnace operation which would be a hard job. Any mistake of slag composition would lead to quality failure.

Based on the above analysis, the BF process could divide into three states: furnace body, regular performance and heat state. The train of thought for specific operation, adjustment and control are shown in Fig. 1.2.

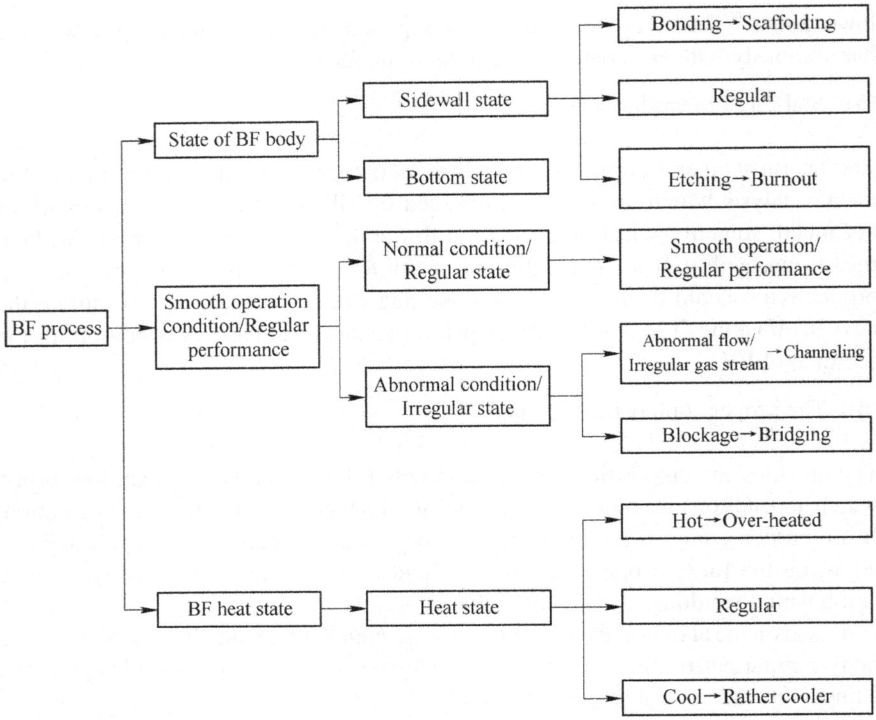

Fig. 1.2 Operation classifying of BF process [3]

References

1. China Metallurgical Construction Association, *Design Specification for Blast Furnace Iron-making Process.* (China Planning Press, Beijing, 2008)
2. Z. Xiang, X. Wang et al., *Blast Furnace Design—Theory and Practice of Iron-making Process Design*, vol. 2 (Metallurgical Industry Press, Beijing, 2014)
3. T. Yang, J. Xu, Y. Liu, et al., Development and application of smelting expert system for No.2 BF of Capital Steel. Iron-Making **6**, 31–34 (1998)

Chapter 2
To Activize the Hearth

Abstract In Guxing, Zhengzhou, Henan Province, there are two sets of ancient (in Han Dynasty, 2000 years ago) BF remainder, near which 15 pieces of salamander were discovered. The largest one weighs about 20 t (Fig. 2.1) (Zhang in Metallurgical Iron Ruins of Guying Town in Han Dynasty. Guangling Book Society, Yangzhou, 2009). Because of the low capacity of blast facility, the hearth was not active then the hot metal would freeze easily. The ancient smelting workers had recognized that the blast could not penetrate into the burden of larger hearth, so they changed the hearth into oval shape to promote blast blowing to the center. Actually, this was the earliest adjust for BF blast.

According to the long-term practice, it is convinced that BF could not obtain high productivity by enlarging the size without active hearth. In the past 1000 years, due to the limitation of blast capacity, the BF volume was smaller all the time and never appear large BF with hearth diameter over 3 m (Fig. 2.1).

Hearth is the "engine" of BF production, where coke and other fuel burn to produce reducing gas CO and H_2, ascending through the stock column to reduce iron oxides and produce metallic iron. At last, the hot metal accumulates in the hearth and then flows out. The space created from combustion facilitates the burden descending. Hearth is the beginning, also the ending of BF smelting process, which means well-balanced hearth is very important, otherwise, production must be influenced critically.

2.1 Combustion Zone of Coke and Blast Kinetic Energy

2.1.1 Combustion Zone of Coke

Terms of Iron-making

Combustion zone, a phrase in iron-making, means the coke combustion at front of the tuyeres,through which blast enters into the zone to roll up and burn up the coke. Also, because of the existence of oxygen, it is called the oxidizing zone. The main

Y. Liu, *The Operation of Contemporary Blast Furnaces*,
https://doi.org/10.1007/978-981-15-7074-2_2

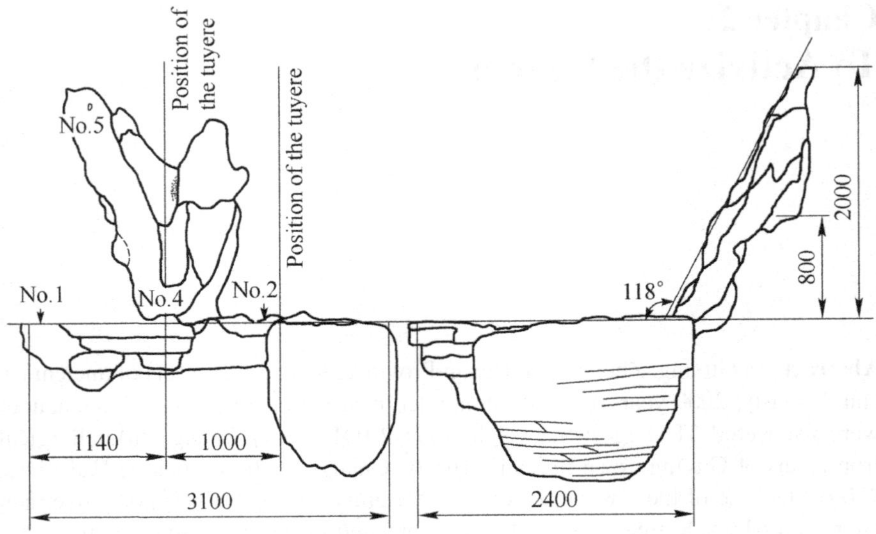

Fig. 2.1 Size of salamander in No. 1 BF pit of Guxing

chemical reactions in the combustion zone (belt) are: (1) $C + O_2 + CO_2$; (2) $CO_2 + C = 2CO$.

When the oxygen and carbon dioxide use up, the combustion would finish at the edge of the zone, which actually is indicated with CO_2 of 1–2%. During combustion, the coke is rolled up by blast in the zone, which is so-called raceway. Both combustion zone and raceway are used in this book.

As early as nineteenth century, the combustion process in BF was measured through tuyeres [1], two young experts from US Bureau, Sir Lowthian Bell and J. E. Johnson first measured and discovered the coke combustion zone before tuyeres [2], which was quoted in textbooks for many years. Figure 2.2 shows the vertical profile of combustion zone. They took the samples from normal tuyeres and special ones in the lower part of the bosh.

A research team headed by M. A. Pavlof did a lot of measurement and research for BF process with tremendous contribution to combustion zone researching from 1934 to 1947. It is indicated from measurements that deferent blast indexes influence combustion very much [1].

In 1952, J. Elliott, R. A. Buchanan and J. B. Wagstaff measured for combustion zone before tuyere with high speed camera at 800–3000 frames per minute, first obtaining the kinetic pictures of coke combustion zone (Fig. 2.3) [3]. Figure 2.4 shows the trail of coke moving in the raceway found by C. Burns. Figure 2.5 shows a diagrammatic sketch of raceway [4], the upper part of which is vertical profile. The blast blows through the tuyere to push coke move, roll and burn and produce high-temperature gas, the composition of which is shown in the lower part of Fig. 2.5. The oxygen content of blast decreases rapidly and converts to CO and CO_2 for the combustion. The temperature peak (about 2000 °C) appears near the end of raceway,

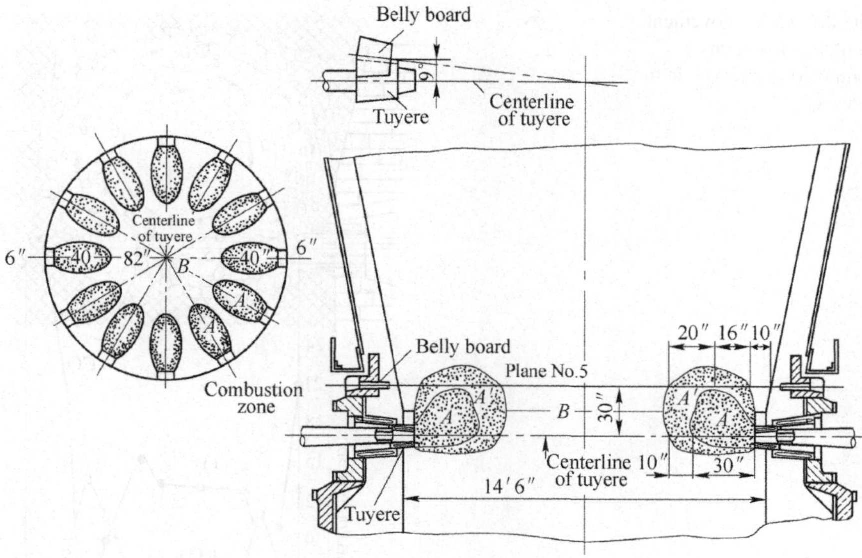

Fig. 2.2 Shape and size of combustion zone

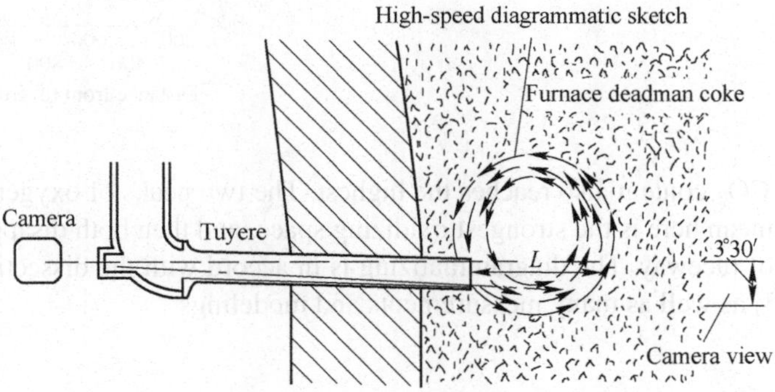

Fig. 2.3 Diagrammatic sketch of raceway before tuyere taken with high speed camera [3]

Fig. 2.4 Trail of coke moving in the raceway [6]

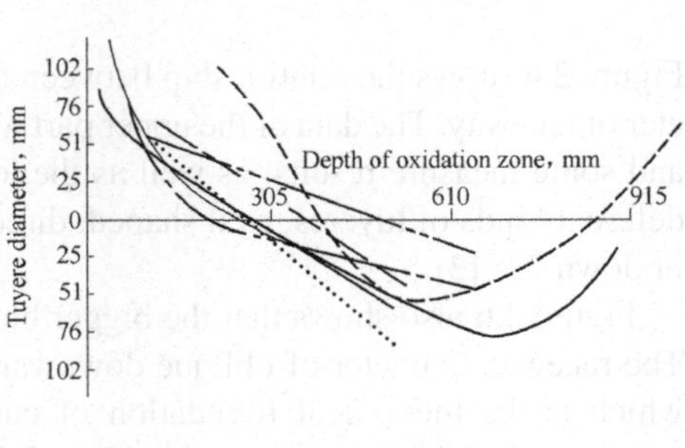

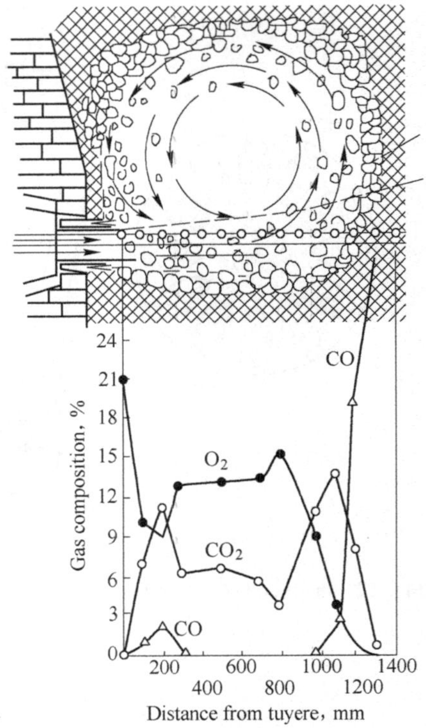

Fig. 2.5 Coke movement, combustion and gas composition changes in the raceway

where the CO_2 content also reaches the highest. The two peaks of oxygen and CO_2 in the gas mean here is the strongest oxidizing space, and then both disappear at the boundary of raceway. The diagrammatizing is in accord with the dissection for BF raceway [5] as well as many measurements and modelings.

2.1.2 Blowing Kinetic Energy and Raceway

Figure 2.6 shows the relationship between the blowing kinetic energy and the diameter of raceway. The data of the upper part were collected by many American authors and some measure results, as well as the lower part include the results tested with deferent kinds of tuyeres, oval shaped, direct blowing, oblique blowing with up 22° or down 22° [3].

Figure 2.6 also shows that the bigger blowing kinetic energy the larger raceway. The raceway diameter of oblique downward tuyere is longer than the upward one, which is the theoretical foundation of current application of downward oblique tyuere. In Fig. 2.6, the results of cold model experiments are matching with productive BF and the principle for different BF is the same no matter what the tuyere form is.

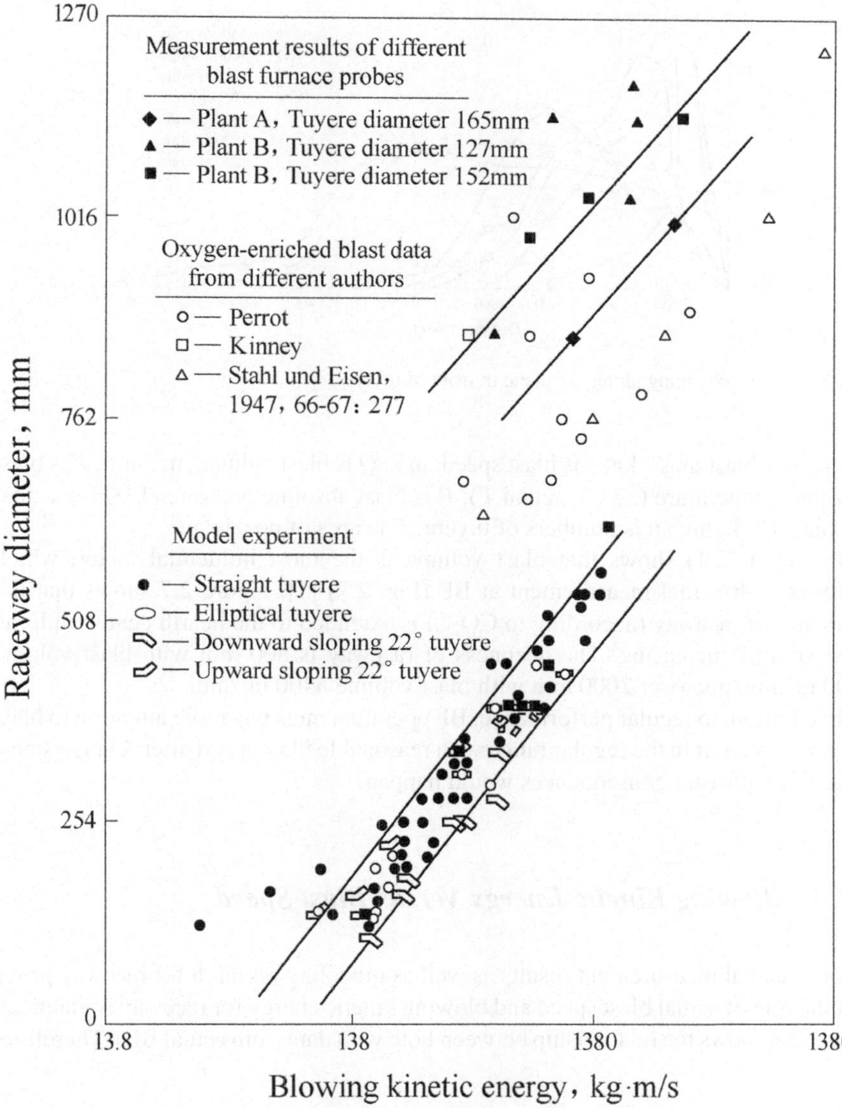

Fig. 2.6 Blowing kinetic energy versus raceway

Blowing kinetic energy (E) is the driving power to form the raceway Eq. (2.1):

$$E = \frac{1}{2}mv^2 = \frac{1.293 \times 1.033^2 \times Q^3 \times T^2}{2 \times 9.81 \times 60^3 \times 273^2 \times n^2 \times f^2 \times P^2} = 4.368 \times 10^{-12}\frac{Q^3T^2}{n^2f^2P^2} \quad (2.1)$$

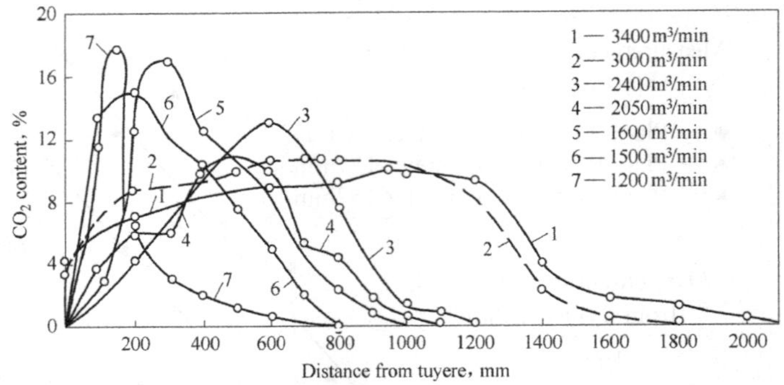

Fig. 2.7 $CO_{2\%}$ changing along the plane in front of tuyere

where m is blast mass, kg; v is blast speed, m/s; Q is blast volume, m^3/min; T is blast absolute temperature (273 + actual T); P is blast absolute pressure (1.003 + actual hot blast P), kg/m^2; n is numbers of tuyere; F is area of nozzle, m^2.

Equation (2.1) shows that blast volume is the most influential factor, which matches with actual measurement at BF (Fig. 2.7) [1]. Figure 2.7 shows that the boundary of raceway (according to $CO_{2\%}$) is extended to the hearth center with the blast volume increasing. The deepness of raceway is 800 mm with blast volume 1200 m^3/min and over 2000 mm with blast volume 3400 m^3/min.

In addition, to regular performance, BF operators must pay more attention to blast volume to keep it in the regular range with reasonable blast speed after relative steps. Otherwise, disaster consequences would happen.

2.1.3 Blowing Kinetic Energy Versus Blast Speed

A lot of actual measurement results as well as modeling research for raceway prove that the role of actual blast speed and blowing kinetic energy for raceway is identical. Figure 2.8 shows the relationship between both with data from actual BFs. Therefore,

Fig. 2.8 Blast speed versus blowing kinetic energy

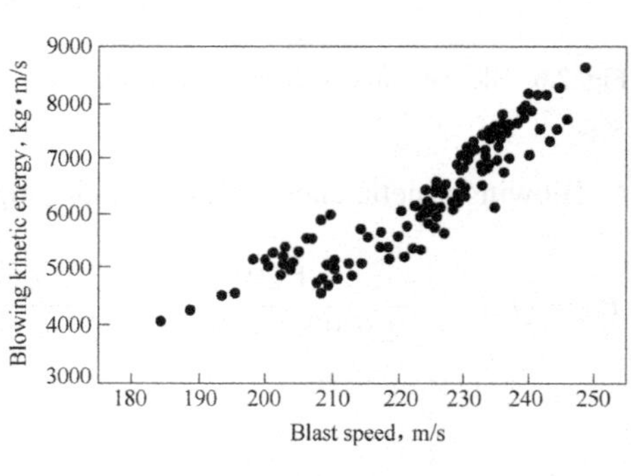

for productive BF, frequently to use blast speed instead of blowing kinetic energy to adjust and control blast volume range is reliable.

2.2 Adjusting at BF Lowe Part

The AN Steel presented BF lower part adjusting in 1961 firstly that was a great creative to influence BF operation very much afterward. During the hard time of economy in China, iron and steel production decreased a lot, including BF operation at serious variety smelting intensity. For BF, regular performance and stability, AN Steel built experience of BF lower part adjusting to obtain proper hearth work by the means of adjusting blast through tuyere to keep the proper blowing kinetic energy and raceway. According to the practice of high, medium and low BF smelting intensity in Anshan Steel for many years, the first "Regulation of BF Lowe Part Adjusting" in the world was formed, the main contents include:

(1) Under certain conditions, the proper blowing kinetic energy is not changeless but decreasing with the smelting intensity increasing the relation between both shows as a hyperbolic curve which is the theoretical foundation of lower part adjusting.
(2) Every BF has its own proper blowing kinetic energy, which is connected with the height of furnace, the higher blast, the bigger kinetic energy. The blast is easier to penetrate into the center for smaller furnace, but hard for larger furnace with longer hearth diameter. So, the bigger kinetic energy is necessary for larger furnace
(3) For shorter and wider furnace with more tuyeres, the resistance of stock is smaller, and bigger kinetic energy could be accepted, vice versa.
(4) The higher blowing kinetic energy could be realized, if the stock porosities improved by the qualified burden with high grade, fewer fines and well-distributed size, vice versa.
(5) It is necessary to keep proper theoretical combustion temperature at 2000–2300 °C (on the high side for large BF), according to the practice on humidified blast in AN Steel for many years [7].

2.2.1 Blast Volume

If the hearth is considered as the "engine" of BF, therefore, the blast is actually the effective power. The blast volume is the most active factor, which is determined by the porosity of stock column as well as influenced by strength and sizing of coke and burden, also the gas stream distribution.

The blast volume (intensity) could be indicated with two idioms:

(1) The acceptable blast volume per every m^3 of BF volume called "blast volume ratio", m^3/m^3.

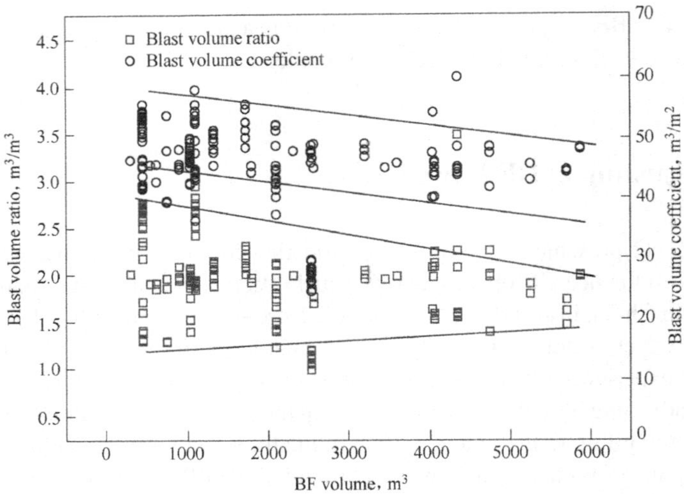

Fig. 2.9 BF volume versus blast volume ratio and coefficient

(2) The acceptable blast volume per every m^2 of hearth area called "blast volume coefficient" m^3/m^2.

The fluctuation of both for evaluating the blast volume is huge, that is the result of BF with a tremendous difference. To compare the fluctuation, blast volume ratio is bigger than the other one, especially for smaller BF. The fluctuations of blast volume ratio and coefficient are 1.0–2.8 m^3/m^3 and about 35–55 m^3/m^3, respectively, from the same BF group (Figure 2.9).

Either blast volume ratio or coefficient does not deal with the oxygen content, which plays an important role actually.

The smelting conditions determine blast volume more or less to a large extent in accordance with burden and production requirement. BF men always say: "the better burden, the more blast," which is very true.

2.2.2 To Determine Blast Speed or Blowing Kinetic Energy

Blast speed or blowing kinetic energy is the critical factor for lower part adjusting. The reasonable raceway is from reasonable blast speed or kinetic energy, which is the important term for regular performance and steady. Figure 2.10 shows the actual blast speed at regular performance BFs, as following:

(1) The larger BF volume, the higher blast speed.
(2) For every BF there is a certain range for reasonable blast speed which could be extended wider under the better burden conditions, vice versa.

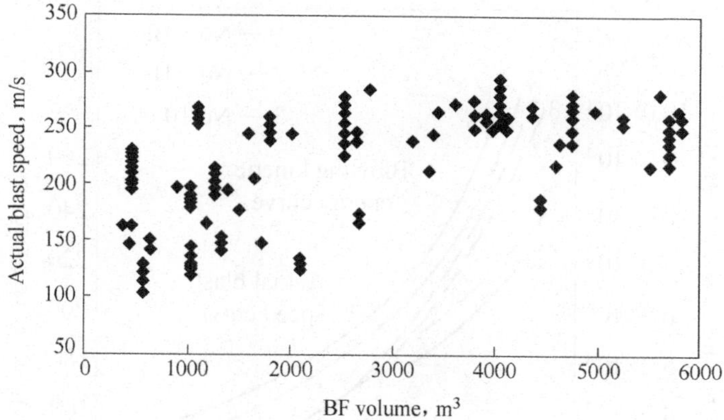

Fig. 2.10 BF volume versus blast speed of actual data

(3) Under the BF premise of regular performance, blast speed should not be changed frequently, especially to change the area of tuyere (nozzle). To change the area must change the tuyere by damping down of BF operation (stopped blast), which would interrupt BF process. The principle of BF operation is that do not stop blast whatever as possible.

2.2.3 *The Relation Among Smelting Intensity, Blast Speed and Blowing Kinetic Energy*

From 1960 to 1962, the BF smelting intensity used to change seriously because of the social economic readjustment. Every steelwork summed their experience at different levels of smelting intensity, among which the BF workers in AN Steel first found out the way to sensitize the whole hearth and the first step of quantificational relationship after long-term exploration (Fig. 2.11).

It is almost a common principle that blast speed decreases with the smelting intensity increasing, but there are some exceptions, for example, No. 1 BF of Capital Steel with volume of 576 m^3 and hearth diameter of 6.1 m (amount to 900 m^3) has the common principle, but opposite under the smelting intensity value less than 1.0 t/(m^3 d) (Fig. 2.12).

This BF has a shorter shaft with 15 smaller tuyeres (the diameter of the largest one is just 110 mm). The smelting intensity decrease begins to shrink the tuyere area slowly, or reduce blast volume without tuyere changing. Because the burden strength was at normal level, the BF still could keep regular performance even at lower blast speed. At the beginning of the operation for the driving changing, there was no convincement for that the proper blowing kinetic energy could be decreased or increased at a wider range so made some losses of production ("circle" in Fig. 2.13),

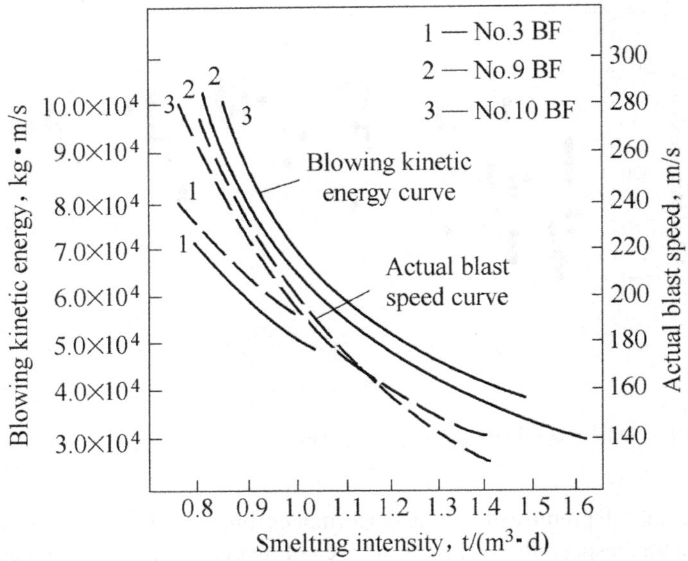

Fig. 2.11 Relationship between the smelting intensity and blowing kinetic energy (speed)

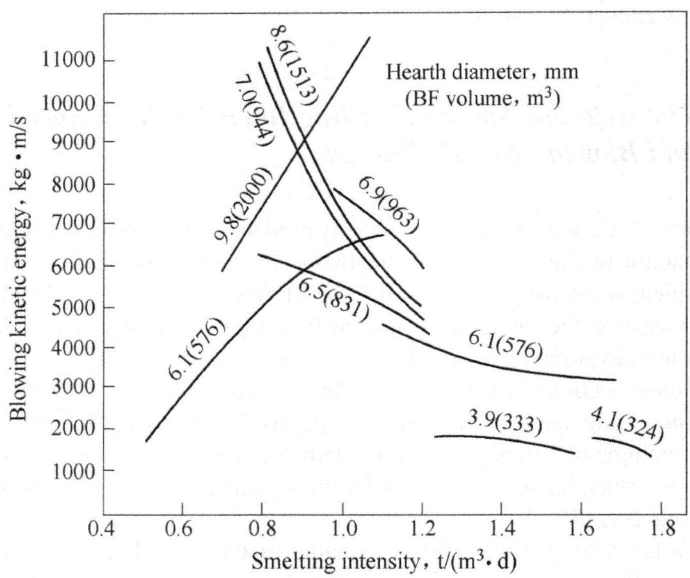

Fig. 2.12 Blowing kinetic energy of different smelting intensity

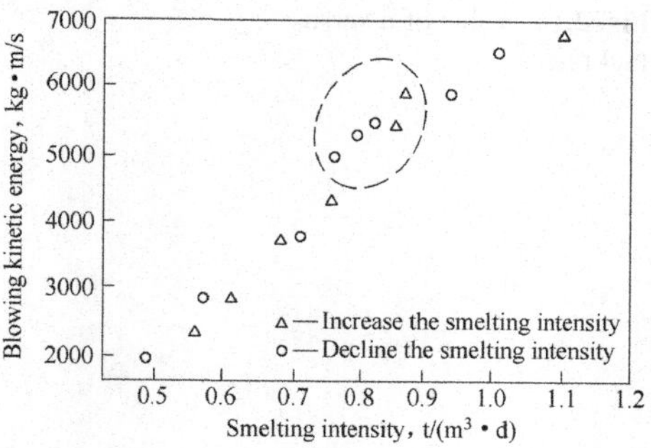

Fig. 2.13 Blowing kinetic energy versus smelting intensity

but it was getting clear during the smelting intensity rising afterward, then the times of tuyeres renewing reduced ("delta" in Fig. 2.13).

2.2.4 The Length (Depth) of the Raceway

The study conclusions of relationship between blast speed (kinetic energy) and depth of raceway in Anshan Steel are presented in Table 2.1. [8] In Table 2.1, A_1 is the area of the raceway and A is the area of the hearth.

There were two ways to measure raceway:

(1) A steel bar (25 mm) was used as a detector to penetrate through the blowpipe as far as possible to raceway bounder.
(2) To analysis the CO_2 content of the gas sample from different points along the centerline of the tuyere and take 2% as the bounder.

Table 2.1 Depth of raceway of different volume BF

Hearth diameter, m	4.7	5.2	5.6	6.1	5.55	6.8	7.2	7.7	9.8	9.4
Depth of raceway, m	0.784	0.949	0.950	0.90	0.902	1.118	1.033	0.965	1.302	1.211
A1/A	0.556	0.596	0.563	0.503	0.541	0.547	0.508	0.44	0.46	0.47
Fuel rate, kg/t	582		680	587	611	562	632	526	513	564
Utilization coefficient, t/(m³ d)	2.16		1.227	1.97	1.822	2.12	1.40	1.863	1.87	1.534
Hearth diameter, m	10.0	11.0	8.8	9.8	9.8	10.3	11.6	12.5	13.4	
Depth of raceway, m	1.11	1.28	1.36	1.20	1.41	1.29	1.45	1.70	1.88	
A1/A	0.932	0.411	0.520	0.43	0.493	0.45	0.438	0.47	0.48	
Fuel rate, kg/t	545	562	505	491	495	520	526	444	431	
Utilization coefficient, t/(m³ d)	1.59	1.56	1.92	1.90	2.342	2.24	1.84	2.00	2.29	

Fig. 2.14 Value of n versus fuel rate

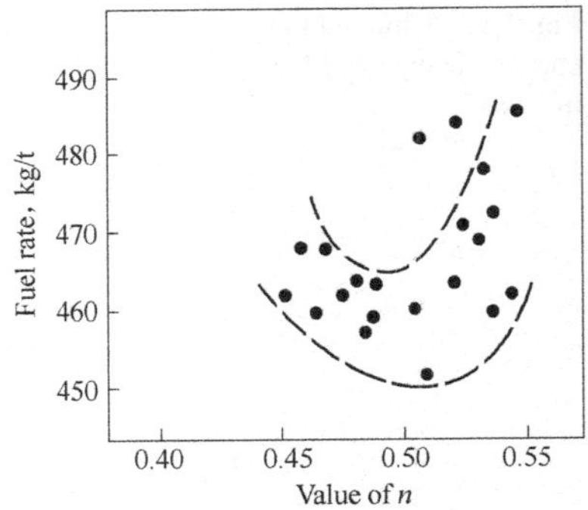

A formula for the relationship between BF hearth diameter and depth of raceway was presented according to the data of Japanese BF [9].

$$n = \frac{d^2 - (d - 2L)^2}{d^2} \quad (2.2)$$

where d is diameter of hearth, m; L is depth of raceway, m.

Figure 2.14 shows the relationship between fuel rate and the value of n that the lowest actual fuel rate corresponds with the value of n at 0.5 [10]. To compare with actual Japanese BF data (Table 2.2), there are some BFs with the value of n close to 0.5 including utilization coefficient, coke rate and others in addition to fuel rate, but for some BFs, the value of n is far from 0.5.

It had been studied by Song Jin and Hongxia Zhang to calculate out the optimum depth of raceway of some Chinese BFs (L) by reference of the Japanese formulae (Eq. (2.2), with value of n is 0.5).

$$L = 0.15\,\mathrm{d}\ (\text{diameter of hearth}) \quad (2.3)$$

It is defined that the optimum depth of raceway is determined by the data from qualified BF, which minus the length of nozzle extension into hearth (setting 0.4 m) is called actual depth of raceway (Table 2.3).

2.2.5 *The Role of the Raceway*

It is not yet known clearly about raceway, the size of which relates closely with intensification of BF. Figure 2.15 is a diagrammatic sketch of coke to enter into raceway [11]. Similar diagrams were affirmed by BF practice that the central coke's main path to the raceway is related to the size of raceway. It has been evidenced in production that the higher the blast speed, the larger the raceway and the better the productivity of BF (Fig. 2.16) [12].

Table 2.2 Depth and other indexes of some Japanese BFs

BF	Hearth diameter, m	Tuyere number	Tuyere diameter, mm	Depth of raceway, m	Value of n	Fuel rate, kg/t	Coke rate, kg/t	Utilization coefficient, t/(m^3·d)	Blowing kinetic energy, kg·m/s
A	6.80	14	150	0.977	0.500	662	662	0.80	1274
B	8.80	18	160	1.36	0.520	505	452	1.92	6974
C	9.35	24	150	1.16	0.435	561	540	1.35	3524
D	9.40	20	120 × 4 160 × 16	1.35	0.490	566	555	1.32	5493
E	9.50	27	149	1.16	0.428	523	495	1.87	3782
F	9.80	27	144	1.41	0.493	495	442	2.34	7469
G	9.80	28	145	1.20	0.430	491	431	1.90	5119
H	10.0	28	140	1.29	0.450	520	464	2.24	6521
I	10.3	28	150	1.26	0.430	529	470	2.28	6361
J	11.5	34	140	1.37	0.420	521	466	2.14	7346
K	11.6	34	140	1.45	0.438	526	489	1.84	7362
L	11.7	28	150	1.51	0.450	489	423	2.86	11,016
M	11.8	34	142	1.75	0.510	515	470	2.37	11,434
N	12.4	36	140	1.17	0.337	548	535	1.28	3402
O	12.5	30	150	1.70	0.470	444	392	2.00	11,281
P	13.4	35	130	1.88	0.480	431	365	2.29	14,110

Table 2.3 Statistics for the depth of raceway from some BFs

BF	Hearth depth, m	Best depth, m	Actual depth, m
Bao Steel No. 1 BF	13.4	1.96	1.56
Bao Steel No. 2 BF	13.4	1.96	1.56
Bao Steel No. 3 BF	14	2.05	1.65
Bao Steel No. 4 BF	14.2	2.12	1.68
Bao Steel No. 2 BF, 2nd generation	14.5	2.12	1.72
Bao Steel No. 1 BF, 3rd generation	14.5	2.12	1.72
Bao Steel No. 3 BF, 2nd generation	14.5	2.12	1.72
AN Steel new No. 1 BF	12.4	1.82	1.42
AN Steel new No. 2 BF	12.4	1.82	1.42
Kashima No. 3 BF, 1st generation	15	2.2	1.8
Oita No. 4 BF, 2nd generation	14.9	2.18	1.78
Chiba No. 5 BF, 2nd generation	15	2.2	1.8
Kimitsu No. 4 BF, 2nd generation	14.5	2.12	1.72
Kimitsu No. 3 BF, 3rd generation	14.5	2.12	1.72
Shang Steel No. 1 Factory	11.1	1.62	1.23
Ji Steel 1750m^{3}BF	9.5	1.39	0.59
MEI Steel 4070m^{3}BF	13.3	1.95	1.55
Ning Steel 2500m^{3}BF	11.2	1.64	1.24

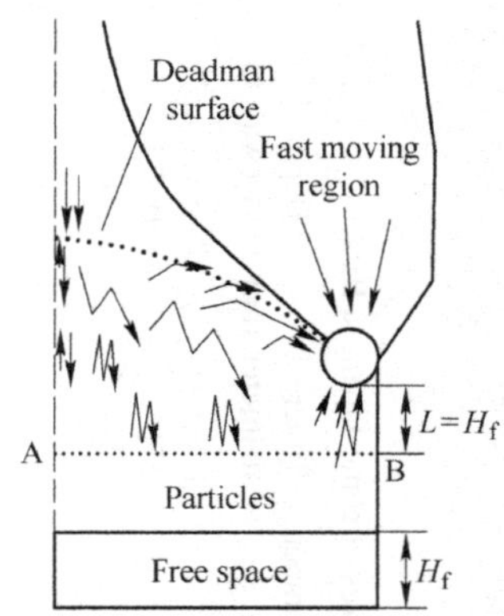

Fig. 2.15 Main area for coke to enter into the raceway

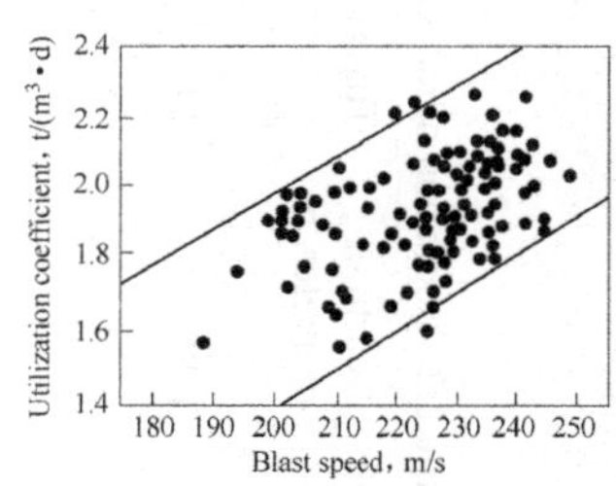

Fig. 2.16 Blast speed versus utilization coefficient

It is obvious that the size of the raceway is controlled by the primary gas distribution in the hearth. If the raceway is too deep, the primary central gas stream would be too strong, then top temperature too high and peripheral burden too heavy, resulting in an irregular performance and unstableness of BF, which is hard to correct, except changing the raceway depth. The same, if the raceway is too short, the primary periphery gas stream would be too strong, then the peripheral channeling unceasing and gas utilization becoming very low, BF could not keep regular performance and stability, even by adjusting burden charging or other methods to change the depth of raceway would be the effective method to improve gas stream primary distribution.

2.2.6 The Length of Tuyere

It is important that the length of tuyere plays role in BF process. When the length is too long, too strong primary gas stream would be introduced to the center; too short, the peripheral gas stream expands too much that used to be the principal reason to make some BFs in China reduce campaign life and lower the gas utilization. Many years ago, one of the famous iron-making experts Liu Qi proposed to increase the length of tuyere for "blowing through" to the center that is effective and easy to be accepted by BF. He introduced that Chinese traditional tuyere length is 5% of hearth diameter, for example, 230–270 mm and 310–350 mm for 300–420 m^3 and 620–750 m^3 BF, respectively. He also introduced that according to the recent BF operation intensification, the range of tuyere length versus the hearth diameter (*d*) should be (5%–5.5%)*d*, (5.5%–6%)*d* and (6%–6.5%)*d* for BF with > 2500 m^3, 1200–2500 m^3 and <1200 m^3, respectively. In recent years, very effective results of BF operation for lengthening the tuyere, to take the BF with 750 m^3 in Laiwu Steel, for example, its utilization coefficient increased over 3.0 t/(m^3 d) whenever the tuyere lengthened from 370 mm to 450–460 mm [13]. The early suggests of Mr. Liu are matched with the results introduced by Jin Song and Zhang Hongxia afterward (Table 2.4) [14].

From the statics by the senior designer Wu Qichang, the relationship between BF volume and diameter of the hearth is shown in Fig. 2.17 [15].

According to the setting in Table 2.4 with the data of Fig. 2.4, the tuyere length of different BF volume had been calculated out (Table 2.5).

In Table 2.5, the calculated values of tuyere diameter are too high for those BFs with >4500 m^3 volume, no one of which used to be successful by the use of larger than 700 mm tuyere. There was a BF used to use 700–730 mm tuyere, because central

Table 2.4 Tuyere length introduced by Mr. Zhang et al

BF volume, m^3	Tuyere length, mm
≥2500	(5–5.5%)*d*
1200–2500	(5.5–6%)*d*
≤1200	(6–6.5%)*d*

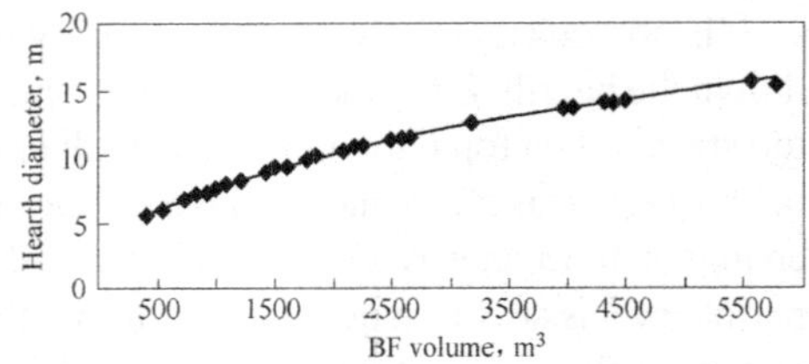

Fig. 2.17 BFs volume versus diameter of hearth

Table 2.5 Calculated tuyere length of different BF volume

BF volume, m^3	500	1000	1500	2000	2500	3000	3500	4000	4500	5000	5500	6000
Hearth diameter, m	5.2	7.6	8.9	10.5	11.2	12	12.7	13.5	14	14.6	15.5	15.5
Tuyere rate[a], %	6		6	5.5			5					
Tuyere length, mm	310	450	534	578	612	660	635	675	700	725	775	775

[a]The ratio of tuyere length to hearth diameter

gas stream was too strong, the tuyeres were changed to 625 mm at last after a period of analysis and observation.

There are a lot of successful examples of tuyere lengthening in China.

In BA Steel, the tuyere length had been changed from 470 to 520 mm at a 2500 m^3 BF [14]. In Baotou Steel, the tuyere length had been changed from 550 to 600 mm at a 2500 m^3 No. 6 BF, the effect was good [16]. In Sha Steel three 2500 m^3 blast furnaces, the tuyere length had been changed from 475 to 580 mm, also very successful [17].

In Ji Steel, the tuyere length had been changed from 550 to 600 mm at a 1700 m^3 BF, it worked very well. Then, for severe erosion parts, used a tuyere with a length of 630 mm [18].

In Hanbao Steel, the No. 1 BF has 32 tuyeres with diameter 130 mm and length 643 mm, the blast speed reaches 263 m/s. In order to "blow through" hearth center, the tuyere length changed in succession to 663 mm for 29 in 32 tuyeres, since February 15, 2012. But the BF operation became irregular and coke rate increased that means the tuyere length too long, then returned to 643 mm except 8 long ones [19].

In Lian Steel, in order to "restrain peripheral gas stream" and "open the center," the tuyere length had been changed to270 mm, but led to peripheral getting hard and wall accreting, then on alternative but to returned to 240 mm to recover the normal furnace operation [14].

At the present level of recognition, the determination for tuyere length would depend on the practical experience and theoretical calculating just for reference.

2.2.7 *The Angle of Tuyere*

Usually, the tuyere is installed at horizontal level, although the modeling tests showed to rise up a few degrees could shrink the raceway (Fig. 2.6), which nobody pay attention to, because no such a tuyere be used before. Since the burden brings too much Zn into make the tuyere rise, then its hazard began to be recognized.

Before some BFs had been installed tuyere with oblique 3°–5° to act hearth and now, many large BF are installed oblique tuyere (5°–7°) but very few > 7°. It is unnecessary to oblique tuyere downward for smaller BF, because the hearth is active enough, especially for that rate of driving (smelting intensity) over 1.5 t/(m^3 d).

2.2.8 *The Operation for Tuyere Plugging*

During some BFs daily operation, the mild blast operation has to be adopted for a long time whenever plugging of part tuyeres is necessary to keep basic blast speed. Sometimes, when the temperature at the part location of hearth wall rises up with the risk of hearth burning through, the relative tuyeres should be plugged. There are a lot of reasons, such as the stock surface decline and channeling; tap hole depth too short, tuyere cooling water leaking to cohere up at the front of the tuyere and so on, to plug the relative tuyeres. When the gas porosity at the front of leaking tuyere becomes worse, so renew the tuyere which should be plugged for 24 h to improve the part situation. Actually, plugging should not be used for temporary treatment, because it requires stopping blast and interrupt production process. In addition, long time for plugging would make the furnace wall accretion at the bosh above the tuyere and extend upward 1–2 m, the thickness of which decreases gradually along the wall with the appearance looks like a "nose." As an example, there was a BF with 1035 m^3 and only one tap hole, which always did not work well because of the unqualified mix clay(for notch closing) in the 1950s, then No. 1 tuyere (above notch) was plugged and notch's percent of pass increased substantially. Hereafter, a piece of small and short wall accretion was found in the intermediate reline, but although no serious influence for production, it was stipulated that any tuyere could not be plugged constantly, but could be alternative in one month to prevent wall accreting, if necessary.

The channeling is easy to appear around the neighbor of plugging tuyere.

2.2.9 *The Theoretical Combustion Temperature*

The range of theoretical combustion temperature (TCT) is usual at 2000–2300 °C, which is very important for keeping the proper temperature of hearth and central coke, especially during coal injection. Figure 2.18 shows the BF volume and TCT influence the temperature of central coke very much [12]. Although the heat transfer mechanism

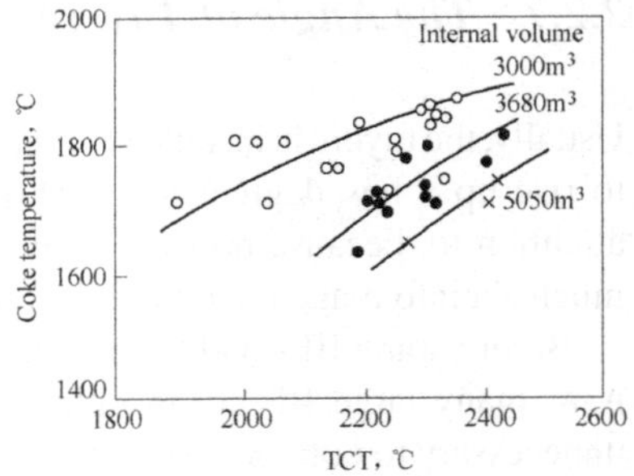

Fig. 2.18 Relationship between coke temperature and TCT

is not clear yet, the temperature of central coke influences the permeability and heat are very important.

Some empirical formulas used to be used to calculate TCT. Now, the computer is very popular on BF, so TCT is easy to get in time and nobody uses the old one anymore.

2.3 The Judging and Adjusting for Proper Blast Speed (Blowing Kinetic Energy)

About BF hearth work, although it has been studied for more than hundred years, some problems still could not be solved quantitatively. The furnace emergency is whether or not from high speed blast. It still depends on the operator to survey and make decision. In addition, the gas distribution also depends on the form of cohesive zone and burden distribution, the role of which is different from blast speed. The presage would not be obvious, when the furnace operation converts from regular to irregular condition.

The inter-reaction between two kinds of adjusting in lower and upper zones will discuss in detail in Chap. 3, here only some key points are introduced:

(1) Hot metal

The hearth work could be reflected from hot metal tapping.

For smaller BF, it taps intermittently. When blast speed is too low, the temperature of hot metal would get cooler and cooler during tapping process. At the same time, the light of hot metal changes from bright to dark and red. But if blast speed is too high, the temperature of hot metal gets hotter and hotter and the light from dark and red to bright.

As for larger BF, it taps continuously, the phenomenon is not clear, but it makes sense during the notch shifting. It is convenient to measure the temperature of hot metal by auto-thermometer.

(2) Burden descending speed and stock bar

The burden (stock) descending speed would not well-distributed with litter slip sometimes, when the blast speed is too low, but during high blast speed, the stock speed increases during tapping and slowdown after notch closing. The stock speed would change with tapping periodically. But there is no obvious change for large BF with continuous tapping and the stock bar work is seminal to small BF with slipping and uneven stock descending.

(3) Tuyere breakage

The "raw" coke particles move up and down often at the front of tuyere that is so-called raw falling when blast speed is too low and blooded tuyere could happen without cooling phenomenon, sometimes, a lot of tuyeres burned out at the upper edge; when blast speed is too high, cooling tendency appears at the front of tuyere under the normal operation, not easy to have blooding tuyere, but at serious situation, many tuyeres might be burned out at the lower edge of the nozzle (Fig. 2.4). Whether the blast speed is high or low, to adjust nozzle's area would be the most effective way to keep the blast volume and temperature at optimum level.

2.4 The Tuyere Burning Out and Preventing

Tuyere burning out is one of the popular accidents, which brings about some indirect losses (even the direct loss is not big) such as stop blast for tuyere renewing, the cooling water leaking and some others. Usually, to renew, a tuyere brings about productivity loss 1/20–1/30. So, the burned tuyere would not be renewed immediately, but reducing the cooling water a litter and waiting for more ones renewing together. The statics for tuyeres breaking in Capital Steel during last 14 years before 2008 is shown in Table 2.6.

In Table 2.7, there are four kinds of tuyere breaking at No. 2 BF Bao Steel including melting damage, break of welded joint, abrasion and buckling, which are from different reasons: The welded joint breaking and abrasion belong to design and manufacture concern, but the melting damage and buckling (from slag skull falling) belong to BF operation, which is to be discussed in detail here.

Of course, the classification is different in every plant. Table 2.8 shows the classification in Baotou Steel for tuyere breaking including chap (from bed manufacture) and tuyere second holder up warping. Most of tuyeres breaking are from melting damage (Table 2.8). Now technical plants have been established for tuyere manufacture to supply qualified tuyeres for BF.

Table 2.6 Statics for tuyeres breaking in iron-making plant, Capital Steel during last 14 years before 2008

Year	No. 1 BF	No. 2 BF	No. 3 BF	No. 4 BF	No. 5 BF	In total
2007	6	113	8		Discontinued	131
2006	2	32	10	5	Discontinued	49
2005	14	20	15	17	1	67
2004	89	55	17	39	10	210
2003	70	28	36	47	13	194
2002	0	2	0	2	17	21
2001	14	7	12	12	13	58
2000	30	4	28	20	8	90
1999	27	33	56	20	1	137
1998	43	32	62	36	13	186
1997	49	55	58	65	27	254
1996	34	74	77	45	21	251
1995	97	27	205	110	15	454
1994	37	73	139	46	17	312

Table 2.7 Classification for tuyere breaking of No. 2 BF, Bao Steel in January–October, 1994 [20]

Project	Melting damage	Break of welded joint	Abrasion	Buckling	In total
Number	21	24	6	8	59
Rate, %	35.6	40.7	10.2	13.5	100
Average campaign live, *d*	183	130	170	103	Ave. 149

2.4.1 The Reason of Tuyere Burning Out

The reason of tuyere burning out could be observed out from the location of damage. If the location is on the head face or exterior margin, that would be caused by the shorter raceway and stronger peripheral gas stream, which flows upwards along the wall to produce peripheral channel and "raw falling," then hot metal drops onto the head face and damage it more or less to form hole or hard spots (Fig. 2.19).

If the blast speed is too high, the pressure would decrease under the jet-blow, then a small amount of hot metal to be collected at under part of inner surface to burn it out (Fig. 2.20). Too strong central gas stream and raceway extended to hearth center would bring about tuyere breaking.

Slag skin fall off is caused in general by strong peripheral gas stream. The falling slag skin impacts against the upper part of tuyere, then to be deformed or even knocked down to hearth, which will be discussed in detail in Chap. 3.

Table 2.8 Number and ratio of tuyere renewing at No. 6 BF in Baotou Steel

Year	Melting damage		Abrasion		Weld		Crack		Medium sleeve bend-up		Length adjudication		Total
	Number	Rate, %	Number	Rate, %	Number	Rate, %	Number	Rate, %	Number	Rate, %	Number	Rate, %	
2007	28	49.1	17	29.8	6	10.5	1	1.7	5	8.8	0	0	57
2008	44	62	4	5.6	8	11.3	0	0	15	21.1	0	0	71
2009	46	73	0	0	3	4.7	1	1.6	10	15.9	3	4.7	63
2010	56	71.8	0	0	4	5.1	2	2.6	6	7.7	10	12.8	78
2011	51	71.8	2	2.8	5	7.0	2	2.8	3	4.2	8	11.3	71
2012	54	78.3	0	0	2	2.9	0	0	5	7.2	8	11.6	69
Total	279	68.2	23	5.6	28	6.8	6	1.5	44	10.7	29	7.1	409

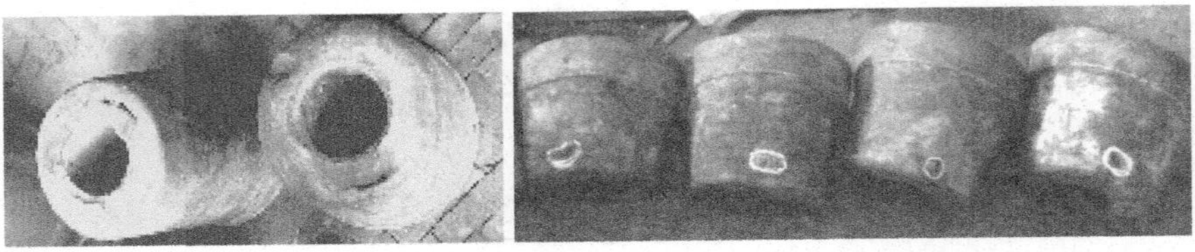

Fig. 2.19 Location of tuyere breaking

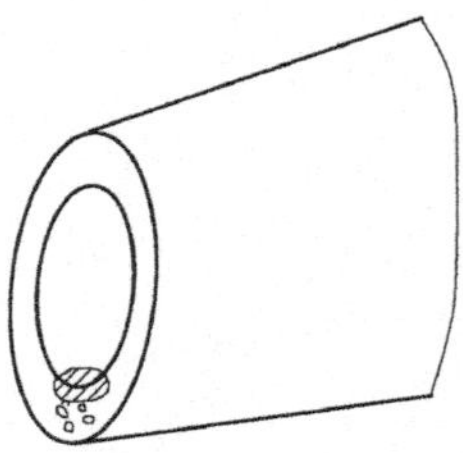

Fig. 2.20 Tuyere breaking at under part of inner surface

2.4.2 The Treatment for Tuyere Breaking

When tuyere breaking, its location should be surveyed firstly to judge the form of raceway and gas stream distribution.

If the breaking comes from too strong peripheral gas stream, the following should be considered:

(1) Low coke strength brings about bed porosity of coke in furnace core and reduction of capacity to receive more blast volume.
(2) Long-term mild blast operation leads to low blast speed.
(3) Incorrect charging program leads too strong peripheral gas stream.
(4) The hearth accumulation, if in condition, should be treated firstly.
(5) Is the tuyere too short?

If the breaking comes from too strong central gas stream:

(1) If the tuyere too long?
(2) Is the blast speed too high?
(3) Is the blast volume too much?
(4) Is the coke to be added in the center too much?
(5) Is too heavy burden at peripheral area?

It should be aimed at different reason to judge and adjust gas stream distribution, which will be discussed in detail in Chap. 3.

The breaking level should be judged to renew the tuyere with serious water leaking in time. Some BF used to control (reduce) cooling water, when leaking, which would be dangerous, especially for small BF to burn out the tuyere and to leak more water into hearth so as to serious accident. Figure 2.21 shows the plight of tuyere burned

Fig. 2.21 Tuyere burned out after overdose cooling water reducing and did not renew in time

out after cooling water reducing too much and not renew in time.

2.5 Hearth Accumulation and Preventing

The most common irregular condition in hearth is accumulation, which would not be harmful very much at the beginning, so it is easy to be ignored, but might become very serious in a short time. A lot of BFs in China had such painful lesson.

2.5.1 The Portent of Hearth Accumulation

The portent of hearth accumulation is represented as following:

(1) The variety of blast pressure, volume and stock descending speed during accumulation.

The blast pressure increases and volume decreases gradually after notch closing, but in contrast after notch opening to produce fluctuation periodically. At the same time, the stock descending speed also varieties that increases at tapping period but decreases after notch closing. Such situation just applies obviously to small BF with batch-type tapping. For large BF with more than one notch, the stock descending speed variety very little during continuous tapping.

The suppression of blast volume and pressure is the common feature of hearth accumulation. At No. 1 BF (380 m^3) of Sha Steel, the blast pressure increases and stock descending speed slow down before notch opening, but in contrast after notch closing. [21] At new No. 1 BF (2500 m^3) in MA Steel, the blast pressure increased suddenly before tapping. The blast pressure did not increase under the regular condition in mid-late of September, 1995 (blast volume 4000 m^3/min, tap to tap time

80 min), but since hearth accumulation and tap-tap time >50 min, the blast suppressed and blast pressure fluctuated from descending to rising suddenly [22].

(2) The features of hot metal and slag.

It is obvious that accumulation stays at the center of hearth, the temperature of hot metal decreases gradually through the tapping while it is difficult to flush from slag notch, but in contrast as the accumulate stays at the periphery of hearth. Whether at center or periphery, [Si] and [S] of hot metal are over the normal level and [S] would be getting higher.

For accumulation, in addition to composition fluctuate of hot metal and slag, the viscous of both increases and some hot metal is brought by slag stream to damage the cinder notch easily. The hot metal temperature at large BF usually keeps 1500 °C with better flowability. No. 1 BF of Sha Steel had a hearth accumulate, the temperature difference between the beginning and end of a same tapping became too big and the notch opened a little bit late, then hot metal and slag burn out the hearth and flow through tuyere cooling holders to outside, consequently to stop blast for 16 h, at the first tapping in the middle shift on December 5, 1999.

(3) The temperature of bottom and hearth decreasing.

One of the predictions of accumulate is bottom temperature decreasing continuously. If the accumulate is at periphery the temperatures of periphery hearth and its cooling plat would be also decreasing at the same time.

The new No. 1 BF of MA Steel had a skull at lower shaft and periphery accumulation leading to frequent slipping and checking with bottom temperature (at 4th level of bottom) decreasing from 545 °C (Oct. 1st–10th) down to 501 °C (Nov. 4th), as shown in Fig. 2.22 [22].

The regular temperatures at upper layer hearth bricks of No. 1 BF of Sha Steel are 584 °C, 506 °C, 461 °C and 489 °C and irregular ones drop to 541 °C, 483 °C,

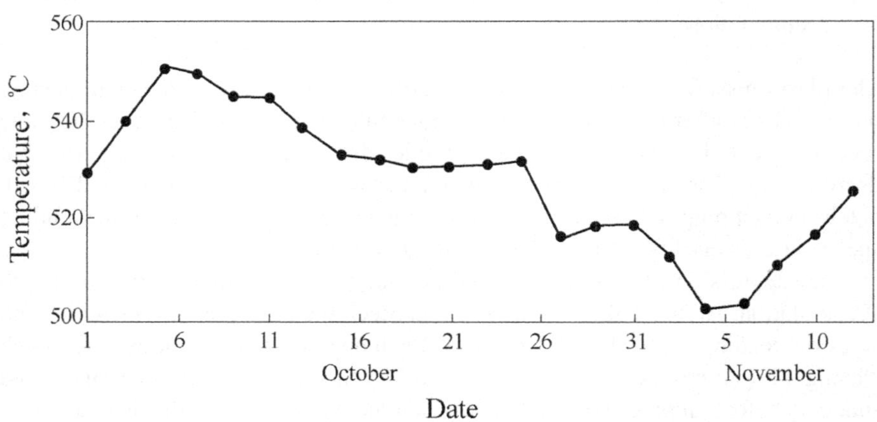

Fig. 2.22 Temperature variation tendency at the 4th layer carbon bottom brick

421 °C and 433 °C. The same drop tendency from regular to irregular condition at lower layers are from 414 °C, 475 °C, 444 °C and 345 °C down to 402 °C, 456 °C, 395 °C and 319 °C, respectively [21].

(4) The feature of gas distribution.

The peripheral gas stream develops that its $CO_{2\%}$ would be very low or temperature very high, when central hearth accumulates. But it would be in contrast as peripheral accumulation that the temperature of peripheral gas stream would be lower, but higher for central gas stream.

(5) The work and burning out for tuyere.

It is not even around all of the tuyeres area under hearth accumulation that "raw falling" (not heated enough) could be found at the front of tuyeres, which could be flooded even burned out the blast pipe, but seldom when accumulate in the peripheral area.

The No. 7 BF (2000 m^3) of Han Steel once had an accumulation to damage 17 tuyeres in 7 days [23]. There were 158 tuyeres, 18 s holders and 1 holder burned out during the recovery period (44 days) for accumulate at No. 2 BF (1200 m^3) of Shui Steel, such accidents were very seldom all over the country [24]. In addition, there were 222 tuyeres in totter burned out during the period to recover the hearth accumulation [25].

(6) Irregular performance.

Because of the hearth accumulation, BF operation could not be regular and channeling, slipping, bridging and slag skin falling happened frequently. But the portent of hearth accumulation is not obvious at the beginning, especially for the BF with qualified coke, so not easy to recognize until it becomes serious that must be treated at budding state in time. Most of hearth accumulates are formed at the center of hearth, especially for those BFs under the condition of low blast. Some of the portents might be easier observed such as bottom temperature decreasing and a bit of hot metal brought by slag. For ultra-large BF with long tapping period (1440 min or more), through one or two notches tapping almost all the time that means the balance of hot metal between producing and tapping with little influence for BF performance. If one of the notches tapes much more hot metal than another (might be only slag tapping), it states hearth work uneven and accumulation could be going to happen. It is the time to do something to prevent it.

2.5.2 *The Main Reasons for Hearth Accumulating*

2.5.2.1 Characteristic of Furnace Stamen

The results of BF dissecting state that the space under the cohesive zone is filled with solid coke and drops of hot metal and slag, which stay at the center of hearth, so-called as furnace deadman or dropping zone. Figure 2.23 is a diagrammatic sketch for hearth working. The coke and injected fuel are burned in the raceway, where a lot of coke drop down to supply the combustion but the coke at central part of hearth are considered no movement, so-called pillaring [26]. But it has been certified that these coke still move slowly from tuyere level to bottom and enter raceway in a week-one month even just a few meters from tuyere to bottom. The dissecting study in Japan shows that "pillaring" floats above the surface of hot metal bath in hearth with some fine particles of coke and limited space for hot metal to move effectively, just around the area of tap hole. [27] The "furnace deadman" under the cohesive zone is filled with coke, which is called "deadman coke," the porosity of which is about 43–50% with some drops of hot metal and slag to flow downwards. The gas from combustion in raceway flows upwards through the "deadman coke." The hot metal and slag are collected in the crucible and a part of hot metal drops to the bottom to float the coke in stamen zone, another part of hot metal and slag are stored amount coke particles and then tap out through deadman coke, which ups and downs along with tapping.

To ensure that hot metal and slag could tap out successfully, must keep them move through the deadman coke thoroughly.

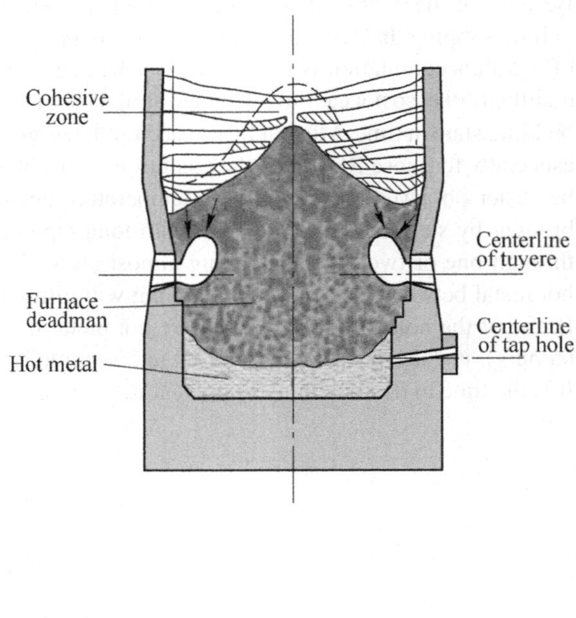

Fig. 2.23 Diagrammatic sketch of hearth

2.5.2.2 The Nature of Hearth Accumulation

The BF dissecting states that the iron ores in BF begin to soften at about 900 °C, soften and melt at about 1000 °C and melt at 1400–1500 °C (Fig. 2.24) with some diversity for different kinds of iron ores. The composition of slag might change, because of the ash from fuel combustion along with which the melting point would divert. According to finishing slag studying, the slag could not flow fluently through the deadman coke under 1400 °C, then the hot metal and slag stay among the coke to form inactive zone in hearth without good liquid penetration leading to hearth accumulation, which is different from furnace cooling as well as chilled hearth.

Hearth accumulation comes from the liquid penetration reducing in part of hearth and resisting gas passing through. Figures 2.25 and 2.26 show the results of temperature measuring and sampling from tuyere area at a Japanese BF under the inactive hearth or accumulation [29].

Figures 2.25 and 2.26 show that utilization decreases when temperature of deadman coke is lower than 1450 °C and some viscous melting thing appears in the samples that means the permeability and liquid penetration be destroyed. [29] The measurement for the resistance of deadman coke at tuyere level by Japanese experts shows that it is related closely to the variation of bottom temperature [29] by the use of a testing tube (diameter 82.6 mm) penetrating through the tuyere (120 mm) [30]. The results are shown in Fig. 2.27 that state the more compact deadman zone, the bigger resistance leading to reduce the permeability and liquid penetration certainly.

2.5.2.3 The Reasons for Hearth Accumulation Forming

There are many reasons to form hearth accumulation, which is a serious one of irregular performance:

(A) The influence of coke quality

Most of the hearth accumulations are caused by unqualified coke, (especially less strength) which bring a lot of breeze into hearth to decrease the permeability and hinder blast into deep hearth, where the temperature goes down to block up by the viscous hot metal and slag. For example, the serious hearth accumulation at No. 3 BF of Hang Steel came from unqualified import coke, including quality fluctuating, too much breeze and smaller size (40% for <25 mm) [31]; the coke reactivity index (CRI) ranges 34.15–36.95% and the strength after reaction (CSR) ranges 53.35–50%, which are far from national standard [32]; A BF (300 m^3) of Liu Steel had a hearth accumulation, when $M_{40} = 68–72\%$; M_{10} as high as 12%; ash contains of coke reach 16–17% and a lot of "ground coke" used to be used [33]; Because of pressure on the money market, the No. 2 BF of Shui Steel had to use unqualified coke ($M_{40} = 72.4\%$; $M_{10} = 11.8\%$), which was made of coal from small mine, then hearth accumulate forming gradually and tuyere burning out frequently to leak cooling water into hearth making the accumulating more serious; [24] Because the fines in burden increased

Fig. 2.24 Temperature distribution at soften zone by the use of temperature measuring tablet [28]

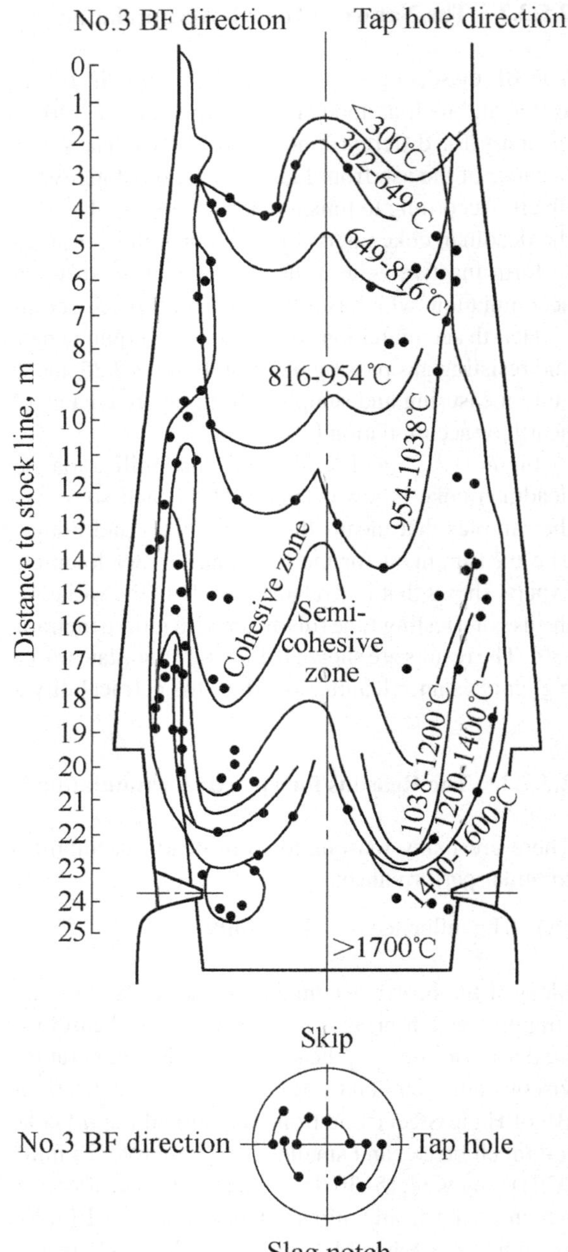

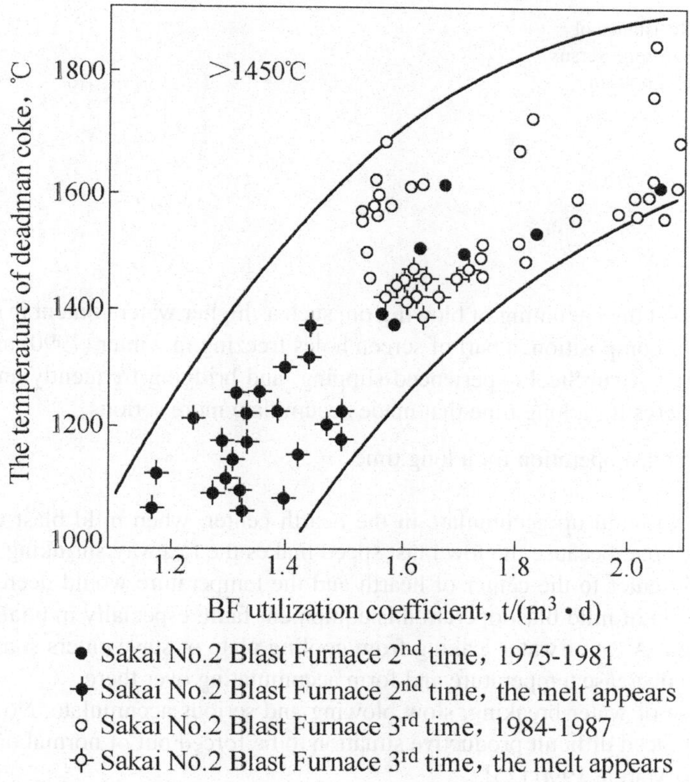

Fig. 2.25 BF utilization coefficient versus temperature of deadman coke

Fig. 2.26 Temperature of deadman coke versus charge index

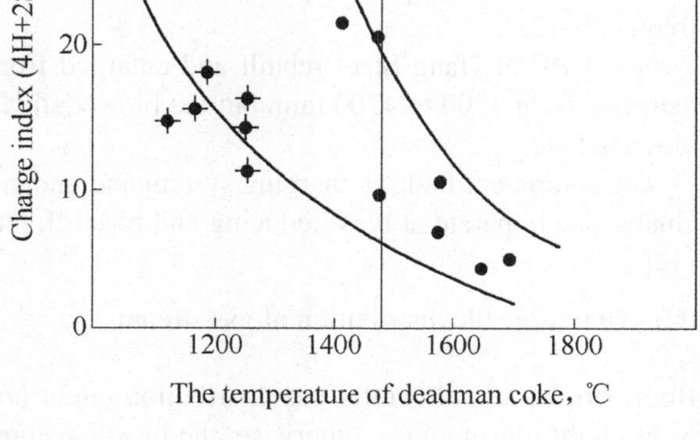

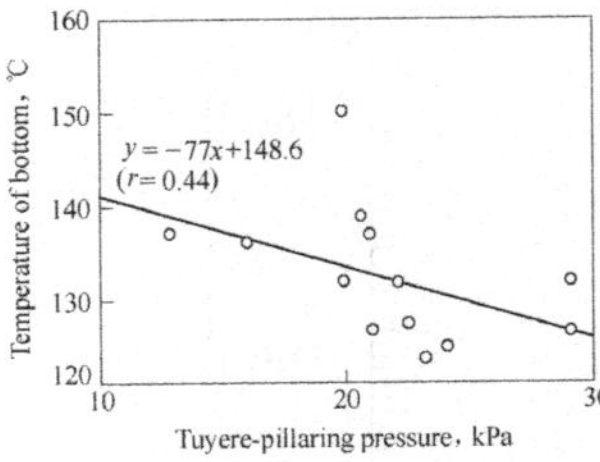

Fig. 2.27 Resistance of deadman coke zone versus temperature of bottom

(Table 2.9) at the beginning of blowing on, such as higher water content in coke, the fluctuate of composition, a part of screen holes freezing in winter (1990) and so on, No. 8 BF of Xuan Steel experienced slipping, and bridging frequently and plug 3 certain tuyeres for a long time that made accumulate more serious.

(B) Mild blast operation for a long time

It is easy to build up accumulate in the hearth center, when mild blast operation for a long time, because the low blast speed makes the raceway shrinking then less blast could enter to the center of hearth and the temperature would decrease. The direct reason of mild blast operation is equipment fault, especially in final phase of a campaign. A lot of water leaking from cooling plate or stave enters some part of furnace to decrease temperature and form accumulating over there.

Because of water breaking, slow blowing and serious accumulate, No. 5 BF of Ben Steel faced difficult productive situation to be forced out of normal order from January to March, 1990 [25].

The temperature at hearth sidewall of No. 2 BF, Tai Steel raised to the historical high (681 °C), then limited the yield to reduce the blast volume from 800 to 750 m^3/min to operate at low blast speed for a long time, so still built up hearth accumulation [34].

Owing to less quality and quantity of sinter, furnace equipment and electricity fault, new No. 1 BF of MA Steel was operated at low blowing for a long time with irregular performance in one and half a year. There were four times of furnace cooling in September 1995 and wall accretion (at lower part of shaft) in October and November leading to hearth peripheral accumulating, slipping and bridging frequently [22].

No. 3 BF of Hang Steel rebuilt and enlarged from 255 to 302 m^3 and hearth diameter from 4200 to 4700 mm, but the blower still 750 m^3/min to operate at low blowing [31].

The equipment fault of charging system and unqualified coke make No. 4 BF of Guang Steel operate at blast reducing and blastoff, leading to hearth accumulating [32].

(C) Unreasonable distribution of gas stream

Both peripheral and central gas distribution ought not to be too heavy (plugging) or too light (developing), otherwise, the hearth accumulation would happen. If the

Table 2.9 Change of burden quality at No. 8 BF of Xuan Steel

Burden	Index	1990				1991				
		September	October	November	December	January	February	March	April	May
Coke	Tumbler index, %	81.56	80.93	80.57	81.27	80.92	80.66	78.80	78.28	79.16
	Water content, %	6.11	7.43	7.08	9.09	9.29	8.44	8.82	7.18	6.20
	<25 mm, %	–	–	6.5	–	–	–	–	5.81	5.99
Sintering	<5 mm, %	–	7.60	8.01	–	5.22	5.68	7.44	7.11	3.57

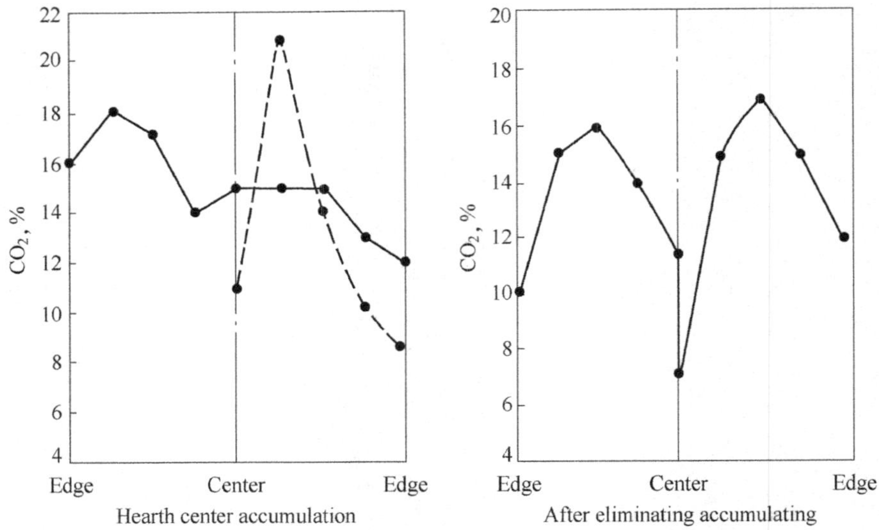

Fig. 2.28 CO_2 curve of No. 3 BF at hearth accumulation

peripheral gas stream is too light, less gas flows to center, where the temperature would be lower; in contrast, if the central gas stream is too light, less peripheral gas could not heat and reduce the burden in peripheral area, where the temperature would decrease. Figure 2.28 shows the change of gas stream distribution before and after the hearth accumulation at No. 3 BF, Hang Steel. The central accumulate was caused by heavy central gas stream actually. The gas stream distribution is along two passes with double curve under regular working of BF after accumulate. The similar situation appeared at No. 2 BF, MEI Steel with two passes before accumulation and heavy central gas stream afterward (Fig. 2.29) [35].

(D) BF irregular working, slipping or slag skin falling off

Fig. 2.29 To compare the gas curves before and after the accumulation at No. 2 BF, MEI Steel (1987)

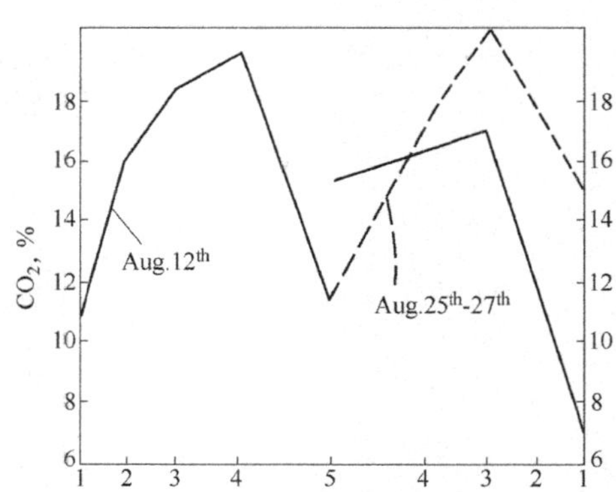

The irregular working, slipping or slag skin falling off lead up to burden descending unstably without enough reduction and heating, then to hearth accumulation and even scaffolding at last. Under the free driving condition destroying and blast reducing sharply, it would be going to accumulate easily.

The situation of BF (2500 m^3), MA Steel had already been stated in this book before, here, No. 5 BF, AN Steel would be described about a lot of reasons for hearth accumulation, charging with chaotic distribution, irregular gas stream distribution, unstable furnace condition, bridging in 12:25 with stock line 800 mm, then checking (blast pressure 50 kPa, stock line not clear), then increase blast volume and charge gradually, bridging again in 14:10, stock line still unknown, after second checking the furnace condition became harder and harder, the yield of hot metal and slag only about 20–30 t and 10 t, respectively, stock line unknown, dull tuyere, small blast volume, low top temperature, less permeability, high pressure difference, low blast volume and pressure. All of these made the furnace operation stayed at lower level with low blast kinetic energy, less peripheral gas stream and no central gas stream opening leading to hearth accumulating at last [36].

No. 3 BF Liu Steel one-sided pursued larger blast volume and higher productivity, ignore the change of burden, to increase the regular charging ratio leading to frequent channeling, bridging, slipping and operating at low stock line then to emergency, hearth accumulation and scaffolding on the north side wall [33].

(E) The slag composition mismatches temperature in the hearth

The fluctuation of slag composition makes its thickness, leading to hearth accumulating. For example, because of the big fluctuation of sinter hard to control, No. 3 BF of Liu Steel had to operate with the slag of higher basicity (higher melting temperature) to guarantee the quality of hot metal, then leading to hearth temperature decreasing and made the hearth accumulating more serious [33].

Owing to long time blowing off and reduced wind, No. 2 BF Meishan Steel had to shift yielding from “steel manufacture iron” to “foundry iron,” then high hearth temperature ([Si] 0.9–1.5%), higher slag basicity (1.15–1.20), low [S] (0.010–0.013%) and high viscous slag lead to hearth accumulation.

2.5.3 *The Principles and Methods for Treating Hearth Accumulation*

The accumulation is the outcome of liquid penetration and permeability decreasing from unqualified coke to worsen the permeability with lower temperature in hearth center zone, where the viscous slag and hot metal drop through hardly. The strength and temperature in the hearth central zone are the deciding factors to treat the hearth accumulating that its original reasons must be analyzed first and then targeted to practice.

2.5.3.1 To Improve Quality of Coke

The "framework" of stock column is consisted of coke which is the important solid materials over there. The quality of coke,especially hot strength, determine the porosity in the central zone of hearth at a large extent. It is very necessary to use qualified coke to prevent and treat hearth accumulation but some iron-making plants did not pay attention and still use bad coke for cost reducing after treatment of accumulation, then it came again. The coke breeze filled the space from tuyere to notch level during the recovery period. The samples from tuyere, south slag hole and notch show that breezes (<10 mm) reach 62.5%–72.9%. The tuyere breaking and renewing came into a vicious circle to leak a lot of water into hearth, which made the recovery more difficult. Such situation never appeared in past 30 years. The way to recover normal condition is to use qualified coke in place of bad ones to improve the liquid penetration and permeability in coke layer for more blast.

2.5.3.2 To Deal with Hearth Accumulating by Upper and Lower Adjusting

The reasons should be analyzed if the peripheral or central gas stream is too light (developing) or too heavy (plugging) at the beginning of hearth accumulating, which could be forecasted:

(1) No tuyere burning out continuously;
(2) No sharp decline for bottom temperature;
(3) Regular performance did not worsen seriously without frequent slipping and channeling.

The process of treatment for the early hearth accumulation is as following.

The first step, to compare the real blast speed and normal level, if close to each other, the tuyere could not be changed. But it could be reduced, if the blast speed is too low and its volume close to normal level, if the volume is much lower than normal, some tuyeres should be plugged temporarily to speed blast, if the blast volume is hard to reach normal level the tuyere should be adjusted.

The second step, the charging program should be adjusted immediately, if not deal with tuyere, to analysis and determine real controlling factors for gas stream distribution (very few BFs equip such apparatus) by the use of cross-beam temperature measuring technology (temperature measuring along redial direction of throat), top temperature, throat temperature, infrared scanning (image) and so on, as well as observing the "raw falling" distributing at the front of every tuyere, the shaft wall temperature distributing (temperature difference of cooling water income and outcome the staves or plates). On the base of these parameters analyzing, the adjustment for charging program might be determined carefully.

It must be very careful to enlarge the sizes of charge (burden). Whether central or peripheral (under lower blast volume) accumulation, it would be hard to increase blast volume. If it is necessary to deal with tuyere (in first step), the upper adjusting should wait for 16–24 h to observe, and then could continue. If the regular performance could be seen, the observing period should extend another 15–24 h. If the furnace driving condition worsened seriously, it must return to the original status to meet an emergency.

If the upper adjusting could not improve the furnace conditions over one day, the other ways should be explored in time without losing the opportunity to avoid hearth accumulation at the beginning.

When facing hearth accumulation, 9 tuyeres (No. 2–10) were plugged at one side to decrease the tuyere area 50% (from 0.2856 to 0.143 m^2) that is aimed at the serious hearth accumulate around northwest part and burned tuyeres concentrated over there [35]. Also facing hearth accumulation at No. 2 BF Tai Steel, the No. 2 tuyere area decreased from 0.101 to 0.097 m^2 to ensure enough blast momentum to penetrate the hearth center [34].

When peripheral hearth accumulation was found in advance at new No. 1 BF MA Steel, the closed tuyeres were reopened before slug (flushing) and top adjusting started to loosen peripheral burden by changing the chute angle [22]. For No. 2 BF MEI Steel, the ores charge size increased from 15 to 23 t and chute angle panned from 29.6º/26º (ores/coke) to 33.6º/33º time by time, then the peripheral and central gas stream developed gradually and each $CO_{2\%}$ were 16.5% and 4.9%, respectively, that is close to normal level of top gas distribution curve.

2.5.3.3 To Decrease Reducing Blast, Blastoff and Water Leaking

There are different reasons for reducing blast, mainly:

(1) During the last period of one campaign life of a BF, the equipment's faults, such as water leaking of cooling system are frequent.
(2) Some BFs start-up in a hurry without enough test run, then more equipment's faults leave behind.
(3) There are some hidden dangers in the equipment's, such as hearth burning through, charging system faults, less burden applying and so on.
(4) The production system is not perfect, such as how to balance hot metal and steelmaking requirement, burden supply falls short of demand because of the weather or other things.
(5) Irregular performance, stiff operation, scaffolding, all of these factors influence BF to accept more blast volume and operate at full blast.
(6) Unqualified burden, no matter of sinters or coke, might force BF operate at slow blast.

Whether or not, every fault should be managed in time or gradually to keep enough blast speed to prevent the temperature too low in the core zone, even could not operate on full blast, and to prevent hearth accumulating, which might become more serious because of water leaking. The tuyere must be controlled and renewed or reducing cooling water at once, when breaking is found out. The operation of BF must be studied fully and resolutely. The blast volume is the secondary factor of BF, just after burden quality something like the life of production. There is no priority without blast volume which is the first important thing for BF operation.

2.5.3.4 Slugging (Flushing Furnace) with Manganese Ores When Serious Hearth Accumulation

It must be slugged if tuyeres burning out continuously. A lot of tuyeres used to be burned out at No. 1 BF Capital Steel in the 1979s, more than 40 a month, then the manganese ores were used to slug with excellent effect. Table 2.10 shows the hot metal composition in that time.

The points of slugging operation with manganese ores at BFs Capital Steel are as following:

(1) To control [Mn] in hot metal at 1.0–1.5%.
(2) To keep hot metal [Si] 0.8% (<1.0%). If [Si] too high, the hot metal would be viscous, but too low might make the manganese recovery rate decrease a lot. The normal manganese recovery rate is 50% or a little more at [Si] about 0.8% with 50% for burdening. The manganese containing in slag depends on slag ratio, 2–3% for slag ratio 500 kg/tHM. According to the practice of Ben Steel, it would be very important to keep proper [Si]% in hot metal, otherwise, the hearth would agglomerate and damage more tuyeres [25].
(3) To tap slag in time for reducing the line eroded by slag with high MnO%.
(4) It is the time to stop adding manganese ores when tuyeres not be burned out any more, blast volume increases or the permeability index improves automatically.

Some iron-making plants have such experience also to slug with manganese ores or fluorite.

At new No. 1 BF of MA Steel, the slug was from November 4–7, 1995 with manganese ores 1.13 t per charge (ores charge 50 t), whenever the [Mn] was 0.75%, binary basicity (CaO/SiO_2) reduced to 1.03 or so, ternary basicity ($(CaO + MgO)/SiO_2$) reduced to about 1.29, [S] in hot metal is 0.024% and average [Si] is 1.20%. [22].

There was a hearth accumulation at No. 7 BF Han Steel after January 24, 2002 with a lot of tuyeres burning out, and then 17 tuyeres were plugged, just the rest 11 tuyeres

Table 2.10 Hot metal composition dealing with the hearth accumulation at No. 1 BF of Capital Steel

Date	Hot metal	01:00	03:15	05:33	08:07	10:32	13:05	15:25	17:46	20:20	22:36
27	[Si], %	0.3	0.4	0.34	0.54	0.94	1.11	1.06	1.28	1.28	1.17
	[Mn], %	–	–	–	1.22	1.38	0.63	–	1.96	–	–
28	[Si], %	0.91	0.84	0.72	0.77	0.78	0.72	0.72	0.64	0.51	0.66
	[Mn], %	–	–	1.73	–	1.44	–	–	–	–	–

Table 2.11 Analysis of hot metal and slag at No. 7 BF of Han Steel when hearth accumulation treated

Date	[Si], %	[Mn], %	CaO/SiO_2	Hot metal temperature, °C
January 31, 2002	1.68	0.15	1.05	1418
February 1, 2002	1.33	0.37	0.96	1385
February 2, 2002	1.75	0.68	1.00	1436
February 3, 2002	0.86	0.95	1.03	1459
February 4, 2002	0.82	1.30	1.04	1447
February 5, 2002	0.70	1.02	1.02	1477
February 6, 2002	0.83	1.12	1.06	1468
February 7, 2002	0.73	1.29	1.05	1484
February 8, 2002	0.72	0.83	1.03	1491
February 9, 2002	0.47	0.17	1.07	1480

blowing with blast volume 1600 m^3/min. For recovering normal condition as soon as possible, to slug with manganese ores at first, which came down to hearth in first January, then the furnace condition improved, blast volume increased and opened all of the tuyeres in January 28th blowing to 3200 m^3/min, then stop to charge manganese ores. The [Si], [Mn] both were 0.8–1.0%, slag basicity (CaO/SiO_2) was 1.0–1.05 to realize a successful slugging (Table 2.11). There were three times of slugging with manganese ores to control [Si] 0.7–0.9%; [Mn] 0.8–1.0% and CaO/SiO_2 1.1–1.15. When the manganese ores came down to hearth, the tuyeres burning avoided gradually and the flowability of hot metal and slag improved to keep the temperature of hot metal at 1450 °C or a little more [23].

The fluorspar was used for hearth accumulation treating at No. 4 BF Guang Steel since July 23, 2004 to add fluorspar 100 kg to every charge (ores 8500 kg) up to middle October with better effectiveness. The tuyeres burning out and bridging decreased obviously in November and December. The flowability of slag improved and easy to tap and blast volume increased to full blast. So, stop the use of fluorspar. But the tuyeres burning out and bridging happened again because of the coke pulverizing since January 2005, then 100–200 kg of fluorspar had to be added to every charge. Although the fluorspar could be used in favor of recovering peripheral hearth stickup, but easy to make the tuyere burn out and still could not perished hearth accumulation radically [32].

During the planning period for slugging, it is expected to control the [Mn] at 1.5 $\pm$ 0.2%; [Si] 0.7–1.1%; CaO/SiO_2 1.04 $\pm$ 0.02 and $(CaO + MgO)/SiO_2$ 1.25 $\pm$ 0.02.

The recovery standard after slugging should be as follows:

(1) The (FeO)% in slag might increase during slugging, so it should decrease to <0.5% after slugging.
(2) It should be stabilized basically including stock descending speed, blast volume and pressure.

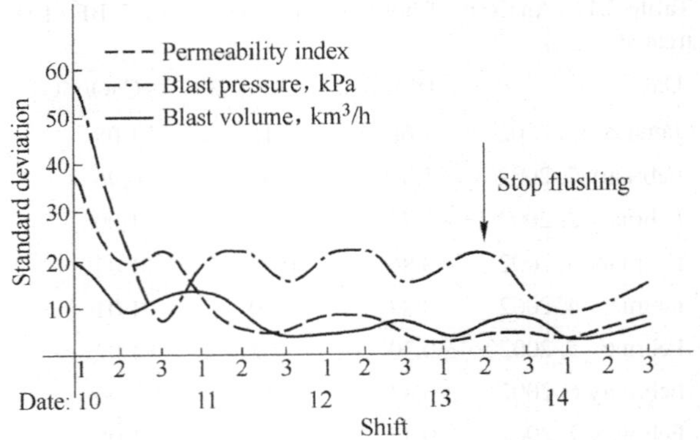

Fig. 2.30 Blast index (shift standard deviation) change during slugging period at No. 8 BF of Xuan Steel before and after flushing

(3) The central gas stream develops well and the central point of cross-beam temperature increases.
(4) The permeability index increases, the ability of accepting blast volume raises and the blast volume increases automatically.
(5) The flowability of hot metal improves and no hot metal stickup at ladle brick.
(6) To stop manganese ores after the above items.

This slugging with manganese ores started since middle shift of April 9th begin to improve performance in 2 days. The central gas stream dredged spontaneously and the gas utilization improved as well as the flowability of hot metal and slag improved. The blast volume increased under the same pressure and both stabilized. The shift standard deviation of blast pressure, volume and permeability reduced gradually to stabilization (Fig. 2.30; Table 2.12). Then, the slugging stopped in the middle shift of 13th to reduce manganese ores gradually to zero in the middle shift of 15th to end the long-term slugging operation.

For No. 8 BF of Xuan Steel, it was successful to use manganese ores for slugging that operated with all of the tuyeres without plugging to stabilize the operation. The slipping and bridging reduced obviously. The blast and charge system improved to increase blast temperature and gas utilization leading to higher productivity and lower coke rate [37].

As a kind of slugging materials, manganese ores are much better than fluorspar, because a small part of Mn join hot metal to dilute it and another big part (MnO) join and dilute the slag. Such hot metal could pass through hearth core to dilute some stick hot metal and slag. But fluorspar could only dilute slag by CaF_2, which is harm for tuyere and furnace wall line and its effectiveness is less than manganese ores.

Table 2.12 Operation index change at No. 8 BF of Xuan Steel (April 1994)

Date	Shift	Blast pressure, kPa	Blast volume, km^3/h	Permeability index	[Si], %	[Mn], %	CaO/SiO2	CO2, %	Slip	Bridging
10	1	191	110	214	1.29	1.08	1.08	17.3	4	1
	2	171	109	209						
	3	214	113	213						
11	1	213	118	221	1.11	1.65	1.05	17.3	1	0
	2	219	128	231						
	3	224	132	238						
12	1	220	127	232	0.60	1.50	1.04	17.2	0	0
	2	228	128	223						
	3	216	130	244						
13	1	217	127	232	0.74	0.88	1.03	17.6	1	0
	2	217	127	241						
	3	211	118	238						
14	1	220	129	238	0.65	0.48	1.05	18.1	1	0
	2	217	132	251						
	3	216	131	250						

Table 2.13 [Si] versus number of tuyere breaking

Date	[Si], %	Tuyere breaking	Slag notch
February 1–10	1.002	38	14
February 11–20	1.032	32	17
February 21–28	1.241	41	11
March 1–10	1.031	38	11
March 11–20	0.724	20	19
March 21–31	0.644	7	11

2.5.3.5 To Improve the Flowability of Hot Metal and Slag

It should be strictly on guard against the high content of Si in hot metal and high basicity of slag, both of which would rise its viscosity to make hearth accumulation more serious.

Because of the sulfur load rising at No. 3 BF of Hang Steel, the high basicity of slag ($CaO/SiO_2 = 1.32$, $(CaO + MgO)/SiO_2 = 1.56$) had to be used. The low flowability of slag made the hearth accumulating [31].

Table 2.13 shows the clear relationship between [Si] and the number of tuyere breaking, which based on the practical data from No. 5 BF of Ben Steel during hearth accumulating [25].

2.5.4 The Measures for Adverse Furnace Condition Preventing

(1) The burden quality must be ensured, especially the strength of coke.

The burden should be screened in stock house before charging to reduce the fine material in the burden.

(2) To keep full blast operation with all of the tuyeres. If reducing blast must be done for some external reason in a longer time, the proper blast speed should be kept.
(3) Pay attention to gas stream distribution to ensure reasonable pass of gas stream.
(4) The first important thing for BF operation is to keep regular performance to realize that the burden quality is geared to the blast volume.
(5) It should be intensified to control the temperature at lower part of shaft and bosh to prevent sculling and hearth accumulating as well as bottom temperature and hearth situation. The operation management and technology should be improved specially to keep hearth thermal system steady.
(6) The equipment maintenance should be strengthened to reduce unplanned blastoff and renew the broken tuyere in time or control the cooling water without leaking into furnace.

References

1. Z. Zhang, *Metallurgical Iron Ruins of Guying Town in Han Dynasty* (Guangling Book Society, Yangzhou, 2009)
2. H. Ralph, H.W. Eetser, in *Blast Furnace Practice* (1968), pp. 143–157
3. J.F. Elliott, R.A. Buchanan, J.B. Wagstaff, Physical conditions in the combustion and smelting zones of a blast furnace. J. Metals, 709–717 (1952)
4. L.M. Qilev, M.I. Ostroukhov, L.Z. Hodak, Combustion process of blast furnace coke. Moscow, 18–67 (1960). (in Russian)
5. B.D. Pandey, U.S. Yadav, U.K. Jha, Control of raceway parameters for improved blast furnace performance. Steel Times Int. 36–37 (1999)
6. M.I. Ostroukhov, L.Z. Hodak, A new study on coke combustion process in front of blast furnace wind, in *Study on Blast Furnace Smelting*, ed. by Wang (Metallurgical Industry Press, Beijing, 1960), pp. 263–276
7. G. Gao, Z. Miao, The law of upper and lower part of blast furnace, in *Formation and Development of Iron-Making Technology in An Steel.* Zhou ed. by (Metallurgical Industry Press, Beijing, 1998), pp. 125–127
8. G. Gao, Lower adjustment. *Blast Furnace Iron-making Process and Calculation.* L. Lanbo, ed. by (Metallurgical Industry Press, 1991), p. 218
9. Original: Gang-Kawagoe (Technical Research): Iron and Steel Association joint research. No. 39th return Milling Department information, 27. Bricks in this book are quoted from [12]
10. S. Asaki, T. Ota et al., The formula for the number of perfect for in the blast furnace. Iron Steel **14**, 23–29 (1979)
11. T. Sanyo, The update of core-filled particles in blast furnace. Iron Steel **5**, 169–175 (2001)
12. Iwanaga, Air vent coke properties and furnace core control. Foreign Steel, **2**, 8–13 (1992)
13. Liu Qi, *Liu Qi Iron-making Essays* (Metallurgical Industry Press, Beijing, 2007), p. 157
14. G. Li, H. Zhou, S. Jin, J. Kunming Metall. Coll. **30**, 16–19 (2011)
15. Q. Wu, Internal type of blast furnace, in *Blast Furnace Design - Theory and Practice of Iron-making Process Design*, 2nd edn. Z. Xiang, X. Wang et al. eds. by (Metallurgical Industry Press, Beijing, 2014), pp. 343–356
16. R. Liang, Z. Wang, Z. Hou, Analysis on the use of blast furnace vents in Baotou steel 6#. Baotou Steel Technol. **6**, 4–7 (2013)
17. Du Ping, Zhao Huotao, Wei Hongchao et al., The causes and countermeasures of burning loss of blast furnace vents of Sha Steel blast furance (2500 m^3). Iron-Making **2**, 47–50 (2012)
18. X. Pan. Treatment of abnormal elevated temperature of the side wall of the blast furnace of Ji Steel No. 2 Blast Furance (1750 m^3). Iron-Making **5**, 27–29 (2010)
19. H. Wei. Thickening treatment of Hanbao Steel No. 1 blast furnace wall. Iron-Making **2**, 37–39 (2014)
20. R. Zhu, Analysis of the damage of No. 2 blast furnace vents. Baosteel Technol. **2**, 7–9 (1995)
21. C. Yu, Treatment of the hearth of blast furnace accumulation of Sha Steel No. 1 blast furance (380 m^3). Jiangsu Metall. **2**, 22–23 (2001)
22. P. Wang, Treatment of the hearth of blast furnace accumulation and thickness of the wall of Ma Steel blast furance (2000 m^3). Iron-Making **3**, 17–20 (1996)
23. C. Yang, Treatment of the accumulation of Han Steel blast furnace (2000 m^3). Iron-Making **1**, 35–37 (2008)
24. L. Wang, F. Wu, Treatment of severe accumulation of No. 2 blast furnace cylinder in Shui Steel. Iron-Making **4**, 18–22 (1997)
25. C. Fan, D. Jin, Causes and countermeasures of the accumulation of No. 5 blast furnace cylinder. Ben Steel Technol. **1**, 16–20 (1992)
26. E. Kangoki, Investigation of blast furnace disintegration in it furnace condition. Iron and Steel **5**, 535–546 (1976); K. Kanbara et al., Discussion of blast furnace and their internal state. Trans. ISIJ, 371–380 (1977)
27. N. Kaoru, et al., in *The 5th International Congress on the Science and Technology of Iron-making (ICSTI' 09)*. Shanghai, China, 20–22 Oct 2009, pp. 1156–1160

28. E. Kangoki, Investigation on the disintegration of blast furnace. Iron-Making Res. 37–45 (1976)
29. C. Shiba et al., Long-term high efficiency operation of the 3rd generation of Nippon Steel No. 2 blast furance. Iron Prod. Res. **333**, 57–68 (1989)
30. T. Koko et al., Development of detection technology for activity degree of blast furnace core. Nisshin Iron Technol. News **75**, 11–18 (1997)
31. Y. Pan, Hearth accumulation reason analysis of Hang Steel No. 3 blast furnace. Iron Steel **2**, 8–11 (1994)
32. X. Cao, S. Jiang, The practice of dealing with the accumulation of No. 4 hearth of blast furnace in Guang Steel. Metall. Ser. **1**, 19–21 (2005)
33. C. Mo, Treatment of the wall scaffolding and accumulation of the hearth of Liu Steel blast furnace (300 m^3). J. Chongqing Iron Steel Coll. **2**, 11–14 (1997)
34. Z. Jin, Cause analysis and treatment of hearth of blast furnace accumulation in No. 2 blast furnace of Tai Steel. Tai Steel Sci. Technol. **2**, 58–60 (2006)
35. J. Liu, Operation practice of the treatment of hearth of blast furnace accumulation in Mei Steel No. 2 blast furnace. Mei Steel Sci. Technol. **2**, 7–13 (1990)
36. H. Wu, Treatment of caking wall of blast furnace and hearth of blast furnace accumulation of An Steel No. 5 blast furnace. Metall. Equip. Manage. Maintenance **25**, 9–11 (2007)
37. C. Zhang, The long-term poor state of treatment of Xuan Steel No. 8 blast furnace. Xuan Steel Sci. Technol. **3**, 17–21 (1991)
38. L. Cheng, Z. Liu, Z. Wang et al., Adjustment of reasonable airflow distribution of blast furnace, in *Formation and Development of Iron-Making Technology in An Steel*. Zhou ed. by (Metallurgical Industry Press, Beijing, 1998), pp. 156–159

Chapter 3
The Structure of Stock Column and the Control of Gas Distribution

Abstract Blast furnace is a countercurrent reactor. The gas rises from the lower part and passes through the charge layers. It interacts with the burden descending from the upper part to complete the smelting process of heating, melting, reducing and carburizing. Smooth burden descend and reasonable gas distribution are essential for stable blast furnace operation. The function of the burden distribution is to change the distribution of gas flow and affect the shape of the cohesive zone through different charging methods. The method of burden distribution with two bells is monotonous. The possibility to change the distribution of burden is limited, and it is quite inconvenient to regulate the gas flow. However, the bell-less top burden distribution method is so flexible that burden can be distributed to any position of the throat. Therefore, it is relatively convenient to regulate the distribution of the gas flow in blast furnace.

3.1 Structure of Blast Furnace Stock Column

The dissection results of blast furnace show that the cohesive zone divides the blast furnace into two parts: above the cohesive zone is the solid stock column called lumpy zone and the solid part below the cohesive zone is basically coke. The structure of blast furnace stock column is shown in Fig. 3.1.

Burden distributing is a kind of operation which changes frequently in the basic operation system of blast furnace. The change of external conditions of blast furnace, or the regulation of production policy of blast furnace, generally requires the adjustment of charging system, which is an important content that the section chief, furnace chief and operation chief should think about and operate frequently, as well as the subject that smelting engineer, factory director often studies and makes decision about.

Y. Liu, *The Operation of Contemporary Blast Furnaces*,
https://doi.org/10.1007/978-981-15-7074-2_3

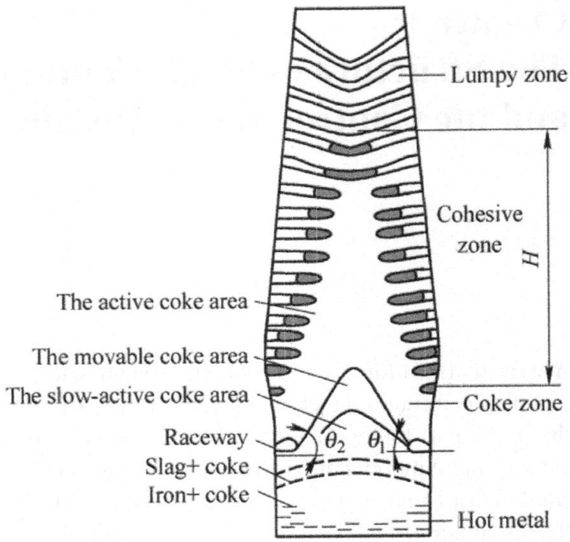

Fig. 3.1 Structure of blast furnace stock column

3.1.1 Characteristics of Burden

The distribution of burdening materials in furnace throat is basically the distribution of ore and coke. The influence of ore layer and coke layer on permeability has been studied by many metallurgists. Table 3.1 shows the characteristics of burden that M. A. Szcefanovich had ever studied before [1, 2].

The ore resistance is 10–20 times greater than that of coke when the burden in Table 3.1 is calculated according to different modes of layered burden distribution at the same gas velocity. Obviously, the ratio of ore layer thickness to coke layer thickness determines the gas flux to a large extent in the layered burden distribution. Taking the equal loss of pressure head on the section of material layer as starting point of analysis, M. A. Szcefanovich simulated conditions of blast furnace throat and analyzes the influence of uneven burden distribution to the distribution of gas under the condition that the temperature was 300 °C, the pressure was 1225 kPa, and the average gas velocity of the section of material layer was 1.75 m/s (see Fig. 3.2).

Figure 3.2 illustrates the important effect of ore layer and coke layer on gas flow distribution. Because the permeability of ore layer is poor and that of coke layer is good, the proportion of ore and coke at various points on the horizontal surface of the throat becomes an important factor affecting the distribution of the gas flow. The adjustment of burden distribution is mainly to change the proportion of ore and coke distribution everywhere in different ways of burden distribution. The ratio of ore and coke to the center and at the wall over the blast furnace throat indicates the basic characteristics of the burden distribution. The most commonly used ore load of the blast furnace actually shows the important characteristics of the permeability of the burden.

Table 3.1 Physical properties of burden

Burden	Bulk density, kg/m^3	Porosity (%)	Calculation of pore diameter, mm	Size distribution, %							
				>80 (mm)	80–60 (mm)	60–40 (mm)	40–20 (mm)	25–10 (mm)	10–0 (mm)	10–3 (mm)	3–0 (mm)
Coke	500	47.0	32	19.4	35.9	36.1	7.4	0.3	0.9		
Sinter	1695	42.6	3.9		3.8	10.2	14.9	40.2		26.9	4.0

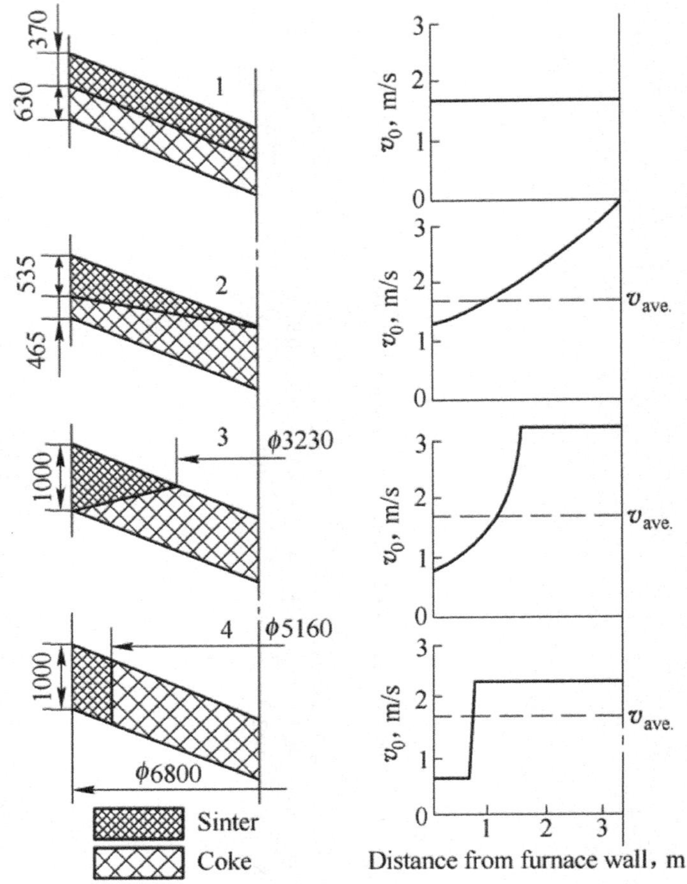

Fig. 3.2 Influence of burden distribution on gas velocity distribution and the loss of pressure head

3.1.2 The Functions of Blast Furnace Burden Distribution

The functions of blast furnace burden distribution are as follows:

(1) The burden distribution could change the output level of blast furnace and improve the smooth operation. The voyage of the stock column in the furnace is about 0.35–0.45. The resistance of the rising gas to the burden is about 40–50% of the effective weight of the stock column. The distribution of gas is uneven; thus, the resistance to the descending burden varies greatly. Utilizing different gas distribution to reduce the resistance of burden can lead to a stable and smooth operation in furnace. With the smooth operation, it is possible to improve the strength of smelting and increase production.

(2) The proper burden distribution can reduce the fuel consumption. It is an important goal of burden distribution to improve the utilization of gas through the burden distribution.
(3) The burden distribution can extend the campaign of blast furnace. If the peripheral gas stream develops excessively, it would inevitably aggravate the erosion of furnace wall. It is the premise of reasonable charging system to control the peripheral gas stream by burden distribution, which not only protects the furnace wall but also improves the gas utilization.
(4) The burden distribution can help prevent and deal with blast furnace upsets, including:
1. Suppressing the wind of blast furnace and rough running of blast furnace;
2. Dealing with buildup on blast furnace wall;
3. Slag crust peeling caused by excessive peripheral gas stream;
4. Increasing removal of harmful impurities, such as alkalis and zinc with the top gas.

The limitations of the action of blast furnace burden distribution are as follows:

(1) Serious hearth accumulation cannot be solved completely by the adjustment of burden distribution.
(2) Effect is small by adjusting of burden distribution alone if the blast furnace wall liner thickening was serious.

3.2 Distribution of Burden and Gas

3.2.1 Distribution of Burden and Cohesive Zone

Terms of Iron-making
Batch weight: The weight of a batch of materials is called batch weight, and some people call it burden batch. In this book, it is called batch weight all the time: The ore batch is called ore batch weight, and the coke batch is called coke batch weight.

Both the model experiment and the dissection study of blast furnace have proved that the distribution of the burden in the blast furnace maintained layered structure of the throat until it is melted. The angle of the coke layer and the ore layer in the throat is usually more than 30°. Data collected from dissected blast furnace show that the layered structure of ore and coke was basically unchanged but the angle was reduced to about 10° in lower part of the furnace [3]. There are two reasons about why the burden angle tends to be flat in the descent process: The first one is that the blast furnace section expands gradually from the throat to the lower part of the furnace and the transverse displacement occurs in the material layer (in the furnace plane perpendicular to the central line of the blast furnace) making the material layer thinner; the second one is that the velocity of burden at the wall is relatively faster than that of the central material due to the coke combustion in the raceway of tuyere. The layered distribution of the burden in the blast furnace is shown in Fig. 3.1.

Obviously, it is not a batch of materials, but the impact of overall stock column on blast furnaces. Thus, it can be seen that the burden distribution plays an important role in the process of blast furnace.

Compared with the solid burden, the cohesive zone has certain plasticity, low porosity, poor permeability and a large resistance to gas. Ono katsuya also established a blast furnace resistance model based on the actual observant state of the blast furnace dissection to conduct a study. Using the model, result of calculation showed: If the permeability of the ore layer is 1, the coke layer is 13; the cohesive layer is only 0.2–0.25. The ratio of the three permeability indexes is: the soft-melting layer: ore layer: coke layer = 1: 4: 52 [4].

The cohesive zone consists of two parts: solid coke and molten ore. Between alternating coke and ore layers, most of the rising gas goes through the coke layer with less resistance. Figure 3.3 shows the lower parts of the blast furnace: hearth, cohesive zone and lumpy zone.

The coke layer in the cohesive zone is the major pathway of gas, so it is also known as "gas window (coke slits)." As can be seen from Figs. 3.1 and 3.3, the higher the cohesive zone is, the larger the area of "gas window" would be and the gas resistance would be smaller.

It can be seen from Fig. 3.1 that the higher the cohesive zone is, the larger the volume of the coke layer in the center of the blast furnace will be and the more central gas flow will develop, which form the type III gas distribution in Table 3.2. If the burden distribution were flat and the area of "gas window" was zero, the cohesive

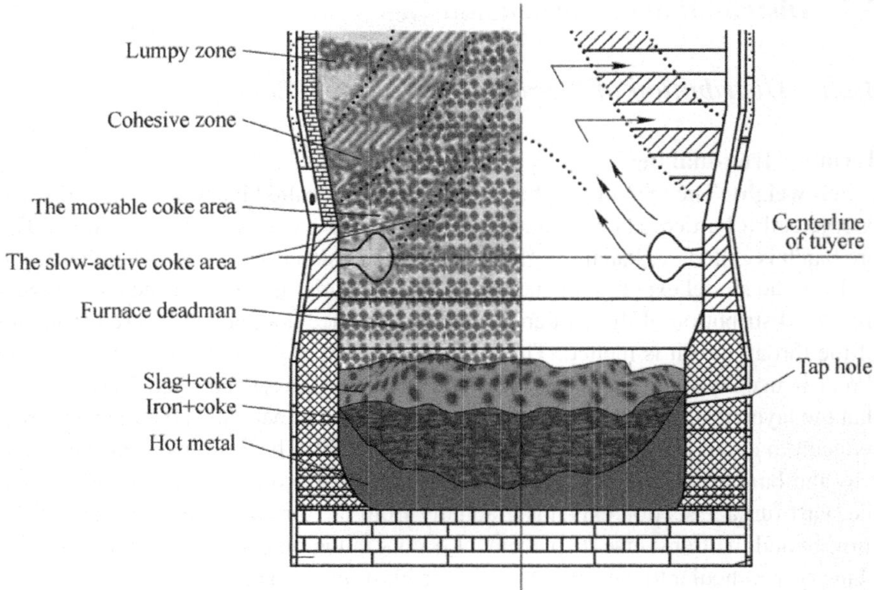

Fig. 3.3 Schematic diagram of hearth, lumpy zone and cohesive zone

Table 3.2 Comparison of operating results of different types of gas distribution

Gas distribution type	Charging system	Curve shape of gas	Gas temperature distribution	Shape of soft-melting zone	Gas resistance
I	Wall development type				Least resistance
II	Double peak type	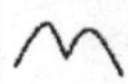	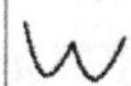	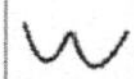	Lower resistance
III	Central part development type				Higher resistance
IV	Flat type				Largest resistance
Gas distribution type	Erosion on furnace wall	Furnace top temperature	Heat loss	Utilization degree of gas	Burden quality requirement to burden
I	Highest	Highest	Greatest	Worst	Lowest
II	Higher	Higher	Greater	Worse	Lower
III	Lowest	Lower	Less	Better	Higher
IV	Lower	Lowest	Least	Best	Highest

zone resistance would reach the highest at this moment, which leads to the difficulty to maintain the smooth operation of blast furnace. The gas distribution is flat then. If the furnace wall above tuyere is mainly made up of coke, peripheral gas is bound to develop the type I gas distribution, namely wall development type. Type II gas distribution has W-shaped cohesive zone, of which the area of gas passage is very large and the edge and center are good gas channels, forming the double-peak-shaped gas distribution. The four types of gas distribution have different functions, which have been compared in detail in Table 3.2. The influence of the cohesive zone on the operation of blast furnace will be discussed in depth below.

3.2.2 Height of Cohesive Zone

It is effective to reduce the blast furnace fuel rate and expand the blast furnace indirect reduction zone. Figure 3.4 shows the result of blast furnace practice [5] that the higher the proportion of the indirect reduction zone, the lower the fuel rate. The indirect reduction zone is generally below 960 °C. The blast furnace cohesive zone is located in an area where temperature is between 1200 and 1400 °C. The dissection of blast furnace revealed that the upper boundary of the cohesive zone is connected to the stack zone, which was about 1200 °C; and the medial side was connected with the

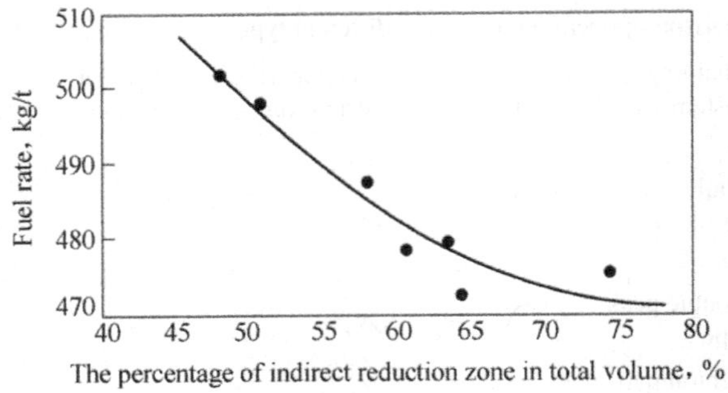

Fig. 3.4 Relationship between the indirect reduction zone and fuel rate of blast furnace

dropping zone at about 1400 °C. In order to expand the area of indirect reduction zone, the height of cohesive zone must be lowered. Figure 3.5 shows the results of the dissection of the two blast furnaces of Nippon Steel Corporation. The operating

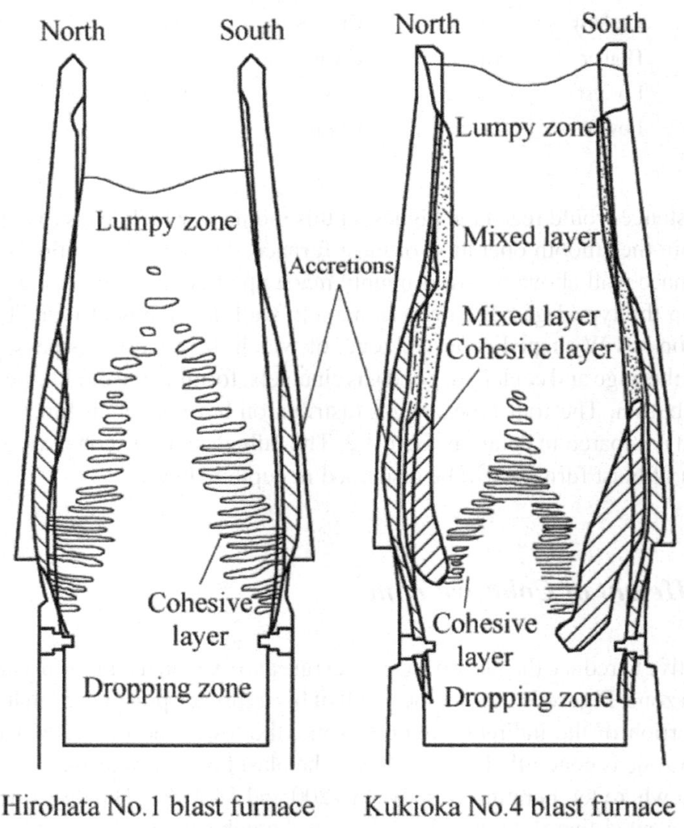

Fig. 3.5 Cohesive zone of blast furnace

data of the two blast furnaces that were collected on the day before shutdown are shown in Table 3.3. Figure 3.5 and Table 3.3 also confirm this point.

It can be seen from Table 3.3 that there is a great difference between the two blast furnaces: Hirohata No. 1 blast furnace boast high output, high top pressure, high furnace temperature, good permeability and high fuel ratio while Kukioka No. 4 blast furnace only had medium output, medium furnace temperature, atmospheric pressure and low fuel rate. The blast furnace dissection found that, as the central gas flow developed in Hirohata No. 1 blast furnace, the height of cohesive zone accounted for 75% of the height of the dissected stock column, and that of Kukioka only accounted for 28% [5–8]. Table 3.3 includes the dissection data of Capital Steel experimental furnace [9].

Another reason for the good permeability of Hirohata No. 1 blast furnace was that the coke layer of cohesive zone was thicker and the area of coke slit was large. Table 3.4 shows the thickness of each layer of the cohesive zone measured from the dissection of Hirohata No. 1 blast furnace. The last few layers were damaged by the melting loss of the ore and the tapping of salamander in blast furnace, and the 17 layers were listed in the table. The thickness of each layer of the cohesive zone in Table 3.4 was gradually thinning from top to bottom. The average thickness of the layer 3–5 of the ore-melting layer (the thickness of the coke slit layer 1–2 was missing in the original report) was 0.28 m, while the thickness of the layer 15–17 was 0.19 m. Thickness of the coke slit at the same position was 0.44 m and 0.17 m, respectively, which accounts for 61 and 47% of the thickness of the soft-melting layers.

According to the dissection results of Capital Steel experimental furnace, the average thickness of the upper three layers of softening-melting layer was 0.16 m, and the average thickness of the layer 5–7 was 0.08 m. The average thickness of the upper layers of coke slit was 0.12 m, and the average thickness of the layer 5–7 of coke slit was 0.1 m. The average thickness of the upper layers and the lower layers of coke slit accounted for 42% and 55% of the thickness of the soft-melting layer, respectively. The volume of the experimental furnace was 23 m^3, and the dissectional measured stock column height was only 4.28 m [10, 11].

The above data were measured after the cooling of the dissected blast furnace. The shrinkage was much higher than that in the production state.

Table 3.5 shows some data about the cohesive zone of the three blast furnaces. The cooling heights after cooling in the table refers to the height of the burden surface with the upper edge of the throat as the datum after the blast furnace was water-cooled. The cohesive zone structure of the three blast furnaces was clear, and the main body presented as an "inverted V." The height of stock column and cohesive zone of Hirohata No. 1 blast furnace was calculated based on the graphic in the literature, although there may be errors by referring to some relevant reports. The height of the stock column given in the table refers to the height measured before dissection after the furnace shutdown, not the height of the stock column in the production state.

The resistance of the blast furnace stock column was mainly in the high-temperature zone of the lower part of the blast furnace, as shown in Fig. 3.6 [12]. The

Table 3.3 Operating data collected on the day before furnace shutdown

Plant	Date	Yield		Operation coefficient						Load
		Daily output (t)	Utilization coefficient, t/(m^3 d)	Blast volume, (m^3/min)	Blast temperature, (°C)	Blast pressure/Blast volume	Moisture, (g/m^3)	Oxygen enrichment (%)	Top pressure, (kPa)	
Hirohata	22	3289	2.34	2300	941	0.9	32	0.96	900	2.97
Kukioka	24	2268	1.77	2039	980	0.61	16.5	1.99	58	3.94
Plant	Fuel, kg/t			Hot metal, %			Slag		Gas, %	
	Coke rate	Heavy oil	Fuel ratio	Si	S	Mn	Slag volume, kg/t	CaO/SiO$_2$	CO$_2$	CO
Hirohata	471	31	502	0.80	0.038	0.62	265	1.12	19.2	23.4
Kukioka	387	78	465	0.52	0.039	0.60	267	1.22	19.0	22.7

Table 3.4 Thickness of each layer in cohesive zone of Hirohata No. 1 blast furnace and Capital Steel experimental furnace

Blast furnace	Hirohata No. 1 blast furnace, m			Capital Steel experimental furnace, mm		
Softening-melting layer	Average thickness of layer 3–5	Average thickness of layer 15–17	Average thickness of layer 3–17	Average thickness of layer 1–3	Average thickness of layer 5–7	Average thickness of layer 1–7
Thickness of ore-melting layer	0.28	0.19	0.26	161.67	78.25	114
Thickness of coke slit	0.44	0.17	0.29	115.00	95	104
Total thickness	0.73	0.37	0.55	276.67	173.25	218
Thickness of coke slit/Total thickness	0.61	0.47	0.51	0.42	0.55	0.49

rising gas was distributed initially through the coke slits of the cohesive zone. Thus, the gas distribution in the furnace was closely related to the shape of the cohesive zone and the area of the coke slits. The area of the lower coke slit was determined by the batch weight of coke and coke distribution. Stable blast furnace process must maintain a stable gas distribution. Therefore, the batch weight of coke should be as stable as possible. In the daily adjustment of the blast furnace, to avoid changing batch weight of coke, burden weight changes should be achieved by changing batch weight of ore.

The height of the cohesive zone was largely influenced by the central gas flow. Because the cohesive zone of Hirohata No. 1 blast furnace was too high, its central gas flow develops abnormally. The blast furnace has excellent permeability and was easy to accept high blast volume; but the main disadvantage of it was that the fuel ratio was too high. The control of center gas flow can effectively control the height of cohesive zone; however, if the cohesive zone were too low, it would be inevitable to lead to the development of peripheral gas stream, which not only eroding the furnace wall, but also greatly increasing the fuel rate.

3.3 Observation and Judgment of Gas Flow Distribution

3.3.1 Velocity and Temperature of Radial Gas

At present, there is no simple and feasible way to directly observe gas flow distribution in blast furnace. As early as 1929, a group led by Kenny and Furnas in the USA carried out actual measurement on the blast furnace of 700t nominal weight. Results are shown in Fig. 3.7 [13]. It can be seen from figure that the radial distribution trend of gas velocity and gas temperature is consistent in the furnace close to the burden surface, and the gas velocity can be described by temperature. In fact, gas temperature

Table 3.5 Height and layer number of cohesive zone in the three blast furnaces

Plant	Thickness of coke layer in throat, m	Thickness of coke layer in belly, m	Height of stock column, m	Height of cohesive zone, m	Number of cohesive zone layers	Average band height of each layer, m	Average band height of the three layers, m		Height of cohesive zone/Height of stock column
							Upper part	Lower part	
Capital Steel experimental furnace	0.31	0.13	4.28	1.7	10	0.17	0.27	0.17	0.4
Hirohata No. 1 blast furnace	1.07	0.51	20	15	20	0.75	0.73	0.37	0.75
Dokaoka No. 4 blast furnace	0.47	0.24	21.5	6	21	0.29	0.47	0.24	0.29

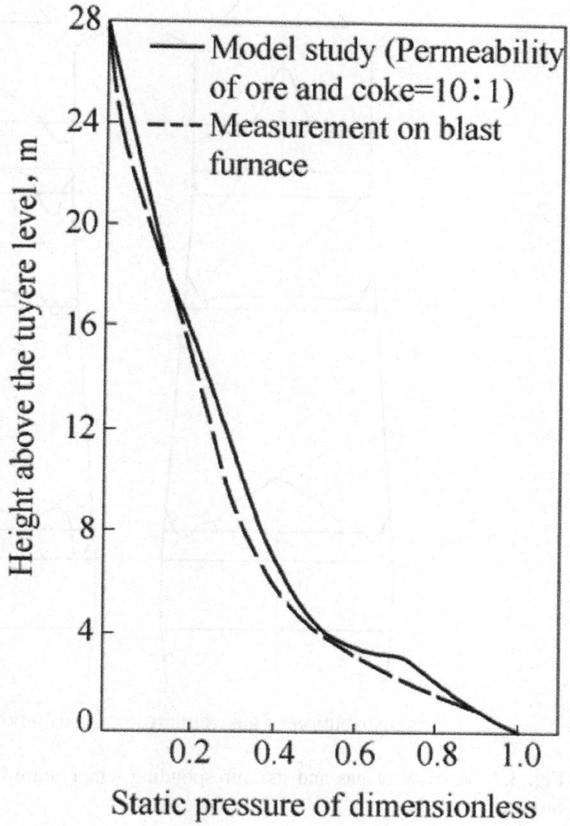

Fig. 3.6 Static pressure in the furnace distributed along the height of the blast furnace

can also describe gas composition, as shown in Fig. 3.8 [14]. So far, to observe the distribution of gas flow mainly relies on the radial temperature mainly. Generally, the blast furnace is installed with the cross-beam temperature measurement (above burden temperature probes) of the throat, which is the main tool for observing the distribution of gas.

In the literature [15], Wang Xiaopeng and his team also used cross-beam temperature measurement to determine the gas distribution and adjusted the charging accordingly. Figure 3.9 shows the results of the change in radial gas temperature at the beginning of their experiment and after a period of experimentation. The comparison of data before and after the experiment shows that the temperature of peripheral gas changed greatly from 260 to 290 °C to 160–180 °C. A result of the ore and coke outward migration is that the center temperature did not decrease, but increased slightly. This is because coke has the effect of improving the permeability of the blast furnace, but its stacking position also affects the rolling of the ore on the burden surface. When the tip of coke pile restricts the ore from rolling to the center of the blast furnace, although the coke impact point is moved outward, the blast furnace center may still be relatively well developed.

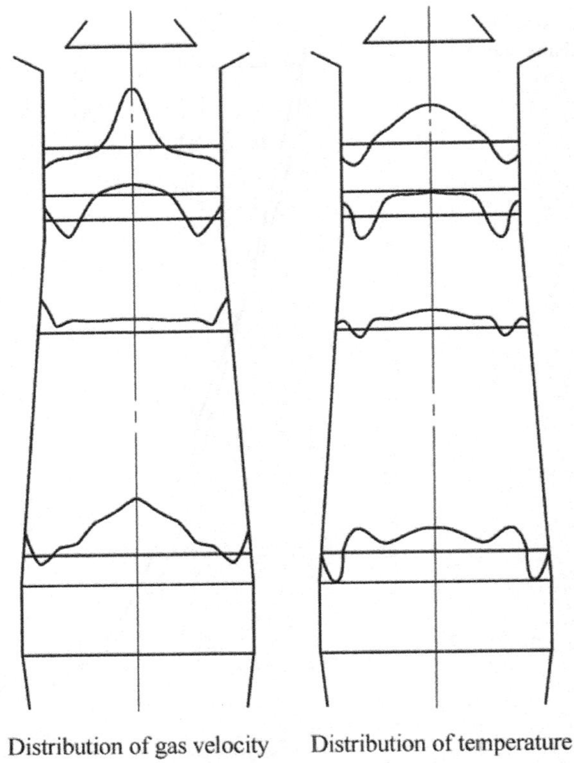

Fig. 3.7 Velocity of gas and its corresponding temperature in actual measurement on the blast furnace

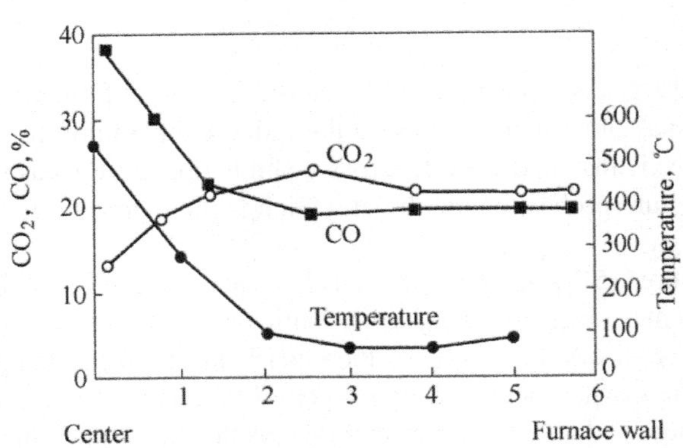

Fig. 3.8 Distribution of gas temperature and composition

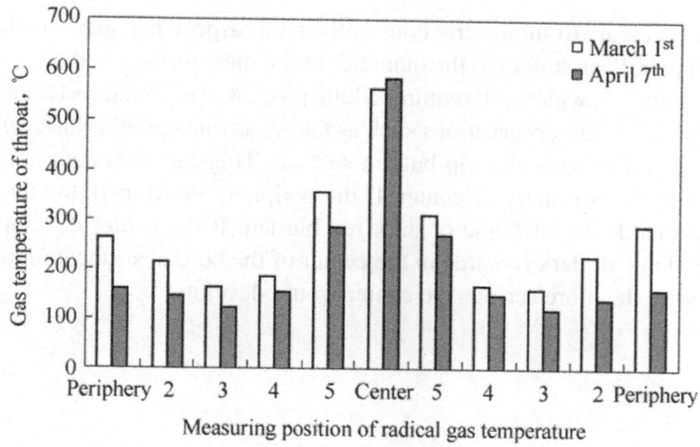

Fig. 3.9 Changes in radical gas temperature (temperature at each point of cross-beam temperature measurement—(above burden temperature probes)) before and after experiment

3.3.2 Direct Observation

3.3.2.1 Furnace Top Imaging

Furnace top imaging is quite common and it can dynamically reflect the gas distribution of blast furnace. The position of the gas flow leaving the burden surface, the size of the stream, and the upward velocity can be seen directly during the production process. Even the direction and position of the channel and the splashing of the burden also can be seen. Figure 3.10 shows the furnace top imaging photographs of the peripheral development, center development and local channeling development of the blast furnace shown by Mei Steel [16].

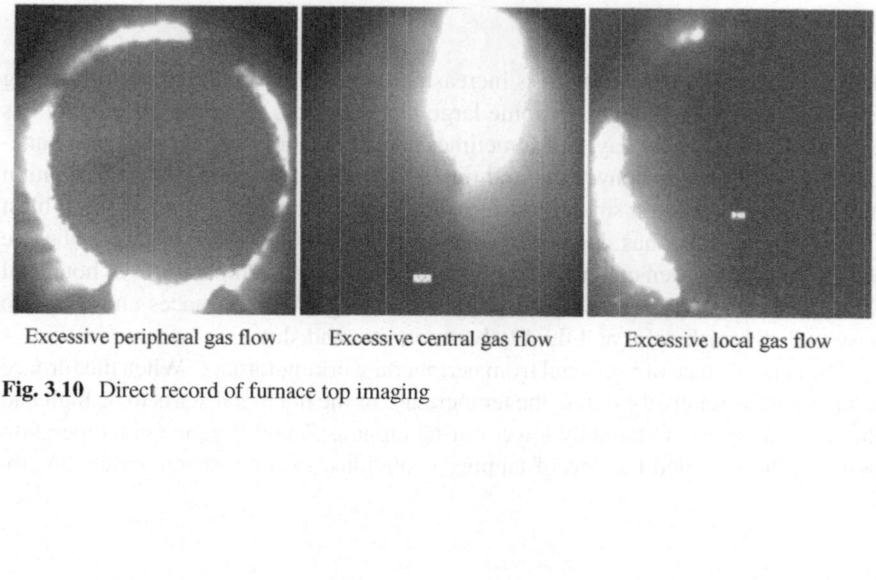

Fig. 3.10 Direct record of furnace top imaging

It is still necessary to summarize constantly in this aspect, in particular, the proportion of the central gas stream to the diameter of the blast furnace and the appropriate smelting conditions, which still requires a long process of research and accumulation.

Other similar direct observations such as taking advantage of a shutdown to open the manhole and observe the top burden surface. This can help determine the gas distribution at the periphery or center. If the periphery developed, the throat might have a peripheral ring of flame or dark red burden. If the center developed, there would be a flame or dark red area in the center of the burden surface; moreover, the wider the area, the more serious the center would develop.

3.3.2.2 Tuyere

Nowadays, many tuyeres have installed automatic facilities displaying the working conditions of the tuyere. It is easier and more frequent to observe tuyere.

When the periphery develops, most front of tuyeres are active and bright, but the circumference is uneven. There is black coke flying in front of some tuyeres, and scab, cold and unreduced burden are observed. When the furnace temperature is low, sometimes slag burned the tuyere and the slag easily enters into the tuyere. When the peripheral gas development is serious, the tuyere is extensively damaged. Due to the development of the periphery, the permeability is poor, and the bad part is mostly at the upper part or the end surface of the tuyere.

When the center develops, the tuyere works unevenly. Although the furnace temperature is normal, the tuyere zone is cool. Although the slag irrigates some individual tuyere when the furnace temperature is low, there is no phenomenon about slag entering into the tuyere. In severe cases, there are also damages in tuyere. But different to periphery development, damages in center development are mostly at the bottom or at the bottom of inner surface.

3.3.2.3 Hot Metal

The volume of the blast furnace is increasing day by day, and the time for tapping is getting longer and longer. In some large blast furnaces, the time for tapping has exceeded 1440 min per day, and sometimes the two tap holes are tapping simultaneously. Therefore, it is convenient and timely to observe hot metal. Gas distribution has a great influence on small blast furnace: Firstly, the tapping time of small blast furnace is short and it has slag notch to discharge upper slag; thus, it is easy to observe the difference between upper and lower slag. Secondly, the temperatures of hot metal from small blast furnace are lower than that of larger blast furnaces and it is also easy to distinguish color and fluidity between iron and slag.

The performance of hot metal from peripheral working furnace: When the furnace temperature is relatively stable, the temperature of the hot metal starts to be high and the temperature gets gradually lower during tapping. The difference in temperature between the start and the end of tapping is obvious. In some severe cases, the hot metal is thick and sticky, and the hot metal ladle would crust. As the physical heat of hot metal is insufficient, in severe situation, FeO > 0.5% in the slag, the upper slag carries more iron and the slag notch is prone to be damaged. If there are two slag notches, the temperature of the upper slag on the two slag notches is inconsistent; compared with the lower slag, the upper slag is hotter and the lower slag is cooler. The upper slag notch is easy to open, easy to discharge slag and the slag rate is high.

The performance of hot metal from central working furnace: The physical heat of hot metal is insufficient. In severe cases, content of silicon and sulfur is high and the desulfurization efficiency is very low, which is about 10% lower than normal furnace conditions. The silicon fluctuated greatly between the two tapping. The upper slag rate is low. When sever, the slag notch is difficult to open, the slag sprays after opening the tap hole and the slag is hard to discharge. As the slag carries iron, the slag notch is prone to be damaged.

The above phenomenon has little influence for large blast furnaces.

3.3.3 Synthetically Judging for Instrument

Observing the instrument displays, such as continuous data and visual image, we draw conclusions through the judgment for a combination of these relevant parameters. For example, to judge the state of periphery, the data collected would be compared with the data during normal furnace condition. These data are: the top temperature, above burden temperature, throat temperature, as well as temperature sat shaft (stake), belly and abdominal layers, and the temperature measurement points at the back of the stave coolers of each layer, these data would be combined with blast volume, blast pressure, permeability index, stock rod and other related data, thus to judge the peripheral gas stream. The specific furnace conditions will be discussed in the relevant sections (Chapters) later.

The same method above can be used to judge the distribution of central gas flow. The edge temperature should be much lower than that of a periphery working furnace. The reflection of various parameters is obviously different. Table 3.6 is the gas distribution judgment.

3.3.4 Judging Gas Distribution from Abnormal Furnace Conditions

In normal furnace conditions, it is easier to judge the distribution of gas; in the case of abnormal furnace conditions, especially in the early stage, the performance of various phenomena and parameters is not very obvious, and it is difficult to determine the trend of the furnace condition. The initial judgment is very important: When the furnace condition has not been seriously dysfunctional, it is easier to change the gas

Table 3.6 Gas distribution judgment

Blast kinetic energy		Much smaller	Normal level	Much larger
Working condition of hearth		Periphery development, accumulation at the center	Active hearth condition	Center working, accumulation at the peripheral area
Direct observation	Off blast and observe the burden surface	Throat may have a ring of flaming or dark red burden	Generally not obvious, sometimes the central part of a light reddish phenomenon	A flaming or dark red area in the center of the burden surface; the wider the area is, the more serious the center working
	Furnace top imaging	Peripheral gas flow is stronger	Peripheral gas is not obvious; central gas is obvious	Stream of central gas is strong and wide
Record displayed on instrument	Hot blast pressure	Low pressure; channeling is prone to occur; rigid curve of blast pressure with constantly sudden rises; spikes exists	Relatively stable wind pressure; slight smooth fluctuation because of the influence of charging and furnace temperature; no zigzag fluctuation except for the change of hot stoves	High pressure before tapping and slag discharging, sometimes 5–10 kPa pressure difference; the pressure lowered after tapping and the fluctuation period is consistent with tapping time; tapping of large blast furnace is continuous; the effect of tapping is not obvious
	Blast volume	It is easy to accept blast volume at the beginning of the periphery development. If the blast volume is small, the long-term development will cause accumulation in central part. The blast volume will not be accepted after the accumulation	The blast volume is stable, and the curve of blast volume fluctuates minutely without spikes. The artificial blast volume changes over a wide range and has no significant effect on the furnace condition	The condition of the furnace is not good, and it will be better after the blast is reduced. It is easy to slip after charging

(continued)

Table 3.6 (continued)

Blast kinetic energy		Much smaller	Normal level	Much larger
	Stock rod	Uneven burden descending; there is often one or two stock rod indicating burden slips. In severe cases, it ends up with a very low stock line	Uniform, stable, gentle changes, no zigzag curve	Burden descending is uneven: It is slow before tapping and significantly faster when tapping. This sign is consistent with the local accumulation of the hearth. We should pay attention to combine with other signs
	Top temperature	The bandwidth fluctuates greatly, sometimes has bifurcation, and the top temperature is high. These phenomena are not obvious when the blast volume is small	Normal, small fluctuations, no bifurcation; temperature difference <50 °C	The strip is narrow, and the temperature difference at each point is small, but the temperature fluctuation is large, which is significantly different from the normal furnace condition
	Throat temperature	Above normal level	Normal; normal distribution	Below normal level
	Throat radial temperature	Higher periphery temperature	Normal distribution	The periphery temperature is low and the center temperature is high

(continued)

Table 3.6 (continued)

Blast kinetic energy		Much smaller	Normal level	Much larger
Gas	Gas Distribution	The periphery is light in burden and channeling is easy to appear; it typically belongs to type I gas distribution. The burden materials after slips are clogged, resulting in difficult and rough burden movement. It is easy to hang when the blockage is serious	The burden at periphery is heavier and the center is light. It is a type III gas distribution	Unstable, generally with heavy burden at periphery and light burden in center
	Gas utilization	Very poor, low CO_2 value of the Gas	Better, higher CO_2 value of the gas; the gas utilization rate is higher and stable	Change a lot (high variation); unstable
Slag	Slag temperature	The two slag notches are inconsistent; the upper slag is hot and the lower slag is cooler	The temperature of the upper and lower slags is close and the temperature is abundant	The lower slag is hotter than the upper slag
	Upper slag rate	High slag rate; easy to discharge slag	The ratio of upper and lower slag is easy to control, and the slag operation is carried out as needed	The upper slag rate is low. In severe cases, the slag notch is difficult to open, and the slag sprays after opening the slag notch, as a result, the slag is hard to discharge
	Composition of slag	With iron; in severe cases, damage slag notch; value of FeO in slag is high, often > 0.5%	Generally without iron, FeO < 0.5%	A large amount of iron is carried in slag, and in severe cases, the slag notch is damaged heavily

(continued)

Table 3.6 (continued)

Blast kinetic energy		Much smaller	Normal level	Much larger
Hot metal	Hot metal temperature	The temperature of the hot metal starts to be high and the temperature gets lower and lower during taping, showing an obvious difference in temperature between the start and the end of tapping	The hot metal temperature is sufficient and the temperature difference is small between the front and the back of tapping	The physical heat of hot metal is insufficient; Si and S are high in severe cases
	[Si] in hot metal	The hot metal is thick and sticky, and the hot metal ladle would crust	Stable Silicon; High Desulfurization Rate	During the two times of tapping, the silicon fluctuates greatly and the desulfurization rate is very low, generally 10% lower than the normal level
Tuyere	Condition in front of tuyeres	Most tuyeres are active and bright, but the circumference is uneven. Cold and unreduced burden appear in front of some tuyeres	Bright; active; uniform	Uneven appearance; relatively cool
	Condition about slag entering into the tuyere	Sometimes slag irrigates the tuyere and the slag easily enters into the tuyere	The furnace temperature is normal; there is no phenomenon about slag entering into the tuyere	Although the slag irrigates some individual tuyere, there is no phenomenon about slag entering into the tuyere
	Damaged part of tuyere	Periphery working furnaces; Poor permeability; Serious damage in the upper tuyere or end face	Rarely damage	Much at the bottom or bottom inner surface of tuyeres

distribution; the blast furnace has not been seriously dysfunctional and has not caused any losses. This is a critical period. Many blast furnaces are seriously dysfunctional, and they are unable to grasp the initial trend and cause major losses. In particular, field operators such as the foreman, the chief of the furnace, and the chief of the work should continuously accumulate data and experience through practice, increase and improve their operational skills.

3.4 Gas Distribution Adapted to Blast Furnace Smelting Conditions

3.4.1 Selecting the Gas Distribution Suitable for the Smelting Conditions

The type of gas distribution depends mainly on the state of blast furnace and burden conditions.

In the four types of gas distribution in Table 3.2, the type I—periphery working furnace—is not desirable. Type IV, the blast furnace is difficult to run smoothly, which does not work either. Type III is the best one between type II and type III gas distribution. When the burden condition is good and the furnace condition is smooth, the gas distribution should be as close as possible to the type IV to obtain the lowest fuel consumption and the highest gas utilization rate; when the burden quality is not good or the furnace condition is not smooth, it should move close to type II (W-shape type) to create a large passage for gas and keep the blast smooth operation. Regarding the advantages of type IV gas distribution, it is imagined that gas utilization might be good, but the passage of blast furnace would be blocked, and the furnace condition should be difficult to go smoothly, which is practically impossible to adopt.

3.4.2 To Determine the Corresponding Burden Distribution

The burden distribution is determined by the selection of the type of gas distribution. The adjustment of batch weight, radial ore-to-coke ratio, and blast system could be used to achieve the required gas distribution.

The radial temperature distribution of furnace with bell-less top is regular. When analyzing the operation of burden distribution of bell-less top, Yu Kun cited the temperature distribution curve of top gas of four stable blast furnaces (Fig. 3.11) [17] and pointed out the characteristics of burden distribution in practice:

(1) The center temperature was high to 500–650°C, and the center gas was unblocked, which would be conductive to smooth blast furnace operation.

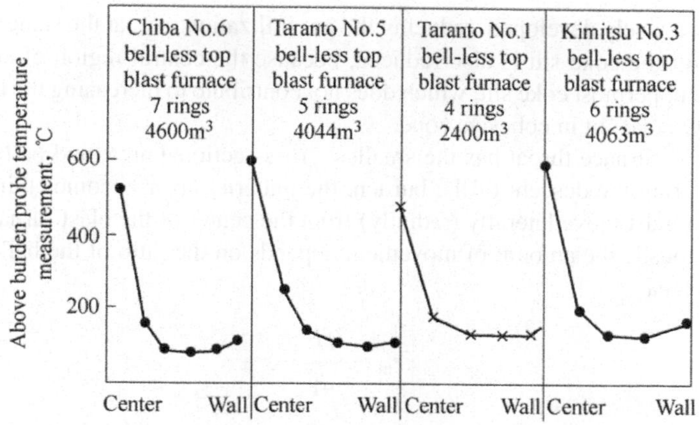

Fig. 3.11 Temperature distribution of top gas

(2) Points 2, 3 and 4 were located at the intermediate area, the temperature of them was close to each other and lower, which would be beneficial to gas utilization.
(3) The periphery temperature was slightly higher than the second point, which would be good for the smooth operation of blast furnace.

These three characteristics are the products of the stable of bell-less top burden distribution. In essence, they are also the characteristic of the type III gas distribution and good standard for production practice. Type III gas distribution, which has both a large gas passage and a large area of cohesive zone coke slits is a gas distribution form of smooth operation and low fuel consumption. Summarizing the characteristics of the bell-less top burden distribution, multi-ring, terrace and top gas temperature distribution is shortcut to achieve good production results.

3.4.3 Stock Line and Batch Weight

The bell-less burden distribution can be smoothly distributed to any position of the throat burden surface by changing the chute angle without changing the stock line.

In the cohesive zone, almost all of the blast furnace gas pass through the coke slits, and the coke batch weight largely determine the thickness of the coke slit. The permeability of the cohesive zone is determined by the height of the cohesive zone and the thickness of the coke slit. To control the height of the high-temperature zone of the blast furnace, the height of the cohesive zone is limited, and the thickness of the coke slit is very important for improving the gas permeability of the stock column. Keeping the thickness of the coke slit of each layer of the cohesive zone, distributing the coke evenly along the radial direction of the throat and maintaining a stable coke batch weight are also an important means to stabilize the coke slit area. Putting a large amount of coke in the center, on the one hand, makes the central gas stream excessively developed, reducing the gas utilization rate; at the same time, the thickness of the coke slit is also reduced, because the central region of coke only affects the uppermost coke slit, which does not contribute to increasing the thickness of majority coke slit in cohesive zone.

The blast furnace throat has the smallest cross-sectional area, and belly has the largest. During the descent of the burden, the material layer becomes thinner, and the burden must move laterally (radially) from the center of the blast furnace to the wall. Obviously, the amount of movement depends on the ratio of the belly area to the throat area:

$$r_{\mathrm{m}} = \frac{D^2}{d_1^2} \tag{3.1}$$

where r_{m} is to the ratio of the belly area to the throat area; d_1 refers to the diameter of throat; D is to the diameter of belly.

Figure 3.12 shows the ratio of the belly and throat area calculated from the actual size of 150 blast furnaces with Eq. (3.1).

It can be seen from Fig. 3.12 that the area of furnace belly is 1.4–2.9 times larger than the area of furnace throat. Therefore, it can be inferred that the thickness of material layer decreases from throat to belly by at least 1.4–2.9 times. When the burden enters the high-temperature zone, its volume changes greatly because of the smelting process, especially the ore, shrinks a lot due to the softening and reduction reactions. The dissectional study of the experimental furnace by Du Hegui, et al. concluded that the average shrinkage rate of the molten layer in the cohesive zone is about 35%.

The blast furnace dissection proved that the thickness of the coke slit in cohesive zone was lower than that of the coke layer at the belly of the furnace, and the thickness of the lower coke slit of the cohesive zone was lower than the thickness of the upper layer. However, the coke slit is the pathway of gas and we have no way to measure the thickness of the lower coke slit of the cohesive zone. According to the existing data of the smooth operation blast furnaces, the lower limit of the coke layer thickness at the belly of the furnace is about 0.2 m, as shown in Fig. 3.13.

In the past, the important of ore batch weight was emphasized. Actually, the coke batch weight has an important influence on the thickness of the coke slit in cohesive zone. The thickness of the coke layer at the belly of the furnace is closer to the thickness of the coke slit in cohesive zone. The ratio of the throat and the belly area

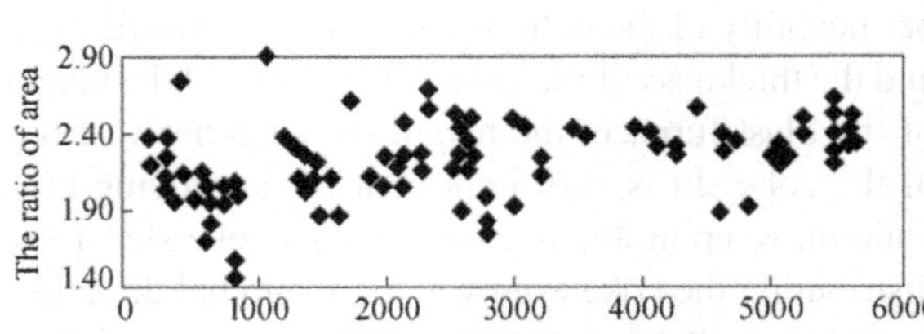

Fig. 3.12 Ratio of the belly and throat area

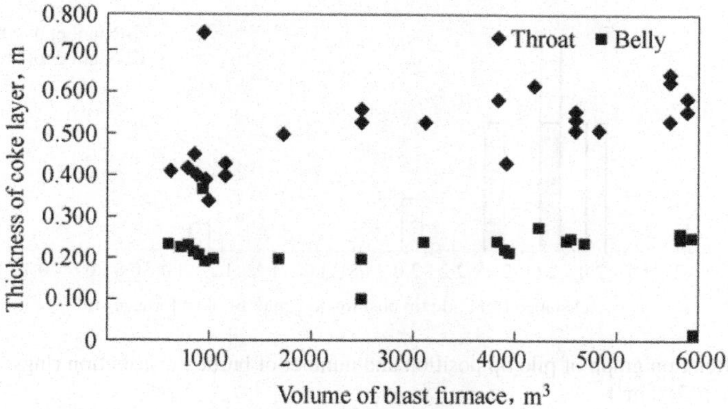

Fig. 3.13 Thickness of coke layer at belly of blast furnace

of different blast furnaces varied greatly from each other (Fig. 3.12). It is easy to make mistakes to measure the thickness of the coke slit of the cohesive zone by the throat area.

From this, the following conclusions can be drawn:

(1) The height and shape of the cohesive zone affect the distribution of the blast furnace gas, thereby affecting the smooth operation, fuel rate and blast furnace campaign.
(2) Controlling central gas flow is an effective way to regulate the height of the cohesive zone. The radial gas temperature in the center of the blast furnace throat should be controlled at 450–600 °C.
(3) In order to keep the thickness of the coke slit at cohesive zone (refer to the vertical direction of the blast furnace) stable, the coke batch weight should be kept as stable as possible, and the daily burden weight adjustment should be done mainly by changing the ore batch weight.
(4) The lower limit of coke batch weight should provide enough thickness for the coke slit in the lower part of cohesive zone. According to the existing practical experience, the lower limit of the coke layer thickness at the belly of the furnace is about 0.2 m.

3.4.4 Improving the Smooth Operation by Charging System

The blast furnace might be sometimes upset resulting in furnace not operating smoothly. If the reasons are unknown or the burden deteriorates, the charging system should be adjusted in time to keep the furnace operating smoothly. A commonly used method is to implement type II gas distribution, in which the burden should be placed in the middle of the throat then the burden dropping point should be placed in the middle zone with a certain distance from the center of the blast furnace and the furnace wall. Figures 3.14 and 3.15 show examples of burden distribution for this

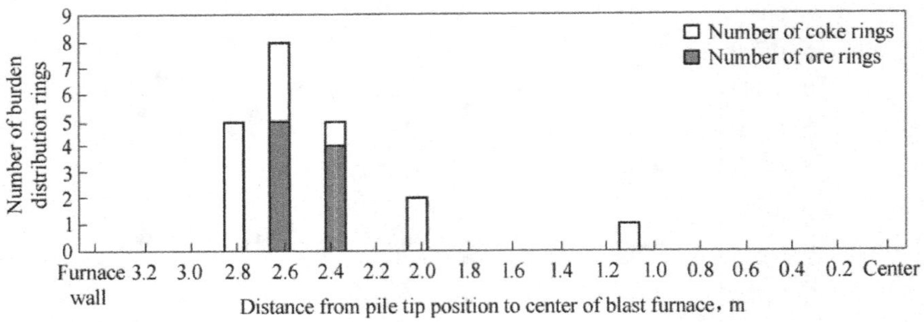

Fig. 3.14 Relation graph of pile tip position and number of burden distribution rings. (volume of blast furnace: 1780 m^3)

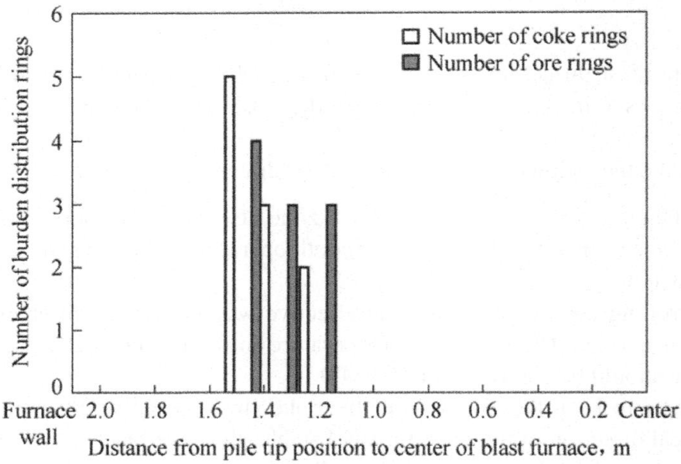

Fig. 3.15 Relation graph of pile tip position and number of burden distribution rings. (volume of blast furnace: 480 m^3)

type of operation.

Referring to the 1780 m^3 blast furnace in Fig. 3.14, in order to cope with fluctuations in the strength of the burden, coke and ore were distributed in the middle of throat radius, and ore was distributed in the middle of coke, forming two strong gas flows at the edge and center of blast furnace, which were essentially type II (W) gas distribution. Due to the development of the peripheral gas stream, the lining was gunned every one to two years. Gas was poorly utilized. Generally, the utilization rate of gas was about 46%, and its fuel consumption was high. But it had strong adaptability to the deterioration of furnace burden.

In the 480 m^3 blast furnace shown in Fig. 3.15, the lining was gunned every two years. It should be said that the gunning repair life was generally only lasted for one year, and it was difficult to maintain for two years. The utilization coefficient of this blast furnace was about 4.2 t/(m^3 d), with a high output and good performance.

Because of the poor utilization of gas, the consumption of fuel was high, and the fuel rate was about 540 kg/t.

These blast furnaces represent a type of blast furnace operation that sacrifices blast furnace campaign and consumption to achieve blast furnace smooth operation and high yield. There are some reasons for this type of operation. When the external production conditions deteriorate, especially when the coke strength deteriorates, the blast furnace could maintain operating smoothly and pass through the difficulties safely, so as to avoid major losses. After all, the smooth operation of the blast furnace should be the first consideration in the process.

3.4.5 To Improve Gas Utilization Rate—Changing from Type II Gas Distribution to Type III

Capital Steel No. 2 blast furnace had a volume of 1780 m^3 and a throat diameter of 6.8 m. It was blown in on May 23, 2002. After the start-up, the condition of the furnace was stable and the production performance was quite brilliant. After that, under the condition that the slag volume per ton was about 330 kg and the blast temperature could reach more than 1200 °C, the utilization ratio of blast furnace gas was only about 46%, and the top temperature of blast furnace was often above 200 °CC. In order to cope with the fluctuation and deterioration of burden and keep the blast furnace operating smoothly for a long time, the gas distribution was controlled to type II, and the expected effect was achieved. The burden distribution and gas distribution before and after the experiment are shown in Table 3.7; Fig. 3.16. It shows that both coke and ore are distributed in the middle of the throat radius, and the ore was in the middle of the coke, forming two powerful gas streams at the periphery and center of the blast furnace.

3.4.5.1 Process of Burden Distribution Changing

Start changing the charging on March 13th:

(1) One ton of coke was extracted from each batch of burden, and every 5 batches these cokes were put into the furnace from No. 62 batch at the position of 43° rings.

Some blast furnace throats had no partition, and the angle of burden distribution had no fixed corresponding ring position. Adjusting charging was realized by changing angle and ring number, so the angle changed a lot in the experiment of burden distribution, which increased the difficulty of the test. Because the angle had no corresponding relationship with the throat area, the comparability between different furnaces was poor, and the distribution test mainly depended on self-exploration. Experience shows that reasonable burden distribution should be divided into 5–11

Table 3.7 Changes of burden distribution and gas distribution before and after the experiment

Date	Batch weight W_k, t	Burden distribution		Gas distribution (above burden temperature), °C						Top temperature, °C	Utilization ratio of gas, %	Load (K/J)
		α_K	α_J	Periphery	2	3	4	5	Center			
March 13th	40	35 38 35	40 38 35 32 26	194	163	136	151	289	505	206	45.5	3.7
		3 4 2	3 3 1 2 1									
April 24th	45	8 7 6 5 4 5	8 7 6 5 4 3 4	132	143	127	176	301	590	175	50.3	5.5
		3 2 2 1 1 1	3 3 2 1 1 1 1									

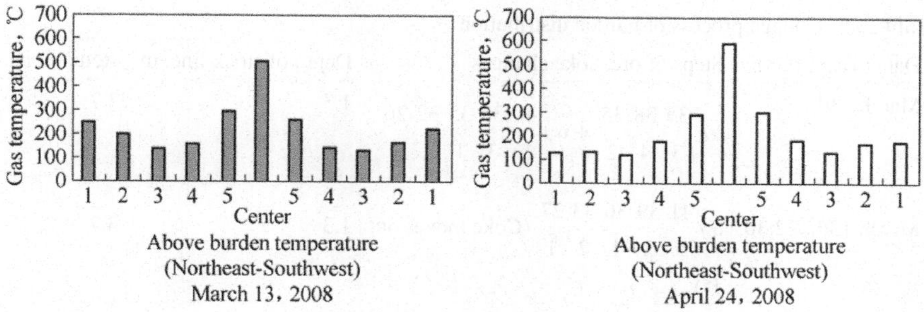

Fig. 3.16 Radial gas temperature distribution at throat

equal areas according to the throat diameter of blast furnace, and the corresponding angles of each zone should be fixed so that each practice could be carried out at a limited fixed angle, so as to reduce the number of tests and improved the quality of practice.

The ring position of burden distribution in Capital Steel No. 2 blast furnace throat was divided into 9 zones according to the principle of equal area (Table 3.8).

(2) From the No. 146 batch of burden, the ring position of ore and coke moved out by 1 degree, which began to change the basic process of burden distribution. At 9:30 on March 14th, the coke extraction was canceled after 24 h in batch No. 68, and simultaneously, the coke ring position was generally moved out by 2°. At 4:00 on March 15th, the coke continued to move out, and it entered the fixed ring position of burden distribution at 11:00. The specific change process is shown in Table 3.9.

(3) After 4 days of adjustment, the smooth operation was completely normal, the gas distribution was obviously improved, the top temperature was lowered, the gas utilization rate was increased, and the temperature of each point on the edge of the above burden temperature measurement decreased significantly (Fig. 3.16). Increase in ore to coke ratio began at 7:00 on March 17th and continued on March 16th to 19th. Each time, the ore batch weight increased by 1t. Total ore weight had increased from 40 to 43t, and the O/C ratio increased from 3.7 to 4.02. On March 19th, the ore continued to move out and the coke remained unchanged. On March 20th, the O/C ratio continued to increase, and the ore batch weight increased to 45 t on April 24th. The O/C ratio increased from 4.02 to 5.5.

Table 3.8 Comparison table of ring position/angle of Capital Steel No. 2 blast furnace

Ring position of burden distribution	9	8	7	6	5	4	3	2	1
Angle of distributing chute	47.5 °	45.5 °	43 °	40.5 °	37.5 °	34.5 °	31.5 °	27.5 °	22 °

Table 3.9 Change process of burden distribution

Data (Year 2008)	Steps of ore, coke changes	Depth of stock line, m	Actual load
March 13th	$\alpha_K\begin{matrix}35 & 38 & 35\\ 3 & 4 & 2\end{matrix}+\alpha_J\begin{matrix}40 & 38 & 35 & 32 & 26\\ 5 & 3 & 1 & 2 & 1\end{matrix}$	1.3	3.7
March 13th, 22:30	$\alpha_J\begin{matrix}41 & 39 & 36 & 33 & 27\\ 5 & 3 & 1 & 2 & 1\end{matrix}$ (Coke moves out 1°)	1.3	3.7
March 14th, 9:30	$\alpha_J\begin{matrix}43 & 41 & 39 & 36 & 33 & 27\\ 3 & 3 & 3 & 1 & 2 & 1\end{matrix}$ (Coke moves out 2°)	1.3	3.7
March 15th, 4:00	$\alpha_K\begin{matrix}37 & 40 & 37\\ 3 & 4 & 2\end{matrix}+$ $\alpha_J\begin{matrix}44 & 42 & 40 & 37 & 34 & 28\\ 3 & 3 & 3 & 1 & 2 & 1\end{matrix}$ (Ore and coke moves out)	1.3	3.7
Entering into fixed ring position			
March 15th, 23:00	$\alpha_K\begin{matrix}6 & 5\\ 4 & 5\end{matrix}+\alpha_J\begin{matrix}8 & 7 & 6 & 5 & 4 & 3\\ 3 & 3 & 3 & 1 & 2 & 1\end{matrix}$	1.3	3.8
March 17th	$\alpha_K\begin{matrix}7 & 6 & 5\\ 1 & 4 & 4\end{matrix}+\alpha_J\begin{matrix}8 & 7 & 6 & 5 & 4 & 3\\ 3 & 3 & 3 & 1 & 2 & 1\end{matrix}$	1.3	3.9
March 19th	$\alpha_K\begin{matrix}7 & 6 & 5 & 4\\ 2 & 3 & 3 & 1\end{matrix}$ (No change in coke angle)	1.5	4.02
April 4th	$\alpha_K\begin{matrix}7 & 6 & 5 & 4 & 5\\ 3 & 3 & 1 & 2 & 1\end{matrix}+\alpha_J\begin{matrix}8 & 7 & 6 & 5 & 4 & 3 & 4\\ 3 & 3 & 2 & 1 & 1 & 1 & 1\end{matrix}$	1.5	5.06
April 24th	$\alpha_K\begin{matrix}8 & 7 & 6 & 5 & 4 & 5\\ 3 & 2 & 2 & 1 & 1 & 1\end{matrix}+\alpha_J\begin{matrix}8 & 7 & 6 & 5 & 4 & 3 & 4\\ 3 & 3 & 2 & 1 & 1 & 1 & 1\end{matrix}$ (Same angle of ore and coke)	1.5	5.5

3.4.5.2 Actual Result

As a result of the smooth operation improvement, the coal injection rate and blast temperature increased rapidly, and the small size coke which had been discontinued for several days was put into operation on March 26th. Specific indicators for each month are shown in Table 3.10; Fig. 3.17.

Table 3.10 Monthly operation indicators of Capital Steel No. 2 blast furnace

Month	Daily output, t	Coke, kg/t	Pulverized coal, kg/t	Small coke, kg/t	Fuel rate, kg/t	Blast volume, m^3/min	Blast temperature, °C	Top temperature, °C	Gas utilization rate, %
1	4125.06	429.9	71.4	0.8	502.1	3565	1044	201	44.68
2	4208.90	414.4	79.9	14.7	509	3660	1097	194	46.27
3	4353.55	426.3	72.3	3.1	501.7	3708	1108	192	46.66
4	4522.6	340.8	136.7	26.5	504	3677	1195	187	49.24
5	4594.77	344.9	130.6	36.2	511.7	3613	1201	182	49.46

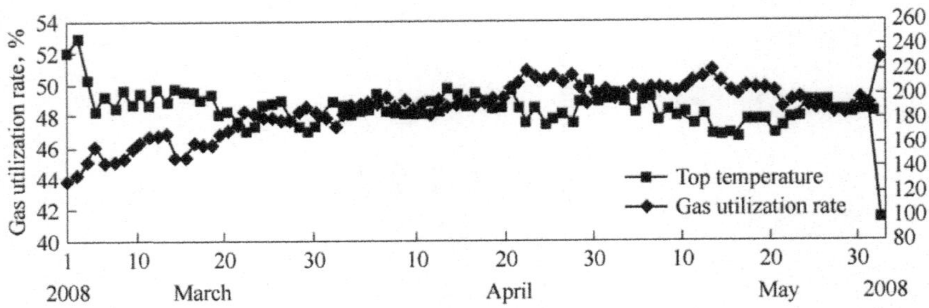

Fig. 3.17 Changes of top temperature and gas utilization rate

The data in Table 3.10, except for the gas utilization rate, were derived from the monthly production report of iron-making plant statistics. The gas utilization rate was taken every 5 min and obtained by the average of the whole month, which was more accurate. Figure 3.17 also shows the monthly average of the data taken every five minutes.

Comparing the results before and after the experiment, the temperature of top furnace was reduced from 200 to 180 °C, a reduction about 20 °C. The gas utilization rate increased from 46% before the experiment to 49%, about 3%. Compared April with February, according to the monthly production data, although the fuel ratio did not change much, the consumption of pulverized coal and small size coke in April was 68 kg/t more than that in February. In fact, the coke ratio not only decreased by 68 kg/t, but also production increased by 7.5% from the average daily output analysis, mainly due to the decrease of fuel rate. According to the estimation of total oxygen consumption and output, the fuel ratio decreases by about 15–20 kg/t, which was generally consistent with the improvement of gas utilization rate, indicating that the data of fuel ratio in the table were not accurate.

The improvement of blast furnace stability was the main reward of the experiment.

Compared with the data before experiment, the angle of ore burden distribution increased from 38° to 45.5°, outward shift 7.5°; the ore peak moved out about 0.7 m; coke moved outward 5.5°, about 0.6 m (see Table 3.7; Fig. 3.18), which completely

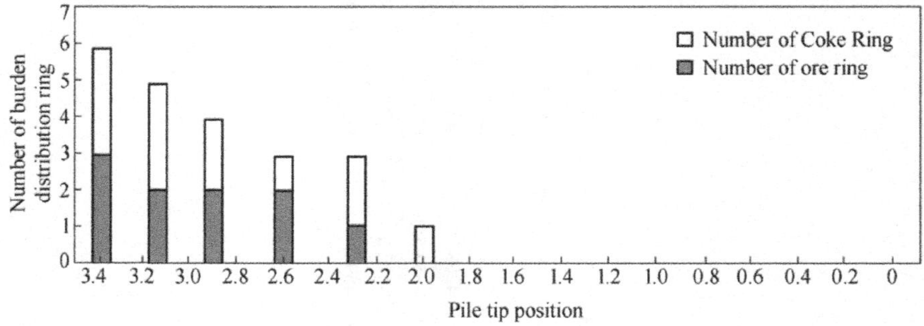

Fig. 3.18 Burden distribution after experiment

changed the adverse consequences of the burden pile tips falling into the middle area of the throat.

3.4.5.3 Conclusion

The distribution of blast furnace gas is affected to a large extent by the burden distribution. Reasonable distribution of burden is very important.

(1) The furnace throat should be divided into 5–11 equal parts based on the principle of equal area and the diameter of the throat. The burden should be distributed according to this specified limited position. Such distribution results can be compared with other different blast furnace practices, as well as with different distribution methods or different periods of the same furnace.
(2) Under normal conditions, the burden peak should not be distributed to the middle area of the throat radius. This kind of distribution can only be used as a treatment means for the poor quality of the burden or the abnormal condition of the furnace. As a means of treatment, it is something like "medicine" and cannot be used as a "meal." Ore is not generally distributed in the middle of coke, as it is easy to have the excessive development of periphery and center gas flow and subsequently forming two gas pathways, which is a poor type II gas distribution.
(3) The long-term use of the burden distribution system requires patience when it needs to be changed. This is because the internal profile of the blast furnace has been adapted to the gas distribution, the pathway of gas is relatively smooth, and the change requires "running-in" and takes time. The experimental process of changing the burden distribution sometimes seriously affects the smooth operation of blast furnace and should not be done frequently. Some experiments change several times a day, so it is impossible to see clearly the real effect of changing distribution of burden.
(4) It is very important to maintain the central gas pathway of blast furnace. The bigger the blast furnace, the more important the central flow. But coke concentrating in the center would not be recommendable. The high-temperature zone in center should not be too wide that would decrease gas utilization and waste fuel. Narrower high-temperature zone in center not only improves smoothness of furnace running but also could save fuel.
(5) Type III gas distribution is the normal operation choice of blast furnace and the basis of high efficiency, low consumption, stability and smooth operation.

3.5 Calculation of Burden Distribution

The distribution of burden in the throat is determined by the landing point of burden, which is the key to the distribution of burden.

3.5.1 Equation for Calculating Landing Point of Distribution of Burden

Calculation of feeder spout outlet velocity (c_0):

$$r = \frac{D - b}{4}$$

$$c_0 = \lambda\sqrt{3.2\,gr} \tag{3.2}$$

where D is diameter of feeder spout, m; b is average particle size of burden, m, ore: 0.02 m, coke: 0.05 m; r is hydrodynamic radius of feeder spout, m; g is the acceleration of gravity, 9.8 m/s^2; c_0 is feeder spout outlet velocity, m/s; λ is burden coefficient, ore: 0.4, coke: 0.9.

Calculation of velocity of burden leaving distributing chute (c_1):

$$c_1 = \sqrt{\frac{2g(l_0 - e/\tan\alpha)(\cos\alpha - \mu\ \sin\alpha) + 4\pi^2\omega^2}{(l_0 - e/\tan\alpha)^2 \sin\alpha(\sin\alpha + \mu\ \cos\alpha) + c_0^2}} \tag{3.3}$$

where c_1 is velocity of burden leaving distributing chute, m/s; l_0 is length of distributing chute, m;e is distance of distributing chute tilting, m; α is distributing chute dip, also current angle of distribution of burden, (°); ω is rotating speed of distributing chute, circle/s; μ is friction coefficient of distributing chute, ore: 0.5, coke: 0.3.

Calculation of landing point (n):

$$n = \left\{l_0(\sin\alpha - e\ \cos\alpha)^2 + 2(l_0 \sin\alpha - e\ \cos\alpha)l_x + \left[1 + \frac{4\pi^2\omega^2(l_0 - e/\tan\alpha)^2}{c_1^2}\right] l_x^2\right\}^{\frac{1}{2}} \tag{3.4}$$

where l_0 is length of distributing chute, m; c_1 is velocity of burden leaving distributing chute, m/s; e is distance of distributing chute tilting, m; α is distributing chute dip, also current angle of distribution of burden, (°); l_x is component of the landing point of burden in the x-axis, m; ω is rotating speed of distributing chute, circle/s; n is peak position, distance between the peak and center of blast furnace, m.

3.5.2 Instructions for the Landing Point of Burden Distribution Calculation Program[1]

This program is used to calculate the distance between the landing point of burden and the centerline of blast furnace in the process of burden distributing. The program consists of three files:

(1) Multi-Ring Burden Distribution Calculation Program. exe;
(2) Config.ini;
(3) Landing Point of Multi-Ring Burden Distribution Calculation Results.txt.

The three files must be in the same folder for the program to run. The "config.ini" is used to store the history information. If it is deleted, the parameters used in the program calculation will be restored to the default settings; the "Multi-Ring Landing Point of Burden Distribution Calculation Results.txt" records the result of the calculation program, and the new result will overwrite the previous one. Therefore, it is recommended to copy the file after each calculation. The main interface of the program is shown in Fig. 3.19.

Landing point of multi-ring burden distribution calculation program (多环布料落点计算程序)		
Parameter setting (参数设置)		Inverse calculation of friction coefficient (反算摩擦系数)
Type of burden distribution: coke (布料种类: 焦炭)	Rotating speed of distributing chute (ω) (溜槽转速ω) 0.15 r/s	Measured depth of stock line (h_1) (实测料线深度 h_1) 1.2 m
Burden coefficient (λ) (炉料系数λ) 0.9	Minimum depth of stock line (料线最小深度) 0 m	Measured burden distribution angle (α) (实测布料角度α) 33.0°
Particle size of burden (b) (炉料粒度 b) 50.0 mm	Maximum depth of stock line (料线最大深度) 4 m	Measured landing point of burden(n) (实测炉料落点 n) 2.84 m
Diameter of feeder spout (D) (导料管直径 D) 0.7 m	Account step of stock line (料线计算步长) 0.1 m	Inverse calculation value of friction coefficient (摩擦系数反算值)
Length of distributing chute (l_0) (溜槽长度 l_0) 2.58 m	Height difference of stock lines (h_2) (料线高差 h_2) 0.3 m	Calculate friction coefficient (计算摩擦系数)
Distance of distributing chute tilting (e) (溜槽倾动距 e) 1.0 m	Friction coefficient of distributing chute(溜槽摩擦系数) 0.53	

(continued)

[1]as the author requested, the calculation program of landing point of burden distribution was compiled and explained by Professor Qi Chenglin and Dr. Li Fengguang of University of Science and Technology Beijing.

(continued)

Calculation result of landing point of multi-ring burden distribution (多环布料落点计算结果)	
Cycle number of burden distribution (布料环数) 5	Calculation of multi-ring burden distribution (多环布料计算)
Angle 1–13 (角度1–13)	

3.5.2.1 Choose the Type of Burden Distribution

Select the type of burden as "ore" or "coke" in the drop-down-list box within the box in Fig. 3.20.

3.5.2.2 Input Calculation Parameter

Input the required parameters within the box marked in Fig. 3.21. In the process of calculation, the burden coefficient is recommended to use the default value, and the friction coefficient of distributing chute can be input with default value or input after inversely calculated based on the measured results. The particle size of burden, diameter of feeder spout, length of distributing chute, distance of distributing chute tilting, rotating speed of distributing chute, and height difference of stock lines are input according to the actual blast furnace parameters. The minimum depth of stock line, maximum depth of stock line, and account step of stock line are determined by the calculation requirements. For example, if you want to calculate the landing point of burden when the depth of the stock line changes within the range of 0–4 m, calculate it every 0.1 m, then input "0", "4", "0.1" each time, the result of landing point will be shown when the depth of stock line is 0, 0.1, 0.2, 0.3, …, 3.8, 3.9, and 4.0 m.

3.5.2.3 Calculation for Landing Point of Burden Distribution

After entering the parameters, you can select the number of burden distribution rings. For example, you want to distribute in 5 rings, inputting 5 into the [Burden Distribution Ring] position in Fig. 3.22, and then the first 5 edit boxes of burden distribution angle will change into importability state. After inputting 5α angles of burden distribution and then clicking on the button of [Multi-Ring Burden Distribution Calculation], you can obtain and display the calculation results, as shown in Figs. 3.22 and 3.23. The results are saved in the program folder entitled "Multi-Ring Burden Distribution Calculation Results.txt", which can be copied and printed later.

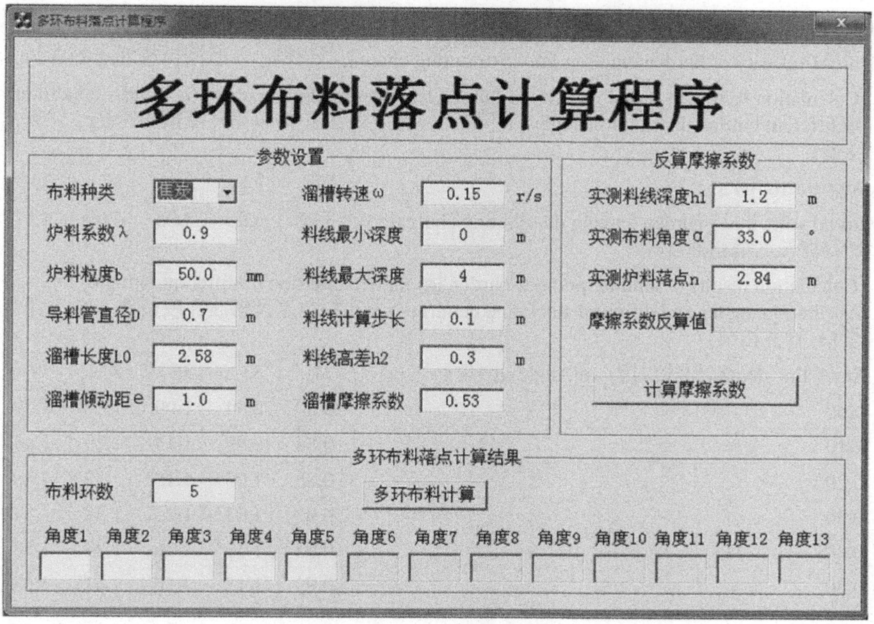

Fig. 3.19 Main interface of the program (landing point of multi-ring burden distribution calculation program)

[Computation Time] (计算时间) 2014–08–08 17:08:15
[Computation Parameter] (计算参数)
Type of burden distribution: Coke (布料种类: 焦炭)
Burden coefficient (λ) (炉料系数 λ)0.9
Particle size of burden (*b*) (炉料粒度 *b*)50.0 mm
Diameter of feeder spout (*D*) (导料管直径 *D*) 0.7 m
Length of distributing chute (l_0) (溜槽长度 l_0) 2.58 m
Distance of distributing chute tilting (*e*) (溜槽倾动距 *e*) 1.0 m
Rotating speed of distributing chute (*ω*) (溜槽转速*ω*) 0.15 r/s
Minimum depth of stock line (料线最小深度) 0 m
Maximum depth of stock line (料线最大深度) 4 m
Account step of stock line (料线计算步长) 0.1 m
Height difference of stock lines (h_2) (料线高差 h_2) 0.3 m
Cycle number of burden distribution (布料环数) 5
[Calculation results of initial velocity of burden entering into distributing chute (c_0) under condition of different burden distribution angles] (不同布料角度条件下 炉料进入溜槽初速度 c_0 计算结果)

(continued)

(continued)

Initial velocity of burden entering into distributing chute (c_0)炉料进入溜槽初速度c_0 2.03					
[Calculation results of initial velocity of burden leaving distributing chute (c_1) under condition of different burden distribution angles] (不同布料角度条件下 炉料离开溜槽初速度c_1 计算结果)					
Angle α (α角)	31	33	35	37	39
Initial velocity of burden leaving distributing chute (c_1) 炉料离开溜槽初速度c_1	3.87	3.98	4.05	4.09	4.11
[Calculation results of landing point of burden (n) under condition of different burden distribution angles and different stock line depth] (不同布料角度及料线深度条件下 炉料落点 (n) 计算结果)					
Stock line depth (料线深度), m|Angle α(α角)	31	33	35	37	39
0.00	0.80	0.94	1.09	1.24	1.39
0.10	0.84	0.99	1.14	1.29	1.44
0.20	0.88	1.03	1.18	1.33	1.49
0.30	0.92	1.07	1.22	1.38	1.54
0.40	0.95	1.11	1.26	1.42	1.58
0.50	0.99	1.15	1.30	1.47	1.63

3.5.2.4 Inverse Calculation of Friction Coefficient of Distributing Chute

In the above calculation, the friction coefficient of distributing chute is taken as the empirical value, which basically satisfies the calculation requirements. The conditional blast furnace can calculate the friction coefficient of distributing chute according to the measurement result of the actual landing point of burden. The specific method is as follows: input the measured stock line depth, the measured burden distribution angle, and the measured landing point of burden in the boxes of Fig. 3.24, click button of [Calculate Friction Coefficient], and you can calculate the friction coefficient of distributing chute, as shown in the area within the box in Fig. 3.25. The calculation result can be used to calculate the landing point after fixing the blast furnace.

3.6 Regulation of Peripheral and Center Gas Flow

3.6.1 Harm of Peripheral Gas Distribution

No. 1 blast furnace of Capital Steel had operated with gas flow preferentially through the peripheral area. 5 years after the start-up of the furnace, more than 50 pieces of cooling stave at bosh and shaft had burned out. As the furnace had operated with a

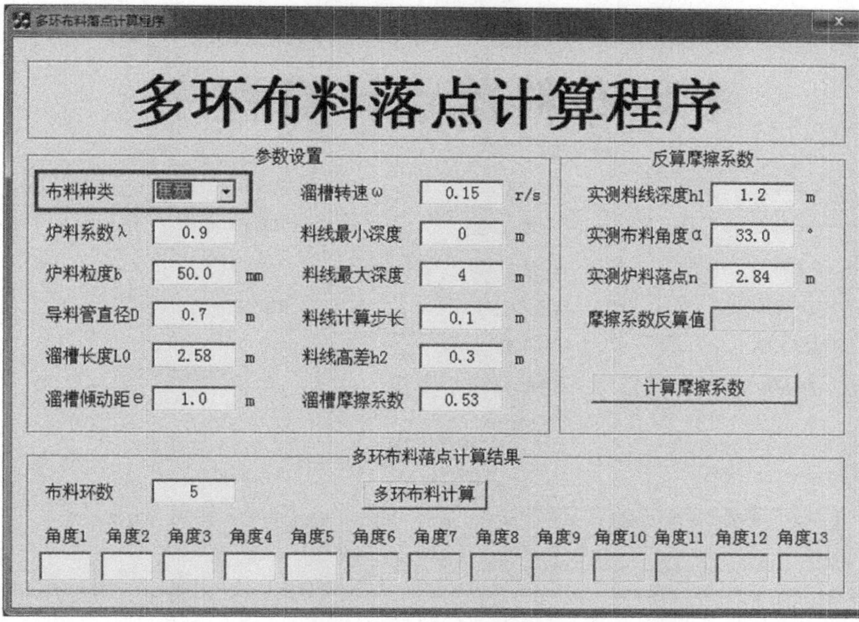

Fig. 3.20 Choose the type of burden distribution (landing point of multi-ring burden distribution calculation program)

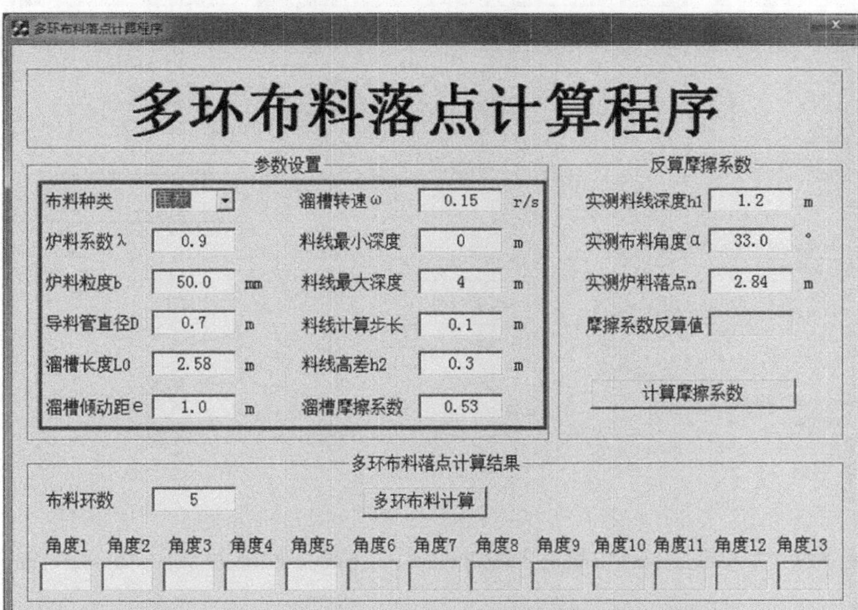

Fig. 3.21 Parameter input (landing point of multi-ring burden distribution calculation program)

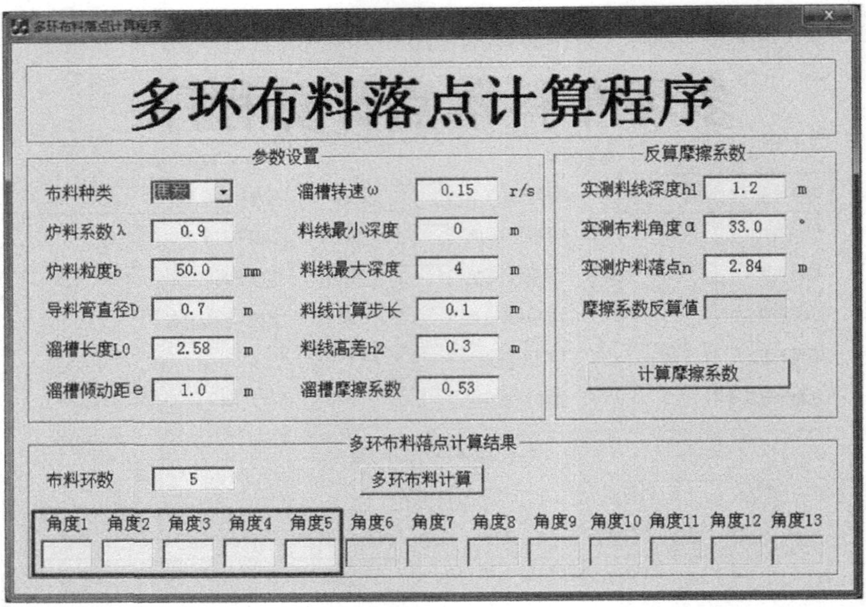

Fig. 3.22 Burden distribution angle inputting and landing point calculating (landing point of multi-ring burden distribution calculation program)

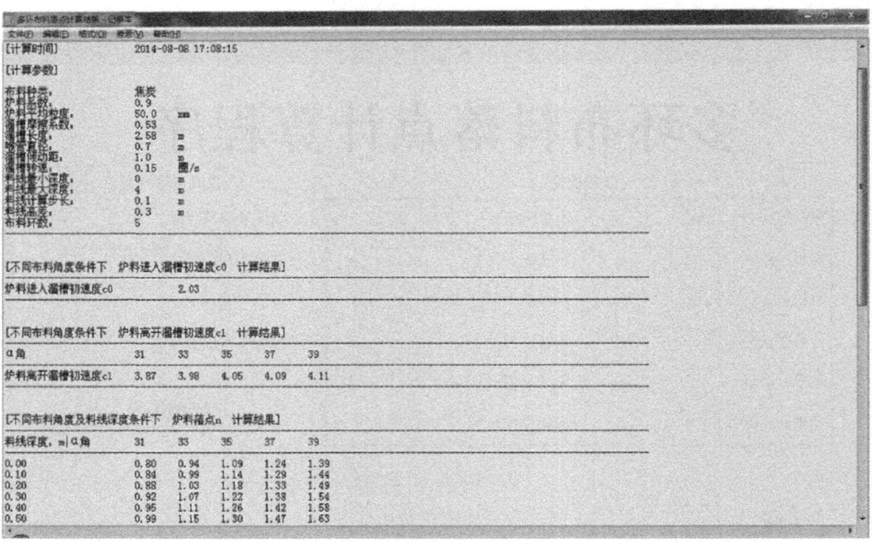

Fig. 3.23 Calculation result (landing point of multi-ring burden distribution calculation program)

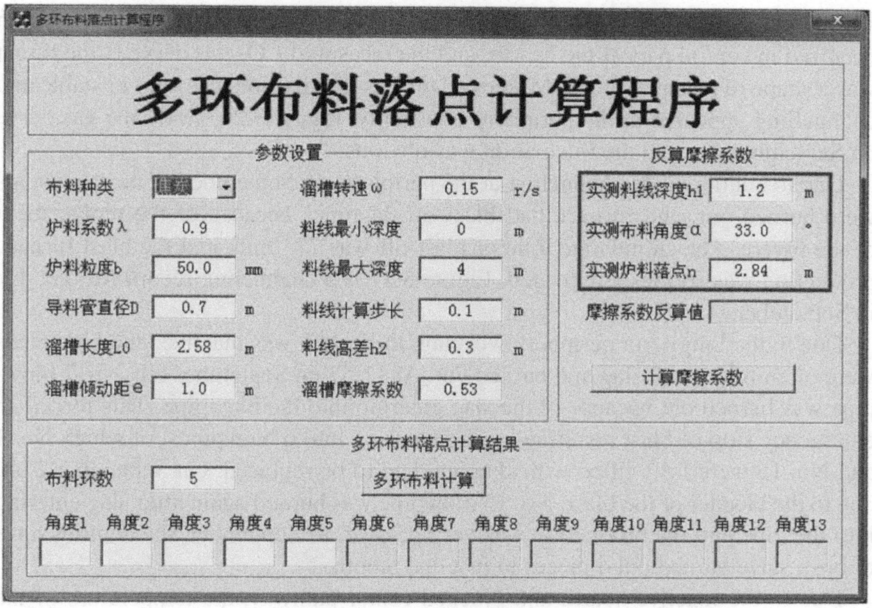

Fig. 3.24 Inverse calculation of friction coefficient of distributing chute (landing point of multi-ring burden distribution calculation program)

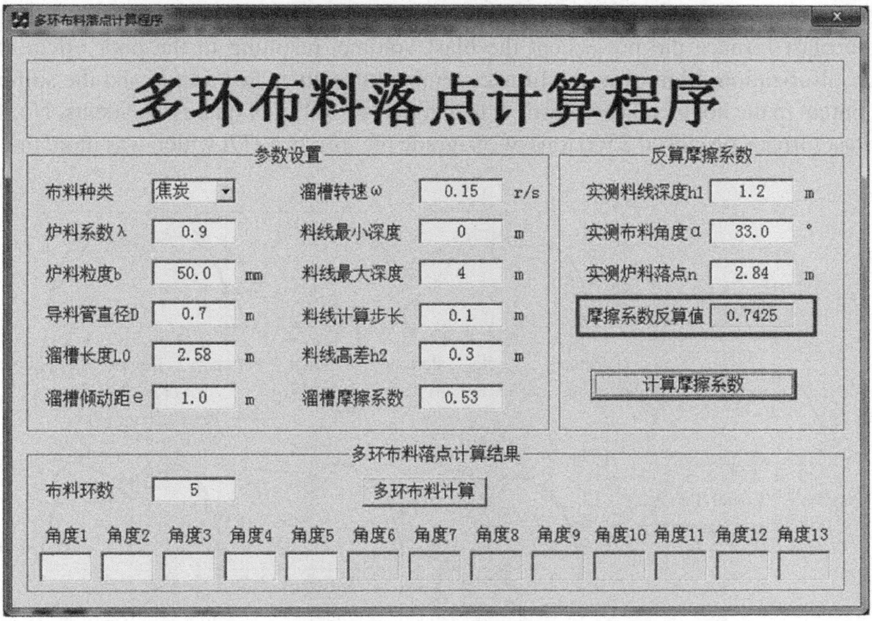

Fig. 3.25 Calculation result of friction coefficient of distributing chute (landing point of multi-ring burden distribution calculation program)

peripheral gas flow for a long period, hearth operation was abnormal. Accumulation occurred in central part of the hearth, and the measured CO_2 gas curve at the throat was a shape of steamed bread. Moreover, the gas distribution was very unstable and channelling appeared at the periphery constantly. Figure 3.26 shows the gas curve on September 1st, and the fluctuation was obvious.

Due to continuously channeling at the periphery, in September alone, 44 tuyeres were burned out, and furnace had blast off 34 times because of the replacement of the tuyere. The cumulative time of blast-off was 713 min, and the blast furnace production was seriously damaged. Figure 3.27 is a channeling record for one day in September.

Due to the long-term periphery working, the hearth was inactive, and blowpipes were often filled with slag and burned out. At 11:57 on September 6th, No. 7 blowpipe was burned out because of the slag entering into the blowpipe. This forced an emergency stop of blast resulting in slag entering into 6 blowpipes, of which No. 1 and No. 15 were fully filled with slag and had to be replaced. On September 27th, due to the bleeder of the blast, No. 13 blowpipe was burned again after slag entering into the pipe and the production was very passive. In September, the cooling stave of bosh stave burned out 4 pieces in just one month.

Due to the inactive hearth and blocked center pathway, the temperature of the burden in the center of blast furnace was low. Therefore, after the cold burden materials entering into the hearth, it caused the central part of the hearth to cool down as the loading was too high for the available hearth heat and semi-frozen liquid to accumulate in the center of hearth. After the accumulation in central part of hearth, the blast furnace did not accept the blast volume, resulting in the poor effect of desulfurization. Moreover, the furnace temperature fluctuated a little and the sulfur content in the hot metal rose, leading to a high probability of quality accidents. No. 1 blast furnace produced 2566 tons of off-grade pig iron in 1977, which was more than

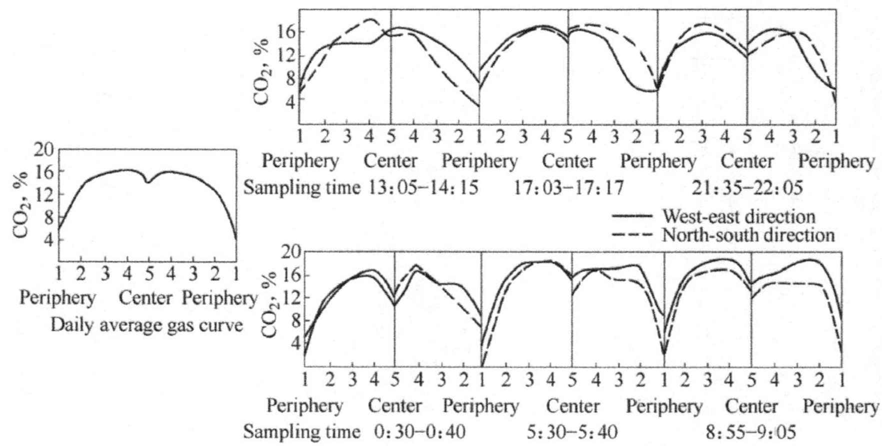

Fig. 3.26 Flow fluctuation in type II gas distribution

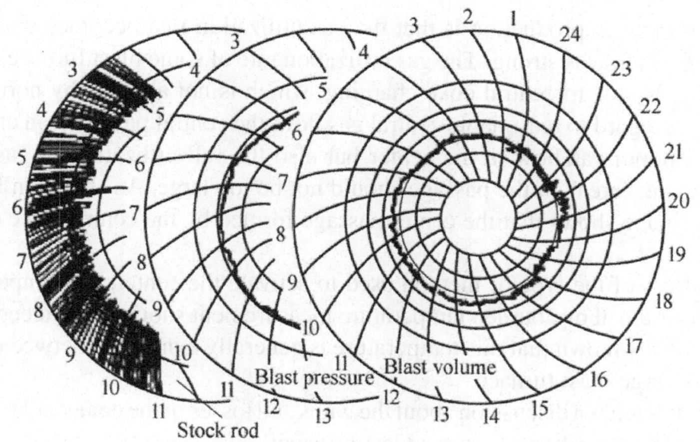

Fig. 3.27 Characteristic of instrument record during the stroke of channel

the total of other three blast furnaces produced. Because of the periphery working, the temperature of the top gas was higher, which affects the life of the top charging equipment; at the same time, the equipment maintenance work was also aggravated.

Figure 3.27 shows a typical instrument record of type I gas distribution, that the blast volume and blast pressure fluctuated frequently. The furnace temperature caused the gradual changes of blast volume and blast pressure. When channeling occurred, the blast volume and blast pressure fluctuated instantaneously and frequently. According to the record of stock rod, the burden descending rate was uneven with fast and slow periods and furnace often slipped. The distribution of type I gas distribution had appeared in many iron-making plants, which is suffered for the plants. Therefore, the type I gas distribution should be eliminated resolutely.

3.6.2 Control of Central Gas Flow

It must be active in the center of blast furnace, especially the large blast furnace. There are two types of methods to open the central pathway: charging the coke to center or changing the depth of the central funnel of the stock column. In 1987, Shimuzu Masakata and others created the central coke charging method, which made an important contribution to the smelting intensity of large blast furnace. Central coke charging facilitated the opening of the center and promotes the development of the central flow. At the same time, it promoted the coke in the hearth deadman to be renewed quickly, which was beneficial to the activeness of the hearth. Because of the active central flow, the peripheral flow was relatively suppressed and the furnace wall was protected. In particular, they pointed out that the volume of central coke charging reached 1.5% of the coke batch weight, which could satisfy the need of replacing the deadman coke [18].

The problem in production is that the gas utilization rate becomes worse when the central gas flow is strong. The gas utilization rate of some blast furnaces reduces more than 2% due to central coke charging, which is not allowed by normal blast furnaces. In regard to reasonable central gas flow, the central passage can ensure not only the smooth gas flow in the center but also the full utilization of gas energy. Therefore, the "area" of the passage should not be too large. As shown in Fig. 3.28 [18], this image shows that the center passage formed by the central coke charging is very narrow.

Regardless of the type of method used to activate the center, the temperature at the center of the above burden temperature measurement should not exceed 700 °C. Experience has shown that this temperature is generally controlled between 450 and 600 °C for large blast furnace.

The following is a discussion about the gains and losses of the center coke charging with the actual operation of a large blast furnace:

The blast furnace volume is 5700 m^3, and the throat diameter is 11.2 m; other specific data are as follows (Table 3.11):

Ore batch weight, t	Coke batch weight, t	Nut coke included, t	Depth of stock rod, m
137	25.1	2.7	1.7

$O^{44.2\ 40.9\ 39\ 37.9\ 32.2\ 29}_{3\ \ 4\ \ 6\ \ 1\ \ 2\ \ 3}$ $C^{44.2\ 43.9\ 39\ 37.9\ 32.2\ 29\ 25.4\ 22}_{5\ \ 3\ \ 2\ \ 2\ \ 2\ \ 1\ \ 2\ \ 3}$

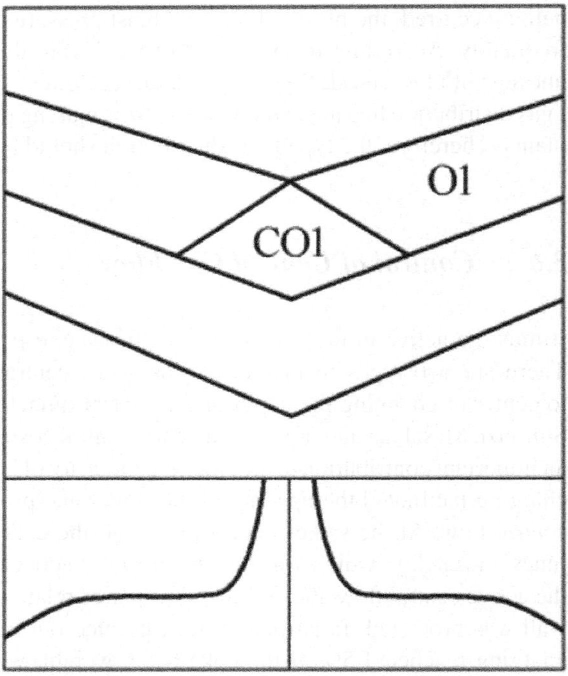

Fig. 3.28 Schematic diagram of ideal central coke charging

Table 3.11 Production index of blast furnace

Index	Blast volume, m^3/min	Blast temperature, °C	Pulverized coal, kg/t	Oxygen Content, m^3/h	Aggregate load	Top pressure, kPa	Top temperature, °C
Daily average	8696	1305	96.9	23,894.38	3.28	2.72	129
Index	Actual blast velocity, m/s	Blast kinetic energy, kg m/s	TCT, °C	Coke ratio, kg/t	Coal ratio, kg/t	Fuel ratio, kg/t	
Daily average	248	13,965	2051	287	161	448	

The 5700 m^3 blast furnace is one of the super large blast furnaces in the world today. The hearth diameter is 15.5 m and requires a sufficient level of center flow. Actually, the center temperature was 440 °C, and the peripheral temperature was between 40 and 55 °C (see Fig. 3.29). When the blast furnace was operating smoothly, the blast temperature was 1305 °C and the fuel ratio was only 448 kg/t. The operation performance was remarkable. The lowest ring position of the coke distribution was 22°, which is 2.06 m away from the centerline of the throat. The width of the terrace was about 1.7 m, and the depth of funnel was about 2.5 m. The huge depth of funnel ensured the central passage of blast furnace gas.

The "expert system" developed by Capital Steel Automation Institute shows the burden distribution, as shown in Fig. 3.30.

As shown in Fig. 3.30, the O/C at the periphery was 3.8 and that in the center was 1.3. This O/C in the center should be sufficient to ensure a smooth gas flow of the center.

The judgment of center flow mainly relies on the furnace top imaging and above burden temperature measurement. The data in the pathway only can be taken for reference since that the central flow is narrow and strong, the flow rate is fast and the

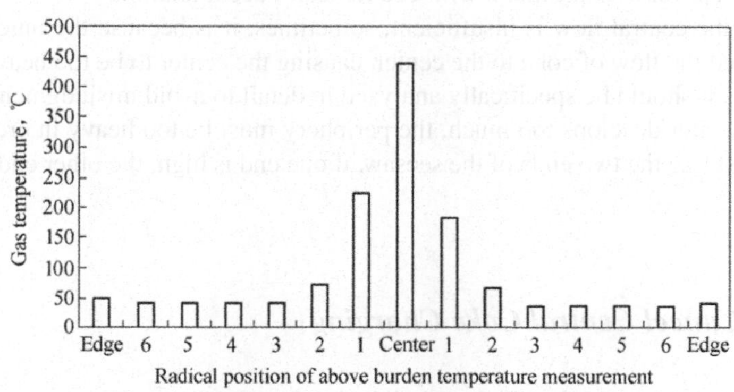

Fig. 3.29 Distribution of radical temperature at throat (above burden temperature measurement)

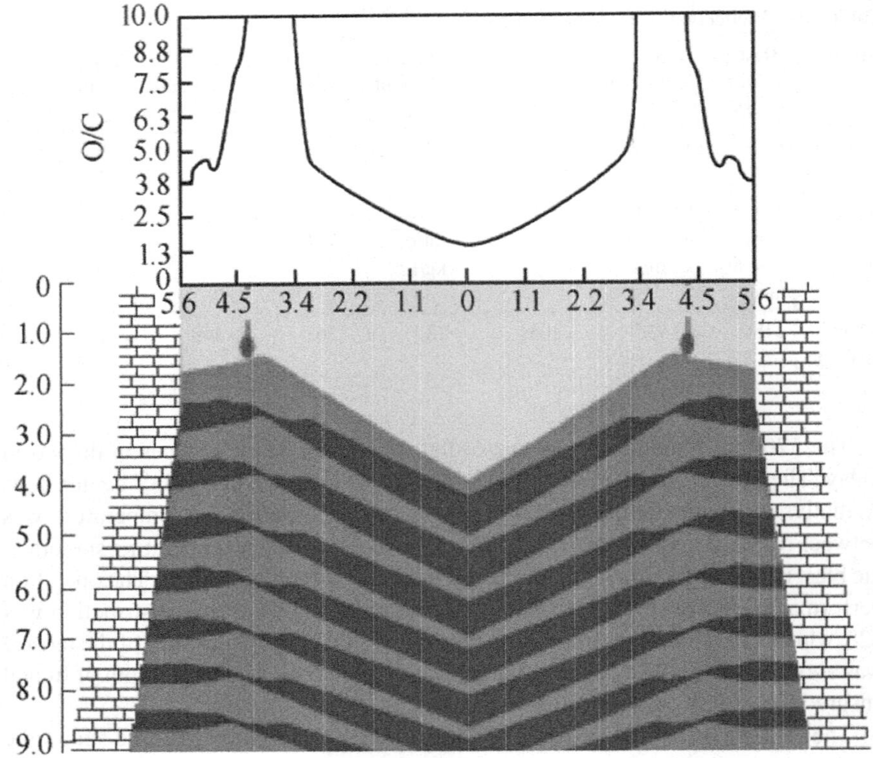

Fig. 3.30 Distribution of burden as "expert system" described

temperature at the central temperature measurement point is higher (the temperature of the large blast furnace center is higher than 400 °C). The specific blast furnace needs experience accumulation.

Insufficient central flow causes the instability of blast furnace. If the central flow is insufficient for a long time, it will lead to hearth accumulation.

When the central flow is insufficient, sometimes, it is because the inner ring of ore blocked the flow of coke to the center, causing the center to be too heavy in ore. Therefore, it should be specifically analyzed in detail to avoid misjudgment.

If the center develops too much, the periphery must be too heavy in ore burden. This is just like the two ends of the seesaw, if one end is high, the other end must be low.

3.6.3 Cancel Central Coke Charging

If the central flow were insufficient and the burden condition poor, the central coke charging could be used to maintain the central passage of gas; if the burden improved or the center was excessively high developed, then the central coke charging should be reduced or canceled. To reduce or cancel the central coke charging, first, there must be enough passage for the gas at the periphery. It has been discussed before that if the center developed excessively in general, the edge would be overweight. Here are some examples:

"Before August 2011, burden surface of No. 2 blast furnace was observed during several shutdowns. It was found that there was no fire at the circumference. At that time, the ore batch weight was 128t, the ore/coke was 4.1, the stock line of ore and coke was 1.7 m and 1.6 m, respectively, and the central coke was 5 circles. When the center temperature was high (1000 °C at high time), the gas utilization rate was low, generally about 46–47%, large amount of top water spray (the highest all day is more than 1000 t), the gas stability was very poor, and occasionally the gas channeling. The judgment should be that the periphery was too heavy and the gas mainly depends on the center passage to maintain the furnace condition " [19]. The matrix of burden distribution at that time was:

$$O\begin{matrix}44 & 42 & 40 & 38 & 35.5 & 32.5 & 35.5\\ 4 & 4 & 4 & 3 & 1 & 1 & 1\end{matrix}\; C\begin{matrix}44 & 42 & 40 & 38 & 35.5 & 32.5 & 25.5 & 13.5\\ 3 & 3 & 2 & 2 & 2 & 2 & 1 & 5\end{matrix}$$

"Later, to ensure the opening of the peripheral gas flow, we reduced the maximum angle for both the coke and ore, raised the stock line, and reduced the ore coke ratio at periphery (i.e., increased the amount of coke at periphery)." In July 2012, using (open periphery) burden distribution matrix, the ore's batch weight was 175 t (including 137 t for Ore A and 38 t for Ore B), and its ore/coke was 5.32. The stock lines were 1.3 m for Ore A, 0 m for Ore B and 1.6 m for coke, respectively. The burden distribution matrix was as follows:

$$O_A\begin{matrix}37 & 35 & 33 & 31 & 29 & 31\\ 3 & 2 & 4 & 1 & 3 & 2\end{matrix}\; O_B\begin{matrix}37 & 35 & 33 & 31\\ 1 & 3 & 2 & 2\end{matrix}\; C\begin{matrix}37 & 35 & 33 & 31 & 29 & 25 & 19\\ 5 & 3 & 2 & 2 & 2 & 1 & 3\end{matrix}$$

"Jingtang Steel No. 2 blast furnace of Capital Steel canceled the central coke charging on June 10, 2012. Initially, the gas stability was slightly poor. Currently (July 2012), the gas was stable, full blast, and the level of furnace temperature was improved. The gas utilization rate rose to over current 51% " [19]. With the increase in gas utilization rate, the temperature of the top decreased dramatically, and no water was needed on the top of the furnace.

Many blast furnaces that cancel central coke charging have generally increased the utilization rate of gas. The above examples have been explored many times and have gone through a long period of time before they have achieved preliminary success. It gives us an important revelation that it is necessary to first create a peripheral gas passage prior to any attempt to restrain the excessive development of central gas flow.

Central coke charging is one of the effective ways to adjust the distribution of gas. Use it if needed. Removed it if you do not need it, it cannot be used or rejected resolutely. There is a blast furnace of 2032 m^3 in Germany. The central coke charging accounted for 28.6% of the total coke. The annual average coke rate was 315 kg/t, including 70 kg/t of small coke, 176 kg/t of pulverized coal injection, 491 kg/t of annual fuel rate, and the blast furnace was operating smoothly [20].

3.7 Slag Crust Falling off

The first slag crust falling off in China occurred in 1982. After Bao Steel No. 1, blast furnace was put into operation, slag crust peeling also appeared. For this reason, Bao Steel organized a special meeting in 1987 to discuss this phenomenon. In recent years, the frequent occurrence of slag crust falling off has become a common problem puzzling the normal operation of blast furnaces. In An Steel, the effect of slag crust falling off on the temperature of cooling stave and the troubles in production had been studied [20–22]. At Bao Steel, the problem of slag peeling was solved in 1987.s Because the second generation cooling stave of No. 3 blast furnace was damaged, the temperature fluctuation caused by slag crust falling off and the measures of stabilizing the temperature of cooling stave and increasing its service life were studied [23–26].

Slag crust falling off often occurs in blast furnaces with copper cooling stave. Considerable progress has been made in the study of the relationship between heat flux intensity of cooling stave and slag crust under different conditions [27, 28].

3.7.1 Characteristics of Slag Crust Falling off

In the smooth process of smelting, the inner profile of blast furnace is stable and the thickness of furnace wall is basically unchanged. Judging from the measured data, the heat flow intensity of each layer of blast furnace fluctuated less, and the relationship between blast volume and blast pressure was stable. However, if the smelting process were unstable, the gas distribution would be abnormal and the channeling appears constantly. As a result, the heat flow intensity of blast furnace fluctuated greatly, and the temperature of brick lining of furnace wall and the temperature behind cooling stave also changed correspondingly.

In the daily operation of the blast furnace, there are two main reasons for the temperature changes of the furnace wall and the cooling stave:

(1) Channeling at furnace periphery;
(2) The formation and falling off of slag crust. If the gas along the periphery is relatively heavy and the furnace wall is easy to slag. When the slag layer reaches a certain thickness, the self-weight of slag plus the frictional force of the descending burden would peel the slag off the lining. Figure 3.31 shows the temperature changes behind the cooling stave before and after the falling of slag crust.

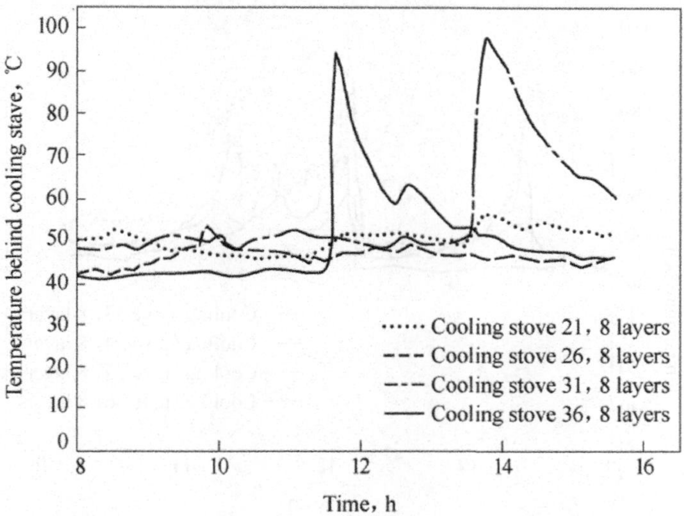

Fig. 3.31 Characteristics of temperature changed behind cooling stave during the slag crust falling off

Figure 3.31 shows the changes in temperature behind 4 pieces of belly cooling stave at the same level (15.2982 m). Figure 3.31 shows that in just a few minutes, the temperature rose sharply and then slowly decreased. The temperature rose as a result of slag crust falling off, and the slow decrease was the result of the process of slag crust re-freezing on the staves. Sometimes it took an hour or several hours for the temperature to approach the original level. In the figure, the slag peeling of cooling stave No. 36 and No. 26 was separated for about 2 h. This characteristic of temperature changes was unique to the slag crust falling off.

Figure 3.32 shows the process of three times falling off and reforming of slag crusts on No. 33–36 adjacent cooling staves. For the first slag peeling, the changes of four cooling staves were different: the slag crusts of No. 34 and No. 35 cooling staves fell off successively, the slag crusts of No. 33 cooling stave melted partly, and the slag crusts of No. 36 cooling stave remained unchanged. The falling and melting slag crusts were bonded again. After more than two hours, they fell off together, and the thickness of peeling was different. The slag crust of No. 35 cooling stave was the thickest, and No. 33 was the thinnest. From the analysis of temperature curve, slag falling and melting sometimes coexisted, sometimes repeated and occurred continuously.

Under the condition of stable smelting process, the temperature of cooling stave was relatively stable. Figure 3.33 shows the temperature change curve behind cooling staves of bosh and belly. In the figure, No. 13–16 were four adjacent bosh cooling staves at elevation of 15.31 m, while No. 17–20 were four adjacent belly cooling staves of layer 8 (at elevation of 16.982 m).

Slag crust falling off and slag crust melting had different effects. The former falls off in an instant, which is very dangerous, while the latter melts over a long period of

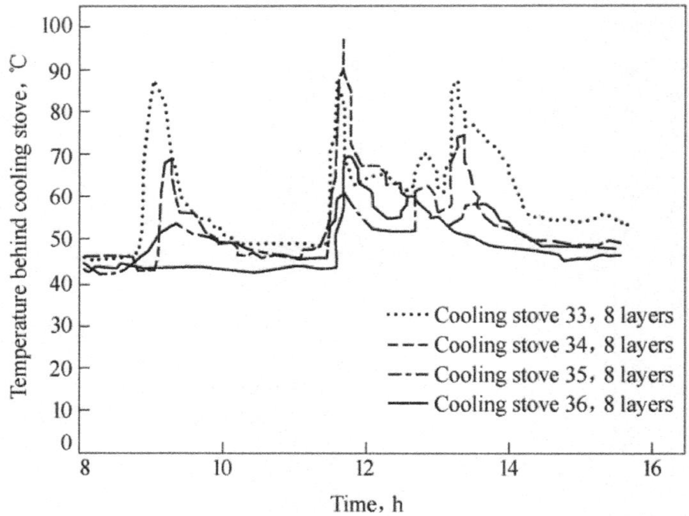

Fig. 3.32 Slag crusts falling off at adjacent cooling staves

time. The molten slag following the burden descent together, which had less impact on the smelting process, Fig. 3.34 shows a typical process of the slag crust formation and melt.

At No. 9 cooling stave on the belly in Fig. 3.34, the thickness of the slag crust changed continuously in the first 4 h (from 8:00 to 12:00), from melting to bonding, to re-melting and then re-bonding; in later 4 h (after 12:00)), it was relatively stable. Adjacent other cooling staves were quite stable. Because slag crust did not fall off as a whole, it had little effect on the furnace condition.

3.7.2 Slag Crust Falling off and Gas Channeling at the Periphery

Because of the deterioration of burden, the gas permeability of the stock column became worse gas distribution was not normal, and burden was too heavy in the center. Often, these lead to gas channeling's along the periphery in blast furnace.

Slag crust fall-off occurs only when the burden at the peripheral area is relatively heavy. In daily operation, it is easy to judge the channeling: Once the channeling appears, first of all, the blast volume increases, the blast pressure decreases, and the permeability index increases obviously. According to the analysis of cause, the two causes are the opposite, so the two treatments are totally different. Table 3.12 lists the differences between slag crust fall-off and peripheral channel.

Figure 3.35 shows the state of the tuyere damage caused by the periphery working blast furnace and/or the frequent occurrence of the peripheral channeling. At the end

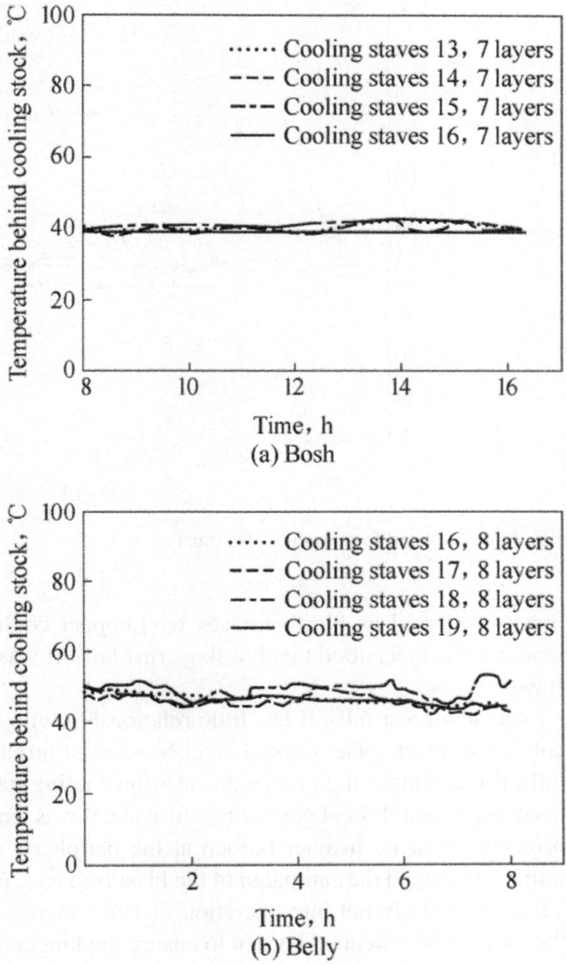

Fig. 3.33 Temperature fluctuation behind stabilized cooling staves

face and upper part of the tuyere, there were pockmarks (bullet holes) and gulfs (deep pit/cavity) eroded by hot metal, and some had been burnt through to the holes already; the tuyere where the slag crust falling off and pressed in was generally intact at the front and end, but the middle part was squeezed and rubbed by the falling slag crust, causing serious deformation or damage. The difference between the two is obvious.

3.7.3 Slag Crust Fall-Off and Copper Cooling Stave

China has not used copper cooling stave for a long time. Little experience has been accumulated. In recent years, slag crust peeling of blast furnace happened more often,

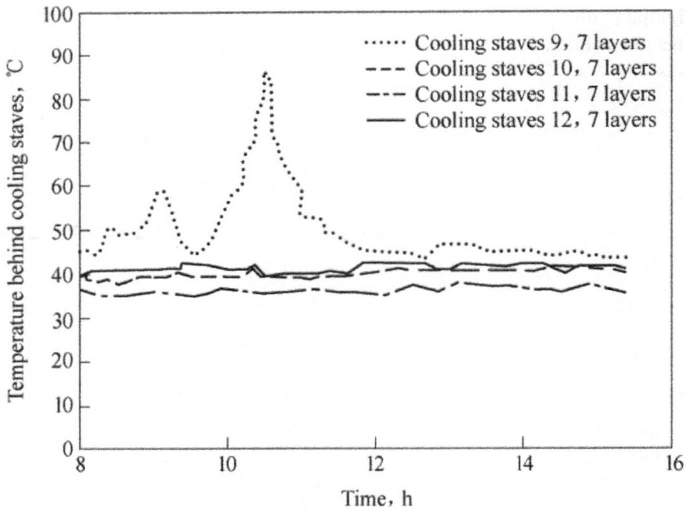

Fig. 3.34 Melt and formation of slag crust

mostly occurred on blast furnaces with copper cooling stave. Some blast furnace workers misunderstood that the slag crust fall-off was caused by the copper cooling stave.

The slag crust fall-off has little relationship with the type of materials used for cooling stave. In 1982, Capital Steel No. 2 blast furnace suffered multiple slag crust fall-off incidents, which was with cast iron cooling stave at that time. After the blow in of new Capital Steel No. 2 blast furnace, it was expected to follow the operating principle of using heavier burden at the periphery, aiming to protect the furnace lining and extend the campaign of the blast furnace. No. 1 blast furnace in Bao Steel, which was newly put into operation in 1987, increased burden at the periphery on the basis of beneficiated burden to ensure the longevity of the blast furnace. During this process, multiple slag crust fall-off occurred, and copper cooling staves were used in Bao Steel at that time. These two events were the earliest slag crust fall-off in China. Placing more burden at the periphery to prevent furnace lining erosion by the gas flow along the periphery was the root cause of the slag crust fall-off. After lightening appropriate burden at the periphery both blast furnaces eliminated the slag crust fall-off problem. At present, although the cast iron or cast steel cooling staves has been used for some blast furnaces, slag crust fall-off still happened [29].

3.7.4 Risk of Slag Crust Falling-Off [21–30]

The slag crust fall-off brings bad effects to the blast furnace production: once the slag crust falls off, the furnace wall and the cooling stave of the falling part would

Table 3.12 Differences between slag crust fall-off and edge channel

Content	Slag crust fall-off	Edge channel
Condition before tuyere	Slag crust sticks to the tuyere and drops slowly. Sometimes can see that the tuyere is blocked and blackened, after the slag crust passing by, it is bright again	In severe cases, a large amount of cold coke descends (black coke) in front of the tuyere, and circulating at high speed in the raceway zone
Temperature at backward of cooling stave	(1) First, the temperature rises sharply (within a few minutes, vertical rose); then the temperature falls slowly, and sometimes it takes an hour or more to fall to the original level or its fluctuation is close to the original level (2) The range of synchronous heating cooling stave along height is small, and sometimes the temperature of adjacent cooling stave at the same level rises more (transversely)	(1) Warming up in the direction of channeling (2) Synchronized heating of cooling staves are mostly confined along the blast furnace height (longitudinal)
Slip	Little	More
Function for tuyere	In severe cases, the tuyere is pressed into the furnace by sliding slag. But the front end of the tuyere was generally not destroyed. When serious, the blowpipe sprays fire and coke. If the blast is not stopped in time, the blowpipe, the intermediate piece and even the elbow would be burned out	The upper part of the front end of the tuyere is eroded into pockmarks and gulfs (bullet holes and cavity); in severe cases, the front or end face of the tuyere was burnt into a hollow

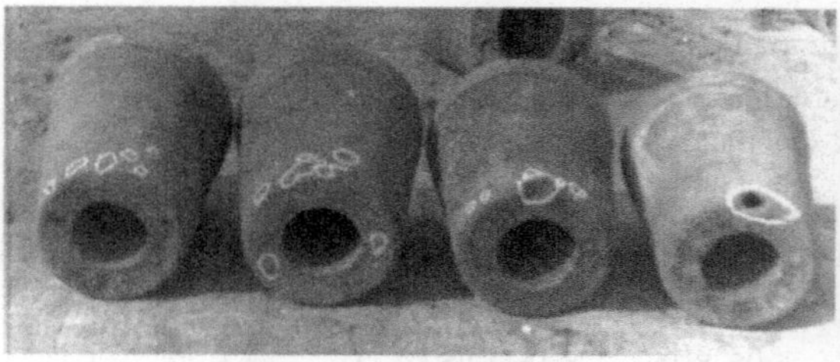

Tuyere burned out by channeling at the periphery

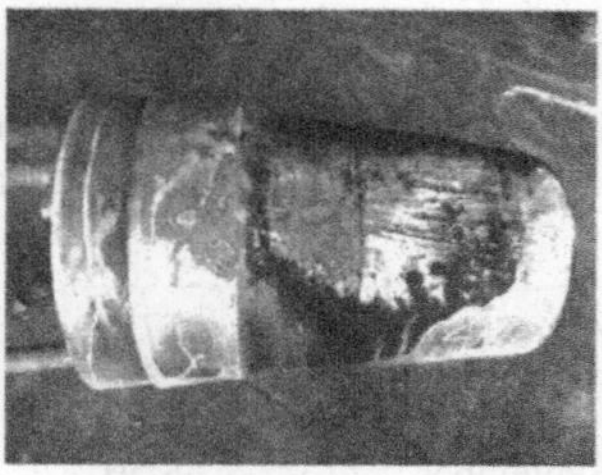

Tuyeres crushed by slag crust fall-off

Fig. 3.35 Damaged tuyeres due to furnace periphery working and slag crust fall-off

be heated sharply, and after the slag crust building up again, the temperature drops also, so that the temperature of the hot surface of the furnace wall and the cooling stave rise and fall sharply, causing a large thermal shock. Subsequently, this causes cracks or erosion of the furnace wall and the cooling stave itself, or even the cooling stave would be damaged.

Severe slag crust fall-off causes the changes of peripheral state of the blast furnace, affecting the distribution of gas, destroying the original distribution, and thus changing and destroying the smelting process.

The descending slag crust makes contact with the upper part of the tuyere that go deep into the hearth. Lighter slag crust peeling slips pass the front of the tuyeres, and the influence is small; in a circumstance of heavier slag peeling, the slag causing tuyere tipping will press the tuyere into the furnace, resulting in the severe blast leakage of blowpipes and the forced shutdown of blast furnace. The tuyere that is pressed into the furnace is blocked in front of the blowpipe. It is difficult to remove from the furnace by general equipment and removal tools. The tuyere must be cut into several pieces with a cutting tool and then be taken out. For this reason, it sometimes requires a furnace shutdown for a few hours.

Due to the tipping of tuyere into furnace, hot blast and coke are ejected from the gap, which not only burn out the blowpipe, but sometimes burn out the tuyere holder and elbow. No. 2 blast furnace which was newly built by Capital Steel that year had suffered damages caused by the slag crust fall-off 5 times in 1982; see Table 3.13. Except for the first treatment, because of the failure to grasp the main point, it took a long time to deal with the damages. For the 4 subsequent treatments, the total time of blast stopping was 18 h and 51 min.

After the slag crust entering the hearth, the temperature of the hot metal dropped. When it was serious or not be handled properly in time, off-grade products would be produced due to the high S content of the hot metal and even lead to serious furnace cooling.

To sum up:

(1) Slag crust fall-off is caused by heavy ore/coke at the periphery.

Table 3.13 Losses caused by slag crust fall-off

Date	Tuyere number	Starting time of blast leak	Time of normal pressure	Time of wind stopping	Damage equipment		
					Tuyere	Tuyere holder	Elbow
August 31st	22	17:18–18:50			1		
	22	22:28	22:45–23:58	23:58–4:13	1		1
September 1st	22	5:50	6:05–8:15	8:15–12:52	1	1	1
	22	15:55	16:07–17:46	17:45–21:56	1	1	
September 2nd	18	4:08	4:05–7:33	7:33–11:49	1	1	
Total			7 h 20 min	18 h 51 min	5	3	2

(2) When the peripheral burden is too high, the operator could develop the appropriated peripheral gas flow to eliminate the slag crust fall-off.
(3) Slag crust fall-off has a little to do with the material of cooling stave. Reasonable gas distribution is the key.
(4) Because the copper cooling stove has an obvious advantage, the operator should continue to accumulate operation experience to develop the equipment advantage of copper cooling stave.
(5) The designer of blast furnace should install additional temperature measurement device at the cooling staves where the slag formed, to observe and measure the slag forming and falling off.

3.8 The Adjustment of Furnace Upper and Lower Parts to Suit to Each Other

Expert in Iron-making
Fan Zhekuan (1932–1995): Based on the long-term production analysis of several blast furnaces in Anshan Iron and Steel Company, he first proposed the relationship between burden batch weight and blast kinetic energy and smelting strength in the field of iron-making and given the quantitative calculation equation, which points out the direction for blast furnace operation and smelting strengthening.

As early as in the 1950s, the famous Chinese iron-making expert Li Jingcun proposed the "combining the upper and lower parts" according to the production practice of the Ma Steel. He pointed out that the furnace top adjustment must be combined with the blast system [31]. At that time, the blast volume and the blast temperature were increased, and at the same time, the batch weight of the ore was reduced. As a result, the gas flow in the center of the blast furnace was excessively developed and furnace slipped constantly. After then they expanded the ore batch weight, changed the reverse unit charging and suppressed the central flow from the upper part. As a result, the collapse was eliminated and the blast furnace ran smoothly.

Cheng Lanbo summed up operating experience of many years in Anshan Iron and Steel Plant and pointed out that before1958, blast furnace had been limited to the upper adjustment and often struggled to cope with the abnormality of the blast furnace. Later the plant paid attention to the lower adjustment and strengthened the smelting. "The practice also shows that the various operating systems have an impact on each other, but a reasonable blast supply system and the charging system can unify the contradiction from the countercurrent flow motion between gas flow and the burden, so that the flow is distributed reasonably and the furnace stable plus operating smooth. Therefore, it is more important to choose a reasonable blast supply system and charging system." [32] Fan Zhekuan, an iron-making expert in China [33], first gave the relationship between ore batch weight and blast kinetic energy of blast furnace in the world. He gave the following equations:

$$\left.\begin{aligned} W_k &= V_k \rho_K A \\ V_k &= 0.09t + 0.21 \\ W_k &= (0.09t + 0.21)\rho_k A \end{aligned}\right\} \quad (3.5)$$

where W_k is ore batch weight, t; V_k is volume of ore per square meter of throat area, m^3/m^2; ρ_k is bulk density of ore, t/m^3; A is throat area, m^2; t is intensity of smelting, $t/(m^3\ d)$.

Fan Zhekuan also gave the calculation equation for the relationship between blast kinetic energy and intensity of smelting in Reference [33]:

$$E = (764t^2 - 3010t + 3350)d \quad (3.6)$$

where E is blast kinetic energy, kg m/s; d is diameter of hearth, m; t is intensity of smelting, $t/(m^3\ d)$.

The relationship between blast kinetic energy and intensity of smelting could be established by Eqs. (3.5) and (3.6).

It can be seen from the equation that when the intensity of smelting increased, the batch weight should be expanded. Under the conditions of Anshan Iron and Steel Plant at that time, it can be calculated by the formula. These formulas are empirical ones based on the smelting conditions of Anshan Iron and Steel Plant at that time. The methods can be used to perform statistics based on their own empirical data to obtain reference results.

The charging system should be compatible with the blast supply system. When the blast velocity was low, the raceway of the hearth tuyere would be small and the initial gas distribution would be more at the peripheral area of hearth. For large- and medium-sized blast furnaces, charging system should not excessively block the peripheral gas flow. The charging system should be adjusted to properly control the peripheral area and to open the center. The focus was to guide the gas flow to center and to prevent the peripheral gas flow from being suddenly blocked and destroying the smooth operation of blast furnace. If gas flow of lower part in center developed, as long as the blast furnace ran smoothly, charging in the upper part should be done properly to open the center and to keep the gas flowing smoothly. If the lower centers were excessively forming a continuous gas channel, the central gas flow should not be blocked at once. Instead, the gas should be properly diverted to the periphery to alleviate the excessive development of the center and keep the furnace operating smoothly. This is the adaptation of the upper and lower parts. In short, no matter upper or lower part of the furnace, it should not form a direct confrontation with each other.

Therefore, when the type I gas distribution was changed to the type III gas distribution, there should be a transition in the choice of the charging system. We first converted the gas distribution from type I into the type II, then increase the blast velocity during the period of type II operation. This would change the initial gas flow distribution and expand the raceway area in front of the tuyere to suit to the upper adjustment. Without the cooperation of upper adjustment, it would be difficult to open the center only by lower adjustment. In August 1977, the gas distribution of Capital Steel No. 1 blast furnace was type I. To activate the gas flow in the center, on August 9th, four tuyeres with a diameter of 160 mm were replaced by small tuyeres of 130 mm, and the area of the tuyere was reduced from 0.217 to 0.199 m^2 blast velocity was increased from initially 100–105 to 120–130 m/s, and the blast kinetic energy was increased from 49 to 68.6 kJ/s. As a result, many tuyeres were burnt out and the gas curve was still type I:

Month	Blast velocity, m/s	Gas (CO_2), %	
		Edge	Center
July	100–105	6.3	13.5
August	120–130	6.4	13.5

The key was that under the long-term distribution of type I gas, the hearth was impossibly active. The center accumulation was almost inevitable. It would be impossible to solve the long-term accumulation of the hearth with either a lower or upper adjustment in one or two days. At that time, Capital Steel No. 1 blast furnace attempted to increase the blast velocity and quickly solved the problem of furnace hearth accumulation, which was the result of insufficient understanding. Due to the reduction of the tuyere size, the effective cross section of the hearth was originally narrow, and the surface of the hot metal rose rapidly. At the beginning, the lower part of the tuyere burned out a lot, especially when taphole openings were late, the number of tuyeres burnt for only the period of August 22–31 reached 8.

Ben Steel No. 5 blast furnace [34] and Wu Steel No. 3 blast furnace [35] had a period of non-accidental experience. In the process of changing the charging practice to make the gas distribution transition from type I to type II, the blast velocity was increased. It was the result of adapting the upper part and lower part to suit each other.

According to the specific conditions of the blast furnace, after determining the type of gas distribution, it is necessary to achieve the intended goals through the charging practice, which parameters are the decision-making links of the charging system, which ones are the first to do and how to achieve them, should be studied carefully.

Expert in Iron-making

Zhang Shourong (1928–): Academician of the Chinese Academy of Engineering. Engaged in the design, construction, production and technology work of steel plant for a long time, he has also conducted research on burden distribution of blast furnace, slag making, injection and longevity of blast furnace.

On the 2nd National Blast Furnace Conference in 1955, on behalf of Anshan Iron and Steel Plant, he delivered a speech entitled "A Report on the Summary of Blast Furnace Top Modulation," and put forward for the first time the concepts of "Upper adjustment" and "Lower adjustment," which had an important impact on the development of China's operation of iron and steel industry.

In 1957, he participated in the construction of Wu Steel and helped the blast furnace which was the first large blast furnace over 1000 m^3 in China put into operation successfully.

Then he put forward the general plan of technical progress of iron-making system.

He is the author of "Selected Works of Technical Paper of Zhang Shourong" and "Longevity Technology of Blast Furnace of Wu Steel." And he is editor-in-chief of "Blast furnace Aberration and Accident Treatment" and "Blast Furnace High-efficiency Smelting Technology and other books."

References

1. M.A. Szcefanovich, *New Research on Blast Furnace Smelting* (Metallurgical Industry Press, Beijing, 1960), pp. 120–158
2. M.A. Szcefanovich, Process of blast furnace process. Metallurgy **5–6**, 85–90 (1960)
3. S. Kantaro, et al., X. Hu Translated. Capital Steel Sci. Technol. Inf. **1**, 78–93 (1978)
4. K. Ono, Reaction and research on cohesive zone of blast furnace. (A Rikun et al Translated). Baotou Iron Steel Cooperation, 57–59 (1980)
5. E. Amadei. Blowing-in and operation of 10.6 m^3 blast furnace, in *Iron-making Proceedings* (1979), pp. 140–148
6. K. Kenjirp, et al., Dissection of blast furnaces and their internal state. Transactions ISIJ, 371–400 (1997)
7. K. Kanbara et al., Inspection of blast furnace anatomy. Study Iron-Making **288**, 37–45 (1976)
8. T. Yiji, et al. Studies on the dissolution of blast furnace cohesive zone. Iron Steel, **10**, 28–35 (1979)
9. H. Du, B. Liu et al., Research on blast furnace cohesive zone. Iron Steel **11**, 40–53 (1982)
10. J. Zhu, Anatomical study of Capital Steel experimental blast furnace. Iron Steel **11**, 1–8 (1982)
11. R. Gao, J. Zhu, Anatomy of capital steel experimental blast furnace. Iron Steel **11**, 9–17 (1982)
12. Yang Tianjun, *Analysis and Simulation of Iron-making Process* (Metallurgical Industry Press, Beijing, 1991), pp. 315–344
13. S.P. Kinney, C.C Furnas, Gas-solid contact in the shaft of a 700-t blast furnace. Trans. AIME, 84–97 (1929)
14. J. Kurishara, et al., Low fuel rate operation in a 4500 m^3 blast furnace with bell-less top at Chiba Works, in *Iron-making Proceedings* (1980), pp. 113–122
15. X. Wang, S. Wang, C. Chen et al., Adjustment of blast furnace gas flow distribution. Iron-Making **1**, 8–11 (2009)
16. P. Gu, Application of the top camera system in Mei Steel No. 2 blast furnace. Metall. Collect. **2**, 22–24 (2006)
17. K. Yu et al., *Coal Injection in Blast Furnace* (Northeastern University Press, Shenyang, 1995), pp. 84–88
18. M. Shimuzu, et al., Control of blast furnace cohesive zone and furnace core filling by coke core charging. R&D Kobe Co. Iron-Making Tech. Rep. **4**, 11–15 (1991). Che Chuanren. Discussion of central coke charging and other problems. Iron-Making, **4**, 14–17 (1992)
19. H. Zhang, Y. Chen, Y. Wang, et al., Jingtang No. 2 blast furnace of Capital Steel cancel its practice of central coke charging. Iron-Making, **1**, 54–56 (2013)
20. Y. Liu, *Distribution Law of Blast Furnace Burden*, 4th edn. (Metallurgical Industry Press, Beijing, 2012), pp. 228–230
21. B. Wang, H. Zhang, Y. Che, Heat load management of copper cooling stave of blast furnace in An Steel. Iron-Making **2**, 1–4 (2008)

22. Z. Zhao, J. Tian, Y. Li, Study on slag crust fall-off of new No. 3 blast furnace in An Steel, in *Papers Collection of the Seventh (2009) China Iron and Steel Annual Conference* (2009), pp. 667–670
23. Z. Liu, P. Zhao, Cause analysis and improvement measures of tuyere sinking of 2580 m^3 blast furnace in An Steel. Steel Technol. **6**, 42–44 (2010)
24. C. Cao, Y. Zhou, Z. Ye, Cause and preventative countermeasure of damage in cooling stave of Bao Steel No. 3 blast furnace. Iron-Making, **2**, 1–5 (2000)
25. Y. Zhou, C. Cao, Installation and use of copper cooling stave on blast furnace. World Iron Steel **3**, 1–10 (1999)
26. N. Xu, S. Wu, Study on thermal load measurement of cooling stave of No. 3 blast furnace in Bao Steel. Bao Steel Technol. **4**, 15–18 (1999)
27. Y. Zhou, C. Cao, Application of cooper cooling stave on blast furnace abroad. Iron-Making **6**, 24–28 (1996)
28. Z. Liu, Z. Li, Q. Chai, Q. Lv, Heat transfer analysis of slag crust growth in cooper cooling stave on blast furnace. Iron Steel **8**, 12–15 (2010)
29. Z. Liu, Q. Lv, Heat transfer analysis of thickness of blast furnace slag curst. Panzhihua New Steel Vanadium Co. Ltd. **3**, 51–54 (2008)
30. S. Wang, A. Gu, Cause analysis of slag crustinstability of cast steel cooling stave of No. 6 blast furnace of Xuan Steel. Xuan Steel Technol. **1**, 14–15 (2007)
31. J. Li, Experience in BF process adjustment in Ma Steel, in *Technical Department of Ministry of Heavy Industry. Compilation of National Blast Furnace Production Technology Conference*, ed. by (Beijing: Heavy Industry Publishing House, 1955), pp. 48–275
32. L. Cheng, Z. Liu, Z. Wang, C. Long, The adjustment of reasonable flow distribution of blast furnace, in *Formation and Development of Iron-making Technology in An Steel*. C. Zhou ed. by (Beijing: Metallurgical Industry Press, 1998), pp. 156–159
33. Z. Fan, Research on blast furnace kinetic energy. Metall. Technol. **2**, 29–32 (1963); in *30 Years of Iron-making in China*. ed. by H. Du (Metallurgical Industry Press, Beijing, 1963), pp. 141–150; in *Formation and Development of Iron-making Technology in An Steel*, ed. by C. Zhou (Metallurgical Industry Press, Beijing, 1998), pp. 160–163
34. W. Zhang, et al., Iron Steel **4**, 41–46 (1980)
35. S. Wu, Summary of mass positive partial loading experiments of three blast furnace (1978) (internal data)

Chapter 4
To Stable Furnace Temperature

Abstract Stable furnace temperature is an important condition to ensure the stability of blast furnace. The change of furnace temperature not only affects fuel consumption, but also affects the blast volume and output directly. Stabilizing the furnace temperature and controlling the required furnace temperature level to produce the most economical varieties is the way to maintain and create the best economic benefits of blast furnace production.

4.1 Hot Metal Characteristics

4.1.1 The Melting Point of Hot Metal

Hot metal contains many elements, which related closely to furnace burden and operation. The most important factor affecting melting point of hot metal is [C]. Figure 4.1 shows the iron–carbon phase diagram [1].

Figure 4.1 shows the melting point of pure iron is 1538 °C. With the increase of carbon content in hot metal, the melting point of hot metal decreases. When hot metal contains carbon content 4.2%, the melting point drops to 1153 °C. Blast furnace workers often regard this point as the solidification temperature of hot metal in hearth, which is the lower temperature limit for judging the erosion of hot metal to furnace linings. Carbon content in iron continues to increase along the dotted line in the graph, and carbon in iron precipitates in the form of graphite. As Fig. 4.1 shows, when the carbon content in iron reaches 6.7%, the melting point rises up to 1257 °C. When the temperature of hot metal decreases, graphite-carbon precipitates out. So in the hot metal iron-making process at high temperature, the graphite powder precipitates out along with the runner. The hot metal with high silicon content is particularly obvious, and graphite flies along the runner.

Y. Liu, *The Operation of Contemporary Blast Furnaces*,
https://doi.org/10.1007/978-981-15-7074-2_4

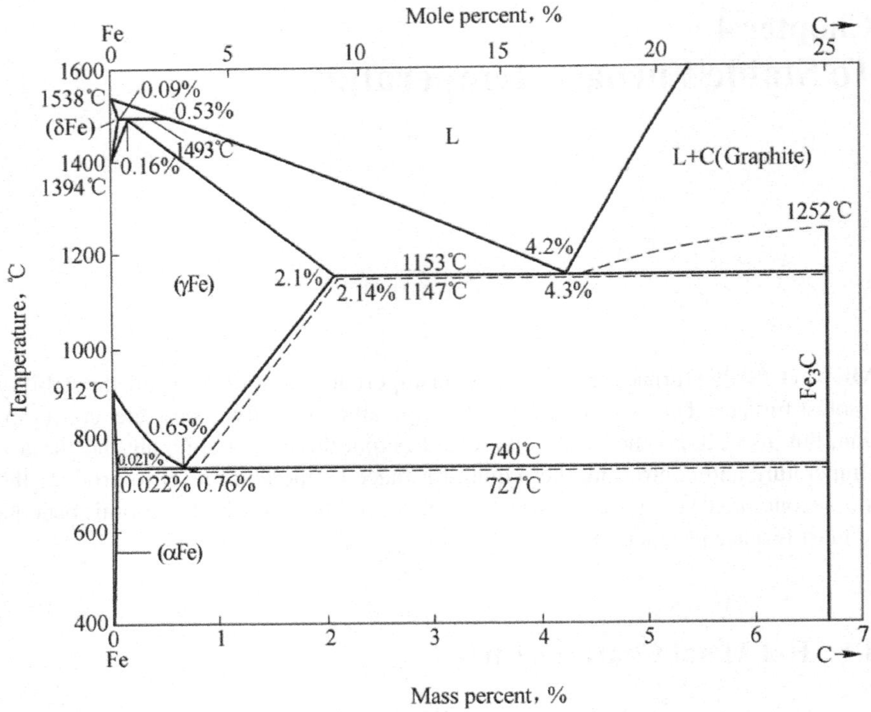

Fig. 4.1 Iron–carbon equilibrium phase diagram

4.1.2 *Temperature Effect on Carbon Solubility*

Furnace burden gradually descends with the temperature increasing and iron oxides are gradually reduced during the descending process. Equation (4.1) shows the relationship between temperature and carbon solubility in iron [2].

$$w[\mathrm{C}]_{\%} = 1.34 + 2.54 \times 10^{-3} t \tag{4.1}$$

where $w[\mathrm{C}]_{\%}$ is carbon content in hot metal, %; *t* is hot metal temperature, °C.

According to Eq. (4.1), when the temperature of hot metal reaches 1600 °C, the content of [C] in iron can reach 5.4%. According to Fig. 4.1, the melting point is about 1200 °C.

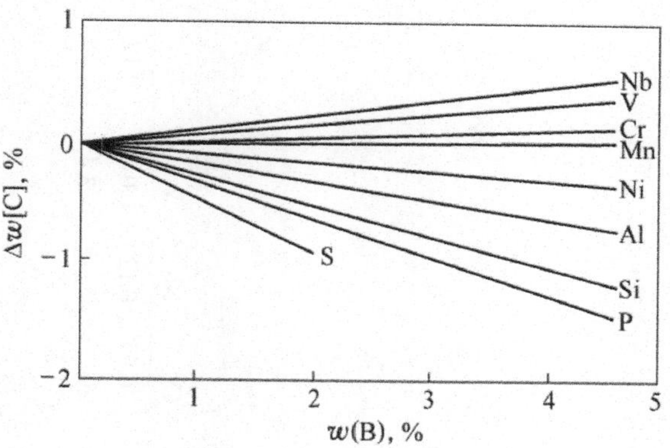

Fig. 4.2 Effect of elements on solubility in iron

4.1.3 The Influence of Some Elements on Carbon Solubility [2]

Carbon solubility is not only related to temperature, but also affected by other elements in iron. Equation (4.2) shows the carbon content of common pig iron:

$$w[\mathrm{C}]_{\%} = 1.34 + 2.54 \times 10^{-3}t + 0.04\,w[\mathrm{Mn}]_{\%} - 0.30\,w[\mathrm{Si}]_{\%} - 0.35\,w[\mathrm{P}]_{\%} - 0.40\,w[\mathrm{S}]_{\%} \tag{4.2}$$

Figure 4.2 and Table 4.1 show the changes in the carbon solubility of hot metal with 1% of each element's percent.

S, P, Si and Mn in hot metal often affect the carbon content of hot metal. These four elements can be regulated and controlled mainly according to the composition of hot metal required by steelmaking, with little regard to carbon content. The influence of other elements is mainly determined by the composition of furnace burden. It is difficult to control the actual carbon content of hot metal.

4.1.4 Carbon Content of Iron for Modern Blast Furnace

There are many factors affecting the carbon content in iron, some of which are less studied and have a wide fluctuation range. In early years, the carbon content in hot metal was sometimes less than 4% because of low blast temperature. In modern blast furnace production, the known carbon content is 4–5.9%. It can be calculated by empirical formulas by I. C. Yusfin and M. A. Arie l [3]:

$$[\mathrm{C}] = -8.62 + 28.8\frac{\mathrm{CO}}{\mathrm{CO}+\mathrm{H_2}} - 18.2\left(\frac{\mathrm{CO}}{\mathrm{CO}+\mathrm{H_2}}\right)^2 - 0.244[\mathrm{Si}] + 0.00143t_{\text{pig iron}} + 0.00278P_{\mathrm{CO}} \tag{4.3}$$

Table 4.1 $w(B) = 1\%$ change in carbon solubility in iron

Element B	S	P	Si	Al	Cu	Ni	Co	Mn	Cr	Nb	V	Ti
Carbon increment, Δ[C], %	−0.4	−0.35	−0.3	−0.25	−2	−0.07	−0.03	0.04	0.09	0.12	0.13	0.17
Applicable concentration, %	<0.4	<3	<5.5	<2	<3.8	<8	<1	<2.5	<9		<3.4	

where CO and H_2 respectively represents the content in top gas, %; $t_{\text{pig iron}}$ is pig iron temperature at tapping; P_{co} is partial pressure of CO in top gas, kPa.

4.1.5 *Trace Elements in Iron*

Carbon in iron comes from fuel. The higher the carbon content in iron, the more consumption of fuel. The increase of carbon by 1% requires 10 kg/t pure carbon, which is equivalent to the consumption of L2 kg/t coke. Due to the high carbon content of hot metal in steelmaking sequence, prolonging decarburization time has a negative impact on steelmaking. According to the need of the whole process of steelmaking, it is necessary to control the carbon content of hot metal at an appropriate level. Hot metal temperature can be controlled, but for feeding scrap steel in steelmaking process, hot metal needs sufficient temperature level. Besides temperature, iron-making process and furnace burden have an important influence on the composition of iron. As shown in Fig. 4.2 and Table 4.1, they also affect the carbon content and other components of hot metal. Nanfen Iron Mine of Ben Steel of China is a world-class iron ore with very few trace elements, similar to Sweden's high-quality iron ores. Sweden is famous for producing high-quality steel, especially ball bearing steel. At that time, Zhang Guangan, a famous expert of Ben Steel, advocated that the high-quality iron ores resources of Nanfen, which lacks in China, should be used to make the high-quality steel and the imported ores or other domestic ores should be used to replace Nanfen Mine to produce ordinary steel. Table 4.2 shows comparisons of hot metal compositions and trace elements between Benxi Iron and Steel Company and other iron and steel companies in China [4].

4.2 State of Hot Metal in Hearth

The hot metal accumulates and releases in the hearth, and its state in the hearth is basically static, disturbed by the tapping, but relatively small. The former Soviet Magnit Golsk Iron and Steel Company took advantage of the relining opportunity to carefully measure the temperature and composition of salamander, so as to "see" the state of salamander [5], which will be discussed in detail in Chap. 8.

4.2.1 *Measurement of Hot Metal State in Hearth*

V. W. Lobyz and others used to blow isotope P^{32} with compressed air into the hearth through the tube inserted the tuyere. The blowing time was 1 or 2 h [6] before the tapping. During the taping, a sample of iron was taken every minute and the intensity of radioactive rays was measured by a counter. The results were inconsistent, and

Table 4.2 Comparison of trace elements in iron in eight iron mills in China

Iron-making plant	Chemical composition, %								
	Pb	Bi	Sn	Sb	As	Ca	V	Cr	Se
Ben Steel No. 1 Iron-making plant	<0.0001	0.0002	0.000108	0.000208	0.00046	0.0008	0.0040	0.0050	<0.0001
Ben Steel No. 2 Iron-making plant	<0.0001	0.0004	0.000111	0.000206	0.00115	0.0025	0.0062	0.0046	<0.0001
A Iron-making plant	<0.0001	0.0003	0.000114	0.000344	0.00120	0.0038	0.00110	0.0086	<0.0001
B Iron-making plant	<0.0001	0.0002	0.00012	0.000428	0.00155	0.0036	0.0079	0.0077	<0.0001
C Iron-making plant	<0.0001	0.0002	0.0123	0.00031	0.00220	0.0021	0.023	0.0043	<0.0001
D Iron-making plant	<0.0001	0.0002	0.000134	0.00067	0.00105	0.0021	0.013	0.0150	<0.0001
E Iron-making plant	<0.0001	0.0004	0.000115	0.00271	0.0022	0.0018	0.015	0.0053	<0.0001
F Iron-making plant	<0.0001	0.0003	0.000134	0.0005	0.0025	0.0094	0.028	0.0047	<0.0001
Iron-making plant	Chemical composition, %							$\sum T$	
	Zn	B	AI	Ca	Ti	Ni	Mo		
Ben Steel No. 1 Iron-making plant	<0.0001	0.00073	0.0018	0.0028	0.052	0.0030	0.00033	0.0620	
Ben Steel No. 2 Iron-making plant	<0.0001	0.00086	0.00086	0.0055	0.059	0.0063	0.00051	0.0715	
A Iron-making plant	<0.0001	0.00074	0.0038	0.0113	0.086	0.0072	0.00093	0.1075	
B Iron-making plant	<0.0001	0.00076	0.0017	0.0059	0.084	0.0052	0.00117	0.1019	
C Iron-making plant	<0.0001	0.00085	0.0008	0.0040	0.138	0.0076	0.0028	0.1803	
D Iron-making plant	<0.0001	0.00073	0.012	0.0075	0.193	0.0052	0.0035	0.2231	
E Iron-making plant	<0.0001	0.00083	0.0012	0.0075	0.0071	0.0023	0.00050	0.0965	
F Iron-making plant	<0.0001	0.00091	0.0055	0.0101	0.091	0.0094	0.00097	0.127	

Note

1. The data in the table is the same as the analysis results of Ben Steel Research Institute in 1979
2. $\sum T = Pb + Sn + Sb + As + Cr + V + Ti + Zn$

there were high peaks. It showed that the hot metal was not stirred and mixed in hearth, although there was little stirring effect on hot metal during tapping, but the effect did not make a difference.

4.2.2 Composition Fluctuation of Hot Metal

There was no large mixing stirring of hot metal in hearth, but every time, the composition of iron fluctuates, because of fluctuations in charge and operating conditions. Table 4.3 shows the result of sampling and determination of cast iron in six blast furnaces of Capital Steel on January 22, 1964. Samples were taken every 2 min from the start of tapping. The composition of hot metal fluctuated obviously. The content of silicon in hot metal is related to the temperature of hot metal. The temperature difference of hot metal in normal furnace condition was less than 10 °C. When the furnace condition was not smooth, the temperature of hot metal fluctuated sometimes to 30°C. The maximum fluctuation of [Si] was 0.32%. It was a group of cast iron with bad furnace condition, with a range of 0.3% and a mean square deviation of 0.11 (No. 12,735 cast in Table 4.3). Figure 4.3 shows the fluctuations in Table 4.4. The fluctuation of iron composition in 12,737 cast was the smallest, the variance was only 0.024, and the range was 0.09. The variance of 12,735 cast was 0.11 and the range was 0.32, which was three times more than that of 12,737. The furnace temperature was rising for 13,735, while stable for 12,737, so the fluctuation was very small.

Because of the low content of [Si], the steelmaking iron amplitude was small. Table 4.4 shows an example measured in No. 3 blast furnace of Capital Steel on July 25, 1975. It was sampled by the same method as the casting iron mentioned above.

The number of casts in Table 4.4 is " No. 16,207 cast," 10:37–11:05, with 204 t iron discharged. The hot metal contained 0.54% Si and 0.037% S. The Si content of this cast was close to the previous cast at beginning and got hotter and hotter. In fact, the Si content of this cast was 0.08% bigger than the previous cast.

4.2.3 Determination of Composition Fluctuation Amplitude of Hot Metal

The natural fluctuation of hot metal composition varies with different plant conditions. The control fluctuation range should be determined according to the plant operating conditions. In some factories, the use of burden was stable for a long time, and the fluctuation range of hot metal composition was very small. In some factories, the fluctuation range of burden composition was relatively large, and the fluctuation range of hot metal composition is also large. Therefore, it was necessary to determine the fluctuation amplitude according to the conditions of the factory. According

Table 4.3 Fluctuation of hot metal composition (%)

Sample number	Tapping number											
	12,726		12,728		12,735		12,737		12,740		12,746	
	C	Si	C	Si	C	Si	C	Si	C	Si	C	Si
1	4.16	2.62	4,19	2.48	4.28	1.9		2.49	4.31	2.22	4.44	1.95
2		2.53		2.5		1.92		2.47		2.2		1.96
3		2.53		2.37		1.94	4.39	2.52		2.24		1.97
4		2.56		2.38	4	1.92		2.56	4.39	2.31	4.54	1.86
5	4.3	2.54	4.39	2.38		1.97		2.56		2.42		2.05
6		2.53		2.42		1.99	4.28	2.53		2.42		2.04
7		2.53		2.42		2.04		2.53	4.39	2.42	4.48	2.04
8		2.54	4.38	2.36	4	2.12		2.51		2.34		1.99

Sample number	Tapping number											
	12,726		12,728		12,735		12,737		12,740		12,746	
	C	Si	C	Si	C	Si	C	Si	C	Si	C	Si
9		2.51		2.36		2.16	4.34	4.34	2.52		2.31	2
10		2.53		2.36		2.3		2.52	4.38	2.34	4.43	2
11	4.43	2.56	4.37	2.35		2.19		2.52		2.34		1.97
12		2.56		2.37	4.38	2.32	4.36	2.53	4.37	2.36		1.99
13		2.67		2.36		2.1						1.99
14	4.28	2.66				2.05					4.46	1.95
15					4.38	2.05					4.46	1.95
16						1.94						

(continued)

Table 4.3 (continued)

Sample number	Tapping number											
	12,726		12,728		12,735		12,737		12,740		12,746	
	C	Si	C	Si	C	Si	C	Si	C	Si	C	Si
17						2.17						
18					4.24	2.15						
Average	4.31	2.56	4.33	2.39	4.34	2.06	4.26	2.52	4.37	2.33	4.47	1.98
Variance(σ)		0.049		0.047		0.11		0.024		0.066		0.056
Pre-furnace analysis (Si)		2.62		2.46		2.1		2.53		2.38		2.05
Hot metal Temperature, °C	1400–1401		1390–1400		1370–1400		1410–1420		1390–1400		1380–1400	

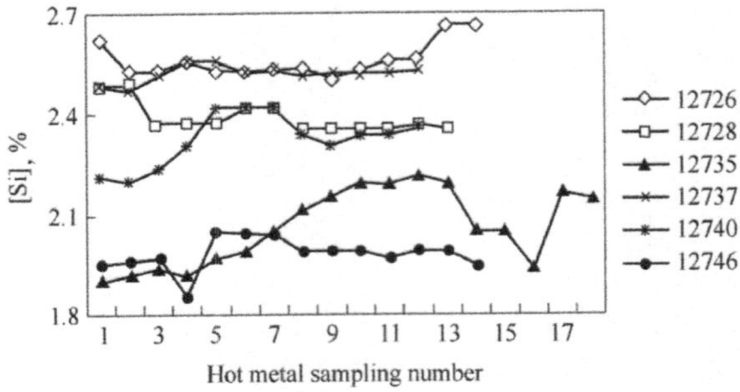

Fig. 4.3 Casting iron wave diagram

Table 4.4 Composition fluctuations in steelmaking iron (%)

Sample number	1	2	3	4	5	6	7	8	9
Si	0.57	0.595	0.5	0.585	0.545	0.64	0.655	0.68	0.67
S	0.05	0.05	0.028	0.045	0.042	0.042	0.05	0.034	0.035
Sample number	10	11	12	13	14	15	Average	Mean square deviation	
Si	0.615	0.685	0.73	0.715	0.615	0.655	0.63	0.062	
S	0.032	0.032	0.027	0.045	0.029	0.027			

to the results measured in Tables 4.3 and 4.4, the mean square deviation of natural wave amplitude of cast iron was 0.024–0.066 and the maximum was 0.01, which was the result of irregular furnace condition. Therefore, in daily operation, the mean square deviation of components should be controlled less than 0.07. Under these conditions, more stable conditions required the improvement of burden stability. It was difficult to achieve this by operation alone. Some blast furnace mathematical models or expert systems stipulated that the silicon content should be controlled at $[Si] = A \pm 0.03\%$. It was easy to break away from the conditions at that time. It was impossible to achieve 100%. Only by creating stable raw material conditions and operation level could satisfy the above requirements.

4.3 Control Furnace Temperature

4.3.1 Indicators for Judging Furnace Temperature

Although the content of [C] in hot metal has a decisive influence on the solidification temperature of hot metal, it cannot be used as an index to determine the temperature of hot metal in the daily operation of blast furnace, because [C] exists in hot metal in the form of graphite besides Fe_3C. When the furnace temperature is high, graphite will precipitate along the runner and fly in the tapping yard. Its existence in hot metal is more complex than [Si]. When the furnace temperature changes in a small range, the scale of change of [C] in hot metal is not as obvious as [Si]; more importantly, the large range of change of [C] in hot metal is mainly determined by the burden, the burden remains unchanged, and the influence of small range of furnace temperature change on [C] lacks long-term experience accumulation.

[Si] has always been the furnace temperature determination index for daily operation of blast furnace, and we accumulated abundant experience about it. In recent years, hot metal temperature has become an important index for judging furnace temperature because of the widespread measurement of hot metal temperature. Some blast furnaces have replaced [Si] with the "expert system" in particular for judging furnace temperature. It is often more convenient and reliable to use hot metal temperature as a judging index.

4.3.2 *Furnace Temperature Judgment*

Variation of external conditions or production organization would lead to fluctuation of furnace temperature. The task of blast furnace operators should keep the minimum amplitude as far as possible, find out the changing trend of furnace temperature in time and adjust it timely according to the situation. Table 4.5 shows the furnace temperature judgment. The front part of the table shows the instrument display, and the back part shows the monitor observation. In modern blast furnace operation, the foreman seldom observes directly. The judgment of furnace temperature basically depends on instrument measurement and screen display, which is verified by direct observation. This form is for the reference of young foremen.

According to the items listed in Table 4.5, the trend of furnace temperature should be determined and adjusted in time.

4.3.3 *Main Parameters Affecting Furnace Temperature*

Each factory has its own parameters of furnace temperature adjustment and empirical values affecting Si content in hot metal. Table 4.6 is the empirical values of Capital Steel [7]. The following are examples to illustrate the usage. These examples still retain the past records, but only introduce the calculation method, which has no demonstration significance.

Example 1 Steelmaking iron produced at No. 3 BF under the following conditions:

Table 4.5 Furnace temperature judgment table

Project		Overheat	Furnace heat	Normal furnace heat	Furnace coldness	Refrigeration
Instrument indicating instrumentation and recording	Blast pressure	Too high	Gradual increasing	Normal	Gradual decreasing	Unstable, volatile
	Blast volume	No acceptance of blast volume	Gradual decreasing	Normal	Gradual increasing	Rapid drop, no acceptance of blast volume
	Blast temperature	No acceptance of blast volume	Hard to increase blast temperature	Normal	Easy to accept blast temperature	Depending on the furnace surface
	Top temperature		Rise four point dispersion	Normal	Temperature decrease	Fluctuation, mutual overlapping, keeping low
	Stock rod	Irregular		Neat ruler		Irregular
	Stock velocity	Stagnation, collapse, suspensions	Gradual slow	Normal	Gradual acceleration	Stagnation, collapse, suspensions
	Gas utilization	Bad	Good	Normal	Not good	Bad
	Permeability index	Low	Decreasing	Normal	increasing	Large fluctuation
	Top pressure			Normal	Sometimes there are peaks	There is an upper peak, and the top pressure decreases after suspension
	Throat temperature		Increase	Normal	Decrease	Low
	Furnace temperature		Increase	Normal	Decrease	Low
	Furnace bosh temperature		Increase	Normal	Decrease	Low

(continued)

Table 4.5 (continued)

Project			Overheat	Furnace heat	Normal furnace heat	Furnace coldness	Refrigeration
Tuyere		Brightness	Bright	Bright	Bright	Light bright	dark red
		Coke movement	Inactive	Active	Active	Generally active	Inactive, rising and falling
		Sloppy tuyere	Nothing	Nothing	Nothing	Slag in lower part of tuyere	Slag leaks at any time
Slag		Temperature	High	Adequate	Normal	Inadequate	Low
		Color	Bright	Bright	Pale	Normal	Dim
		Fluidity	Sticky	Good	Normal	Stacked in the runner	No flow, no slag
Slag	Slag sample	Color	Yellow-brown	White	normal	Green-black	Black
		Look like	Brown stone	White stone	normal	Green glass-black	Black
		FeO, %		<0.5	<0.5	0.5–1.5	>1.6
Pig iron	Hot metal	Brightness	Bright	Bright	Bright	The darker	Dull red
		Temperature		High	Normal	Low	Low
		Fluidity	Poor, viscous groove at high Si	Poor	Normal	Good	No flow solidification in ditches
		Clip flower	Few, big particles	High, low, particles	Normal	Low, many, broken	
	Iron sample	Surface shape		Big hollow hearts.	Small hollow hearts.	Bulge	Bulge, a dot.
		Fracture color	Black	Gray	Normal	White	White, bright stripes
		Graphite precipitation	There is a layer of graphite	Fine graphite	Normal	Graphite-free precipitation	
		Silicon content	High	Higher	Normal	Low	Very low

Table 4.6 Quantitative adjustment parameter coefficient of Capital Steel

Category	Parameter name	Amount of change	Affecting coke ratio	Influence of pulverized coal
Ore	Iron content Size of <5 mm Sinter % FeO, %	±1% ±5% ±10% ±1%	±1.5% ±1% ±2% ±1.5%	
Coke	Ash content Moisture S	±1% ±1% ±0.1%	±2% Deduction ±1.2–2%	
Pulverized coal	Ash content Weight	±1% ±1 kg	±1.5% ±0.7%	
Auxiliary material	Limestone Iron scrap	±1% ±1%	±0.3% ±0.22%	±0.735% ±23%
Air	Air temperature Humidity Air volume	±10 °C ±1 g/m^3 ±1 m^3/(min·m^3)	±1.7 kg ±1 kg ±1.4 kg	±2.1 kg ±1.25 kg ±1.75 kg
High pressure	Top pressure	(±1 kg/cm^2) ±10 kPa	±6 kg	±7.5 kg
Raw iron composition	[Si] [Mn]	±1% ±1%	±6 kg Si <1.5%, ±40 kg 1.51–2.5% ±60 kg 1.51–3.5%, ±85 kg ±14 kg	

Ore batch weight W_k	Coke batch weight W_j	Limestone quantity	Humidity
17.4 t	5.25 t	0.1 t	23 g/m^3
[Si]	Blast temperature	Batch iron discharge	
0.95%	950 °C	9.7 t	

At that time, slag basicity and [Si] content in hot metal were high, and limestone was ready to be removed, and [Si] was reduced to 0.75%.

Adjusting

According to the data in Table 4.6, the consumption of coke per kilogram of limestone is 0.3 kg, [Si] 1% is equivalent to 40 kg/t of coke, so each batch of burden can reduce coke:

$$(0.95 - 0.75) \times 40 \times 9.7 + 100 \times 0.3 = 107.6\,\text{kg/batch}$$

The actual coke was reduced by50 kg/batch at 22:00, another 50 kg/batch at 6:00 the next day, and the humidity by 23–20 g/m^3, and then, the hot metal contained 0.75% of [Si]. Quantitative adjustments were issued at that time. The foreman was cautious in his operation. The coke was reduced twice, wasted 8 h and coke:

$$64\,\text{batch} \times 50\,\text{kg/batch} = 3200\,\text{kg}$$

This 3.2 t coke is wasted, which is the cost of unskilled operation.

Example 2 If [Si] in Example 1 is improved to 1.2%, how to adjust it?

There are three ways of adjusting:

(1) Lifting blast temperature:
The blast temperature at 100 °C, equivalent to 17 kg/t iron coke, to increase [Si] from 0.95 to 1.2%, requires blast temperature to be raised:

$$(1.2 - 0.95) \times 40/1.7 \times 10 = 58.8\ ^\circ\text{C}$$

(2) If [Si] is raised with pulverized coal:

$$(1.2 - 0.95) \times 40/1.7 \times 10 = 58.8\,^\circ\text{C}$$

(3) After removing limestone, [Si] increases:

$$100 \times 0.3/9.7/40 = 0.077\%$$

Comparing the above three schemes, it is reasonable to remove limestone firstly, which can not only reduce the consumption of limestone but also save coke; if there is still room for higher blast temperature, it is also reasonable to supplement heat to increase [Si] content and give full play to the role of blast temperature; if there is no more room for higher blast temperature, it is also reasonable to use pulverized coal to supplement it. This is also economical and reasonable.

Example 3 On November 17, the ash content of coke was 13.45%, and 400 kg of limestone was added to each batch. On the 18th, the ash content of coke rose to 14.08%, and that of limestone increased to 550 kg/batch. How to adjust and keep the furnace temperature stable?

Solution

At that time, a batch of burden produced hot metal 5.2 t, coke ratio 623 kg/t, coke ash increased by 0.63%, so coke should be added:

$$5.2 \times (0.63 \times 2\% \times 623) = 40.8\,\text{kg/batch}$$

When limestone increased by 150 kg, coke should be added: 150 × 0.3 = 45 kg/batch. Therefore, 85.8 kg coke should be added to each batch.

Modern blast furnaces are no longer fed directly with limestone. The above examples introduce the principle and method of furnace temperature adjustment. At present, the daily furnace temperature control of blast furnaces is mainly to adjust auxiliary fuels and pulverized coal. If the furnace temperature is adapted for a long time, the burden load should be changed.

4.3.4 Action Time of Each Parameter

The volume of blast furnace is very large. Any parameter change will take a period of time to show. If the air temperature is raised by 50 °C, it usually takes two hours for the furnace temperature rising. The foreman used to call this time "inertia." Automation experts study inertia more carefully: When a parameter changes when the system works, it is called "response time" or "dynamic action time." Staib [8] and Wood [9] had done pioneering work in blast furnace dynamic control. Figure 4.4 shows a "dynamic characteristic curve" of humidity and blast temperature.

Humidity decreases by $\Delta\varphi$ g/m^3 from t_0 time, and [Si] in hot metal increases from t_{in} time to 2/3 of the total effect in around 2 h. This period is called "dynamic response time." It takes about 8 h for full effect. This is a very careful consideration. What we can feel in our daily operation is 2 h, that is, 2/3 of the total action time. The response characteristics of blast temperature and humidity are similar.

In terms of practical blast furnace operation experience, blast temperature takes effect about 2 h, so does humidity. Figure 4.5 shows the empirical values of blast furnace operation time. The diagram of control expert changing parameters is called "dynamic response curve." Tables 4.4 and 4.5 show the dynamic thermal curve of humidity and the response curve of common parameters, respectively.

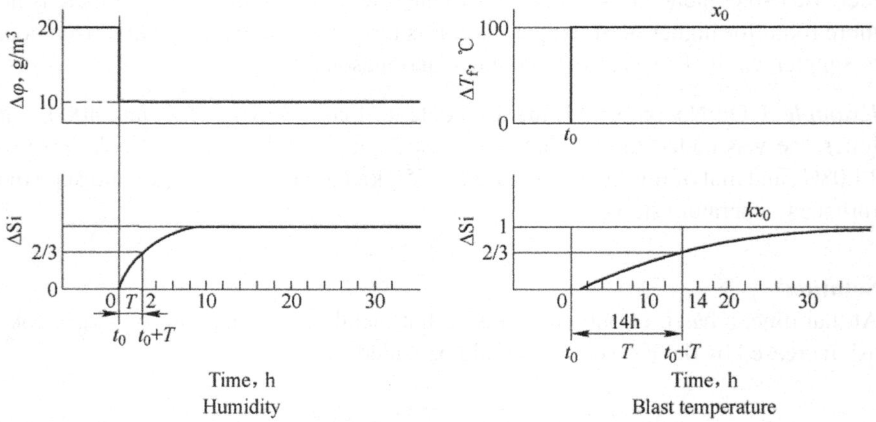

Fig. 4.4 Dynamic characteristic curves of humidity and blast temperature

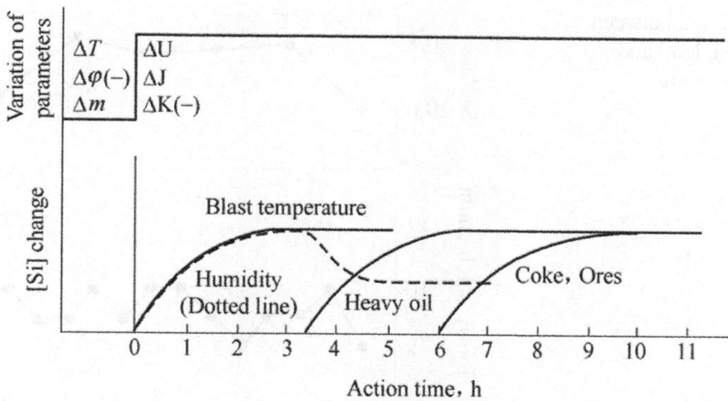

Fig. 4.5 Response curve of common parameters

Because of different operation conditions and parameters adjustment time, each factory should accumulate experience according to its own conditions. The reaction time of ore and coke reaches the hearth through a smelting cycle and then accumulates a secondary iron content to complete the furnace temperature raking. Figure 4.5 shows that the smelting cycle is 6 h. When the burden reaches the hearth and continues to be heated up, it takes over 2 h, that is, 8 h, to really work.

4.3.5 Furnace Temperature Control Method

Stability of furnace temperature is the basic condition for high production and low consumption of blast furnace. Excellent blast furnace workers always keep the furnace temperature near the required lower limit. To stabilize the furnace temperature, there are two main points of operation:

(1) The stability of comprehensive load is the basis for the stability of furnace temperature.
(2) Burden velocity is stable, which is the standard for measuring furnace temperature stability.

Blast furnace foreman should always pay attention to the change of hourly comprehensive load. Generally, the comprehensive load is stable and the furnace temperature is naturally stable. Figure 4.6 shows a production example of No. 1 BF of Capital Steel in March 1978.

If there is a change in the comprehensive load in the first hour, it should be added in the second hour so that the average comprehensive load in the two hours is consistent with the normal load. Experienced foremen find the change of burden velocity, pay attention to observation and adjust in time. If the burden velocity is slow, the furnace temperature should be prevented from overheating; if the burden velocity is found to be too fast, it should also be adjusted in time to prevent the furnace cooling.

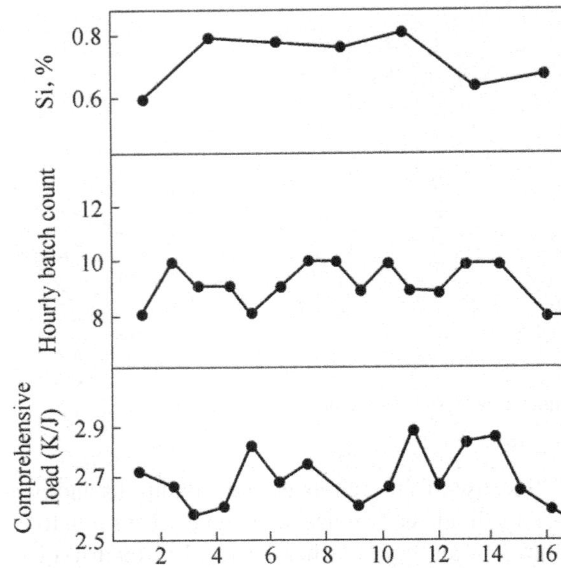

Fig. 4.6 Relation between comprehensive load and furnace temperature

Compared with the consequences of furnace temperature change, furnace cooling is more serious: Serious furnace cooling can lead to malignant accidents such as excessive iron, slag filling, and even chilled hearth. Furnace cooling is the starting point of accidents, and prevention of furnace cooling is an important measure to prevent accidents.

4.3.6 Burden Velocity Change

It is an important test standard to judge the trend of furnace temperature and the change of burden velocity. Usually, when the furnace temperature is normal, burden velocity is stable. Once the furnace temperature changes, burden velocity is bound to change too. When the operating conditions change little, the burden velocity is fast, CO_2 in gas decreases and CO increases, while the burden velocity is slow, the opposite situation will appear.

The change of burden velocity reflects the utilization degree of gas to a certain extent and also reliably reflects the change of furnace temperature. The relationship between gas utilization and burden velocity is determined by the material balance of blast furnace. The change of CO_2 in medium gas depends on the ratio of oxygen to carbon in blast furnace. When O/C is high, CO_2 in gas is high. Oxygen and carbon in the blast furnace come from the top charge and lower tuyere injection. About 70–100% carbon and 40% oxygen are brought into the top charge, and the rest is blown in through the tuyere. When the blast volume and coal injection quantity are constant and the burden velocity is fast, the carbon in the furnace will increase, O/C

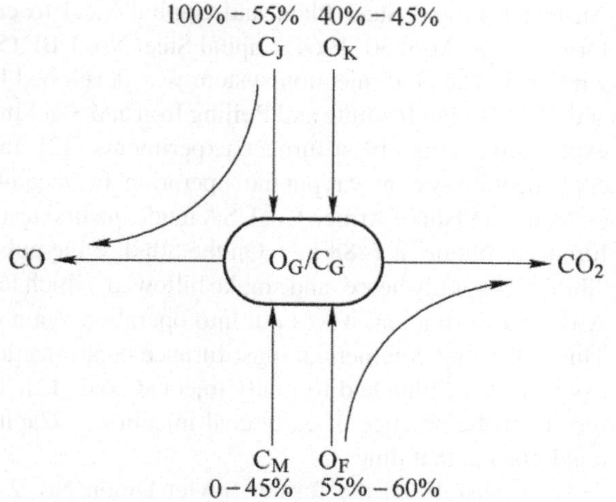

Fig. 4.7 Diagram of burden velocity and gas composition

decreases and CO_2 decreases, while the carbon decreases when the burden velocity is slow, resulting in the increase of O/C and CO_2 in the gas. Figure 4.7 is a schematic diagram of the relationship between burden velocity and gas composition.

Experienced blast furnace foremen always pay attention to the change of burden velocity. In daily operation, when it becomes abnormally fast, they should take immediate measures to control the speed. When burden velocity increases continuously, blast should be reduced and the speed should be controlled.

4.4 Coal Injection Operation

Pulverized coal injection has a long history. As early as 1840, S. M. Banks has been patented and tested it on blast furnace from 1840 to 1846 [10]. In the 1960s, the test of injecting coal into blast furnaces was carried out again. In 1961, Hannah Company of the USA and Sars Company of France, Stanton Company of the United Kingdom and Wilton Company of the USA, Oriental Metallurgical Plant of Democratic Germany in 1963, Karaganda Iron and Steel Company of the Soviet Union in 1964, and so on all did the experiment. However, the test of injecting coal into blast furnaces interrupted to continue because at that time, oil and gas are cheaper in these countries [11].

4.4.1 Coal Injection in China

In his early years, Mr. Shen Yijia, who worked under a work-study program in France, handed over the information of coal injection in the International Iron-making Conference held in Belgium to the Ministry of Metallurgy of China, which arranged

Anshan Iron and Steel Plant and Capital Steel to carry out the practice in blast furnaces. On April 30, 1964, Capital Steel No. 1 BF (576 m^3) was formally injected with coal. The coal injection system was developed by Capital Steel, Beijing Iron and Steel Design Institute and Beijing Iron and Steel Institute, referring to the defects exposed in foreign blast furnace experiments [12]. In June, An Steel blast furnace coal injection system was put into operation. In December of the same year, Bellefont of Ashland Plant of Armco Co. USA made the first test of injecting coal into the blast furnace (volume of 1488 m^3). On the 8th day, the pulverized coal burned itself, and "flutter" was only heard, and smoke billowed, which forced the test to be interrupted. After transformation, it was put into operation again on January 12, 1966 [13–15]. This is the first commercial blast furnace coal injection abroad. At this time, many ironworks in China had formally injected coal [12]. Tables 4.7, 4.8 and 4.9 are the results of the practice of early coal injection at Capital Steel and the raw material conditions at that time.

In August 1965, the former Soviet Union No. 2 BF of Karaganda Plant with 1719 m^3 injected coal. By 1970, a total of 95,000 t of coal were injected. The first coal injection blast furnace in the Soviet Union was No. 2 blast furnace in Donetz Plant. The Soviet Union reported a lot. Only from 1972 to 1976, there were eight papers [16–19] and a monograph [20]. Like the Americans, they attached great importance to propaganda.

After the oil crisis in 1973, coal injection had received widespread attention, coke resources were increasingly scarce, and the pollution caused by coke ovens had been thoroughly studied. It was found that the gas leaked from coke ovens had carcinogenic substances, which caused people to be extremely uneasy. In the 1980s, the world-wide construction climax of blast furnace coal injection was formed. Countries began to inject coal one after another:

Country	The year of beginning coal injection
China	1964
USA	1964
Soviet Union	1968
Japan	1981
Britain	1983
Germany	1985

At that time, coal injection was the most common in China. Japan once organized a group to negotiate with the USA, the former Soviet Union talked over with China to purchase coal injection technology. They talked with Capital Steel and asked to inspect the coal injection facilities. At that time, China had not yet opened up, nor understood the rules of patent negotiation rules. As for the reasonable requirement from Japan, only one circle visit of blast furnace was allowed, and no coal injection facilities were opened. Japanese experts said, "You said that the fish is fresh, but you put it in your pocket and refuse to take it out for us to see, how do we make decisions?"

Table 4.7 Early coal injection operation of capital steel blast furnace

Project		Baseline period 4 days	The first period is 6 days	The second period is 15 days	The third period is 15 days	The fourth period is 12 days
Utilization coefficient, t/(m^3 d)		1.97	2.18	2.55	2.59	2.65
Smelting intensity, t/(m^3 d)		1.160	0.895	0.990	0.887	0.870
Comprehensive smelting intensity, t/(m^3 d)		1.160	1.220	1.485	1.499	1.585
Coal injection quantity, kg/t		0	26.6	35.9	40.5	45.2
Coal injection rate, %		0	26.6	35.9	40.5	45.2
Actual coke ratio, kg/t		587	412	375	344	328
Actual replacement ratio (coke/coal)			1.17	1.01	1.0	1.04
Corrected replacement ratio (coke/coal)			0.94	0.82	0.83	0.90
Air temperature, °C		17.9	3.8	4.6	9.3	13.5
Humidity, g/m^3		17.9	3.8	4.6	9.3	13.5
Oxygen in blast, %		20.5	20.5	21.92	23.72	24.22
Furnace top temperature, %		331	395	415	446	443
Sinter proportion, %		95.6	99.5	100	99.4	100
Iron content in ores, %		51.71	50.20	53.09	52.78	52.82
Limestone usage, kg/t		37.2	8.2	24.2	2.7	0.34
Scrap quantity, kg/t		0	45.7	16.4	57.0	33.7
Pig iron contains Si, %		0.80	0.55	0.61	0.61	0.64
Pig iron contains S, %		0.018	0.026	0.014	0.018	0.015
Slag basicity (CaO/SiO_2)		1.06	1.035	1.03	1.01	1.00
Mixed gas component, %	CO	27.3	25.9	27.8	29.8	30.0
	CO_2	16.0	17.0	16.5	16.5	16.7
	N_2	55.33	54.81	52.02	50.08	49.1
	H_2	1.2	2.1	3.5	3.6	4.0
	CH_4	0.17	0.19	0.18	0.20	0,20
	CO/CO_2	1,71	1.52	1.69	1.82	1.80
Direct reduction degree, %		51.9	47.5	50.9	51.5	51.2
Slag volume, kg/t		640	640	558	571	571

Finally, they introduced coal injection technology from the USA. Originally, we had an advantage in blast furnace coal injection. It was the most common in our country with the highest injection volume and the best economic benefit. We lost the opportunity of technology export because of a lack of negotiation understanding.

Table 4.8 Sinter composition (%) and screening rate before feeding

Test period	Fe	Mn	S	P	FeO	SiO_2	Al_2O_3	CaO	MgO	Residual C	CaO/SiO_2	Sieving rate, %
Base period	49.36	1.29	0.07	0.049	14.09	9.70	2.27	11.60	4.55	0.12	1.19	0
Phase one	49.17	0.30	0.075	0.081	15.10	10.54	1.86	13.03	3.94	0.13	1.14	0
Phase two	52.87	0.35	0.049	0.022	16.01	8.78	1.69	10.63	4.67	0.12	1.23	76
Phase three	52.10	0.30	0.075	0.027	15.70	9.18	1.02	11.48	4.97	0.13	1.17	100
Phase four	60.0	0.12	0.049	0.026	15.45	8.22	1.70	10.30	4.68	0.13	1.25	39.5

Table 4.9 Compositions of coke and pulverized coal (%)

Analytical principle		Base period	Phase one	Phase two	Phase three	Phase four
Coke	Moisture	4.43	4.80	5.40	5.30	5.27
	Fixed carbon	86.92	86.53	87.48	87.74	87.11
	Ash content	11.67	12.04	11.09	10.85	11.50
	Volatile matter	0.82	0.84	0.86	0.85	0.85
	Sulfur	0.59	0.59	0.59	0.56	0.54
	Drum index, kg	326.5	334	324.8	317	/
Pulverized coal	Moisture	0.60	0.77	0.95	0.65	0.59
	Fixed carbon	78.40	77.56	77.11	76.17	76.59
	Ash content	12.39	13.80	14.52	15.05	14.40
	Volatile matter	9.21	8.27	8.03	8.24	8.52
	Sulfur	0.43	4.37	0.34	0.54	0.49

4.4.2 The Role of Pulverized Coal

The function of pulverized coal is mainly to replace "tuyere carbon" (coke in front of tuyere). With the increase of pulverized coal injection rate, the proportion of pulverized coal replacing tuyere carbon increases rapidly. The pulverized coal injection rate reached 45.2%, the pulverized coal accounted for 60.2% of the total tuyere carbon ratio, the heat released by pulverized coal in hearth area accounted for 35.99% of the total iron-making income, and the corresponding heat released by coke in hearth area was only 20.85%, as shown in Tables 4.10 and 4.11 [21].

4.4.3 Pulverized Coal Combustion

Pulverized coal begins to burn in the blowpipe. Figure 4.8 shows the range of pulverized coal combustion estimated by the measured composition of No. 1 blast furnace of Capital Steel. At that time, the particle size of pulverized coal was very fine, less than 0.088 mm, accounted for about 87%; blast furnace blast temperature 1110–1150 °C, oxygen-enriched 2–4%, spray gun outlet was 1.12 m from the front of the tuyere, pulverized coal combustion in the blowing pipe was between 2 and 18%. From the coal gun outlet to the front end of the tuyere, the average passing time of hot blast was 0.0103 s (Fig. 4.8).

With the increase of pulverized coal injection rate, incomplete combustion of pulverized coal tended to occur. In a relatively short period, the pulverized coal injection rate had been raised to over 49% (see Table 4.12). Practice showed that "the phenomenon of incomplete combustion was serious when the coal injection rate exceeded 47%. In the same smelting conditions, the injection rate rose from 45.4 to 49.1%, the fuel ratio rose by 62 kg, the coal injection rate reaches 52.2%,

Table 4.10 Carbon balance at different coal injection rates (kg/tHM)

Different periods				Base period	1	2	3	4
Coal injection rate, %				0	26.6	35.9	40.5	45.2
Carbon income	Coke introduction			511.9	357.0	328.00	309.50	286.0
	Pulverized coal introduction			0	117.50	162.80	179.80	208.20
	Limestone introduction			1.85	1.79	1.79	1.91	1.92
	Sinter introduction			0.38	0.42	0.41	0.39	0.40
	Total			512.34	476.71	493.00	44.50	43.90
Carbon expenditure	Carbon content in pig iron			43.80	43.00	45.40	44.50	43.90
	Carbon in furnace dust			3.08	4.05	6.10	5.16	4.44
	Direct reduction consumption of carbon			109.71	102.08	111.44	105.27	102.18
	Tuyere carbon	Total		356.55	327.58	330.06	329.22	346.00
		Of which pulverized coal	kg	0	117.5	162.8	179.8	60.2
			%	0	35.9	49.2	54.5	60.2
	Total			513.32	476.71	493.00	486.60	498.52

and the fuel ratio rose by 81 kg." "Practice shows that the pulverized coal was fully utilized with oxygen enrichment of 4–5%, blast temperature of over 1100 °C and injection rate of 45%" [22].

4.4.4 Effect of Coal Injection on Smelting Cycle

With the increase of pulverized coal injection rate, the proportion of iron-bearing burden in the burden increases significantly, and the contact time between iron-bearing burden and gas increases relatively. The smelting cycle of burden also prolongs correspondingly when pulverized coal largely replaces tuyere carbon. The influence of pulverized coal injection rate on the treatment cycle can be quantitatively calculated by the following formula:

$$\tau = \frac{24\,V}{\xi\left(\frac{1}{\rho_{\mathrm{J}}} + \frac{\mathrm{K/J}}{\rho_{\mathrm{K}}}\right) i\,V_{\mathrm{n}}(1-n)} \tag{4.4}$$

where τ is smelting cycle, h; V is blast furnace volume from stock line to tuyere center line, m^3; V_{n} is effective volume of blast furnace, m^3; K/J is coke load; P_{K},

Table 4.11 Heat income of different coal injection rates

Different periods Project	Base period		1		2		3		4	
	kcal/kg	%	kcal/kg	%	kcal/kg	%	kcal/kg	%	kcal/kg	%
1. Heat release from coke combustion	764	55.1	485	35.23	389	27.81	319.5	24.32	281	20.85
2. Heat release from pulverized coal combustion	0	0	273.5	19.88	378	27.00	418	31.82	485	35.99
Total tuyere carbon heat release	764	55.1	758.5	55.09	767	54.81	737.5	56.14	766	56.84
3. Hot air brings in physical heat	456	32.80.30	456	33.12	468	33.45	421	33.02	424	31.50
4. Slag forming heat	4.15	0.21	0.915	0.007	2.7	0.193	0.31	0.024	0.02	0.001
5. Heat of formation of methane	2.92	11.57	2.91	0.213	3.03	0.217	2.98	0.226	3.04	0.226
6. Physical heat brought in by furnace charge	160.5		159.0	11.55	158.5	11.33	152	11.59	153.9	11.43
Total	1337.57	100.00	1377.33	100.0	1399.0	100.00	1313.79	100.00	1346.00	100.00

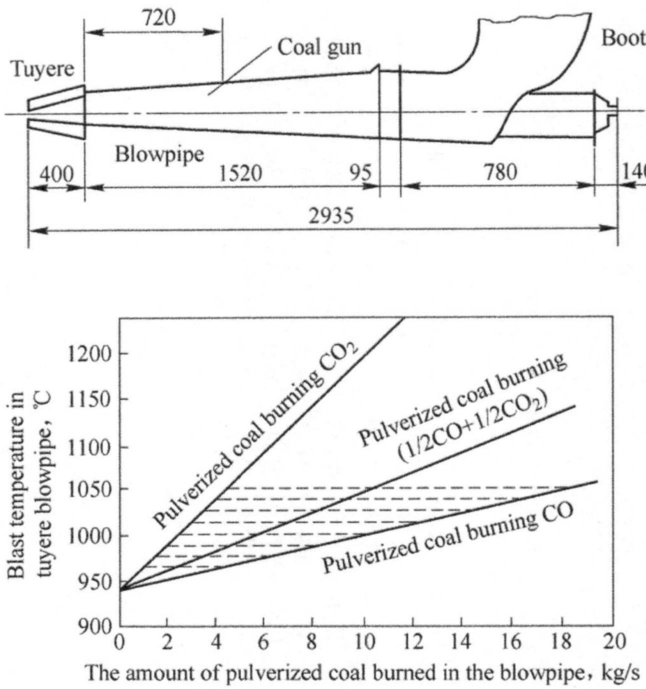

Fig. 4.8 Combustion rate of pulverized coal in blowpipe

Table 4.12 Fuel ratio of different coal injection rates

Time	April, 1966	May, 1966	June 2–8, 1966	June 7, 1966
Utilization coefficient, t/(m^3 d)	2.425	2.45	2.31	2.22
Coke ratio, kg/t	370	336	345	337
Coal injection quantity, kg/t	226	279	332	369
Coal injection rate, %	38	45.4	49.1	52.2
Fuel ratio, kg/t	596	615	677	706
Oxygen in blast, %	23.4	24.9	25.2	23.5
Blast temperature, °C	1144	1150	1150	1150
Furnace top pressure, kg/cm^2	0.58	0.58	0.60	0.60

P_J is bulk density of ore and coke, t/m^3, respectively; i is comprehensive smelting intensity, t/(m^3 d); ξ is compression coefficient of burden; n is coal injection rate.

In Eq. (4.4), the compression coefficient of burden was obtained by the method of measurement: the first measurement: at 8:20 adding 4t manganese ore in burden and another 4 t every 32 min, continuously discharging slag and sampling from

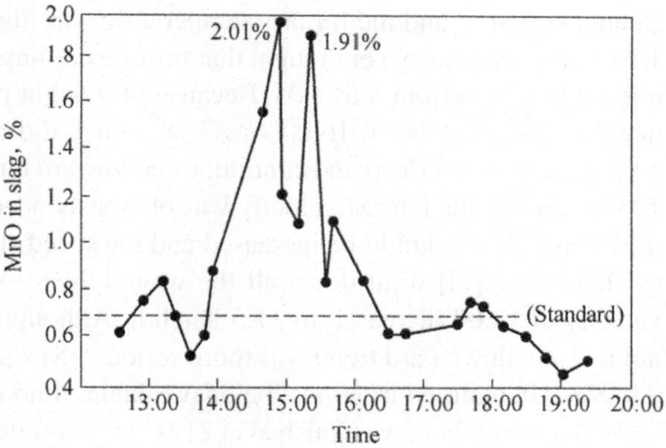

Fig. 4.9 Changes of MnO in slag

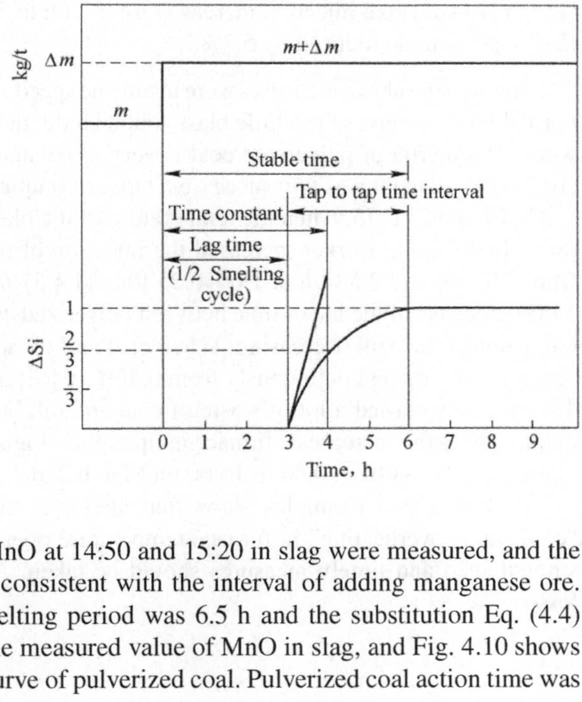

Fig. 4.10 Dynamic characteristic curve of pulverized coal

12:35. The peak values of MnO at 14:50 and 15:20 in slag were measured, and the interval of peak values was consistent with the interval of adding manganese ore. It was obtained that the smelting period was 6.5 h and the substitution Eq. (4.4) was 0.9. Figure 4.9 shows the measured value of MnO in slag, and Fig. 4.10 shows the dynamic characteristic curve of pulverized coal. Pulverized coal action time was about 4 h

4.4.5 Temperature Adjusting with Pulverized Coal Injecting

When the adjustment for furnace temperature inertia (hysteresis) of pulverized coal is large, the adjustment should be started as soon as possible, but often be overlooked. For example, when Capital Steel No. 3 blast furnace (1036 m^3) was making steelmaking iron and the furnace temperature was higher, hot metal containing [Si] 1.24%, the foreman never thought that furnace cooling would occur. At that time, the normal load was from 3 to 3.05. Because of thought paralysis, the average comprehensive load at 10:00 to 16:00 was 3.18, which did not attach importance. Due to comprehensive overload and continuous downward temperature of the furnace, from 12:00 onward, the burden velocity was obviously accelerating, and at that time, the injection volume should be increased and the speed should be controlled. At 18:22, the hot metal [Si] went down all the way: 1.24% → 0.9% → 0.6%. The burden velocity at 18:00 increased to 11.5 batch/h. Although the furnace temperature was not low, the downward trend was more serious. [S] was from 0.017% → 0.027% → 0.048%. Unqualified iron must be unavoidable, if no measures.

When the tapping was finished at 21:00, it was found that the hot metal contained 0.06% of [S]. Until then, measures were taken to try to rise the furnace temperature:

(1) Pulverized coal injection increased from 3 t/h to 3.6 t/h;
(2) Blast volume reduced by 5.7%.

Although the above measures were taken, the speed was not lower than the original normal level because of too little blast volume reduction, indicating that they did not work. The inertia of pulverized coal is very large, and it takes 3–4 h to work. The cost of the accident was two successive tapes of unqualified iron.

On March 2, 1975, when the temperature of the blast furnace went down at 1:00 on night shift, the worker increased the injection of pulverized coal and heavy oil from 5.05 t/h and 2.58 t/h at 1:00–6.25 t/h and 4.57 t/h at 5:00, which lasted until 14:00. Because of the lag of time between pulverized coal and heavy oil, the furnace temperature was still decreasing. When coal and oil acted, the furnace temperature began to rise, rising continuously from 0.42% of [Si] at 8:00 to 0.68% and to 1.0%. This not only caused a lot of waste of coal and oil, but also caused serious loss of output due to the increase of furnace temperature. Figure 4.11 shows a record of the operation process from 0:00 to 16:00 on March 2nd.

The above two examples show that attention should be paid to preventing "cold" and "overheating" in furnace temperature operation. Observation should be strengthened and timely measures should be taken. Any delay would bring huge losses.

4.5 Low-Silicon Iron Smelting

The role of [Si] in hot metal has been discussed in Sect. 4.3, especially its effect on focal ratio:

[Si] content	$< 1.5\%$	$1.5\% - 2.5\%$	$> 2.5\%$
Influencing coke ratio	40 kg/t	60 kg/t	85 kg/t

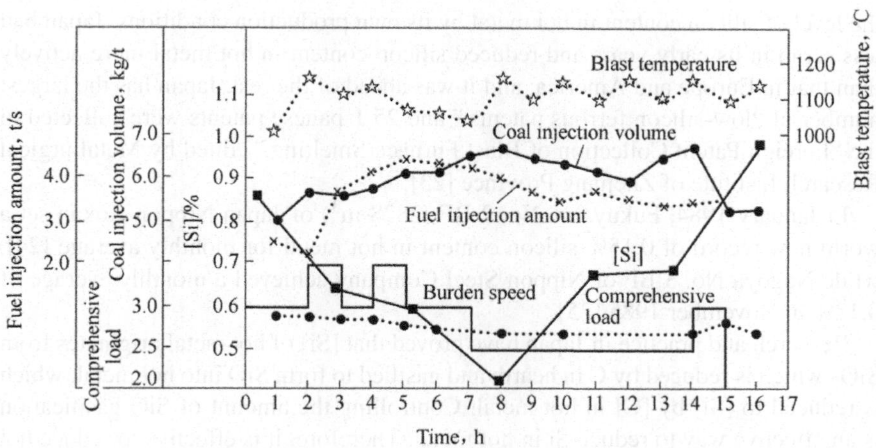

Fig. 4.11 Large fluctuation of furnace temperature caused by operation fault

Fig. 4.12 Silicon content in hot metal of Capital Steel

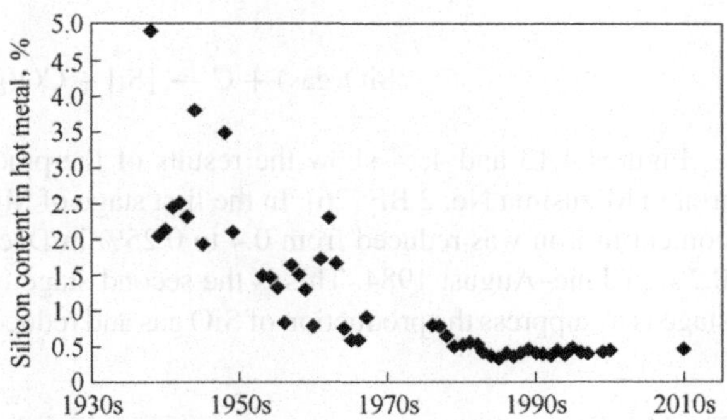

Therefore, reducing the silicon content in steel not only meets the requirements of hot metal for steelmaking, but also is the goal pursued by blast furnace workers. Figure 4.12 shows the change of silicon content in hot metal in Capital Steel over the years. Unlike now, Capital Steel only produced iron pigs for marketing, often producing cast iron and high manganese iron.

4.5.1 Low-Silicon Iron Production

With the development of the times, the quality requirement of steel is increasing day by day. High-quality steel requires very low sulfur and phosphorus. Before desulfurization and dephosphorization of molten steel, desilication is necessary. The content of silicon in hot metal reflects the temperature of hot metal. Steel making with scrap requires sufficient temperature of hot metal, so each factory determines the level of silicon content in hot metal by its own production conditions. Japan had less scrap in its early years and reduced silicon content in hot metal more actively than that in Europe and America, and it was ahead of the rest. Japan has the largest number of "low-silicon ferrous patents," and 25 Japanese patents were collected in the "Foreign Patent Collection of Blast Furnace Smelting," edited by Metallurgical Research Institute of Zhejiang Province [23].

In January 1984, Fukuyama No. 2 BF (2828 m^3) of Japan Nippon Kokan set a world new record of 0.15% silicon content in hot metal for monthly average [24], while Nagoya No. 3 BF of Nippon Steel Company achieved a monthly average of 0.12% in November 1985 [25].

Research and practice in Japan have proved that [Si] of hot metal originates from SiO_2 which is reduced by C in hearth and gasified to form SiO into hot metal, which is reduced to [Si] by [C] in hot metal. Controlling the amount of SiO gasification is an effective way to reduce Si in hot metal. Therefore, it is effective to reduce hot metal temperature or basicity of blast furnace slag and suppress SiO_2.

$$(SiO_2) + C(solid) \rightarrow SiO(gas) + CO(gas) \tag{4.5}$$

$$SiO(gas) + C \rightarrow [Si] + CO(gas) \tag{4.6}$$

Figures 4.13 and 4.14 show the results of the practice of smelting low-silicon iron in Mizusima No. 2 BF [26]. In the first stage of Mizusima No. 2 BF, the silicon content in iron was reduced from 0.4 to 0.25% in December 1983 and to less than 0.2% in June–August 1984. This is the second stage of silicon reduction. The first stage is to suppress the production of SiO gas and reduce the volume of melting zone.

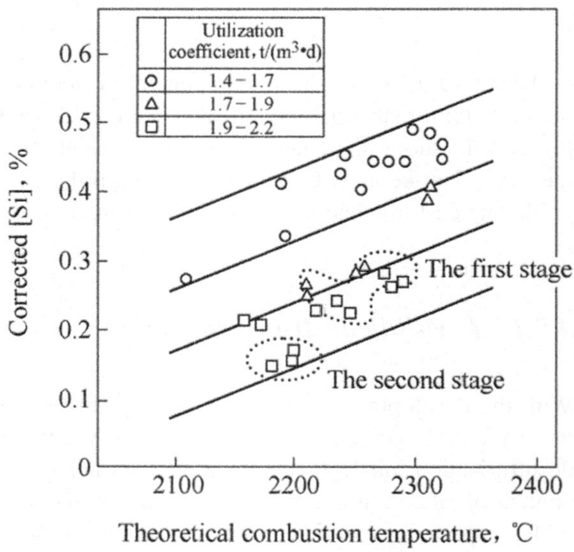

Fig. 4.13 Corrected relation between silicon content and theoretical combustion temperature

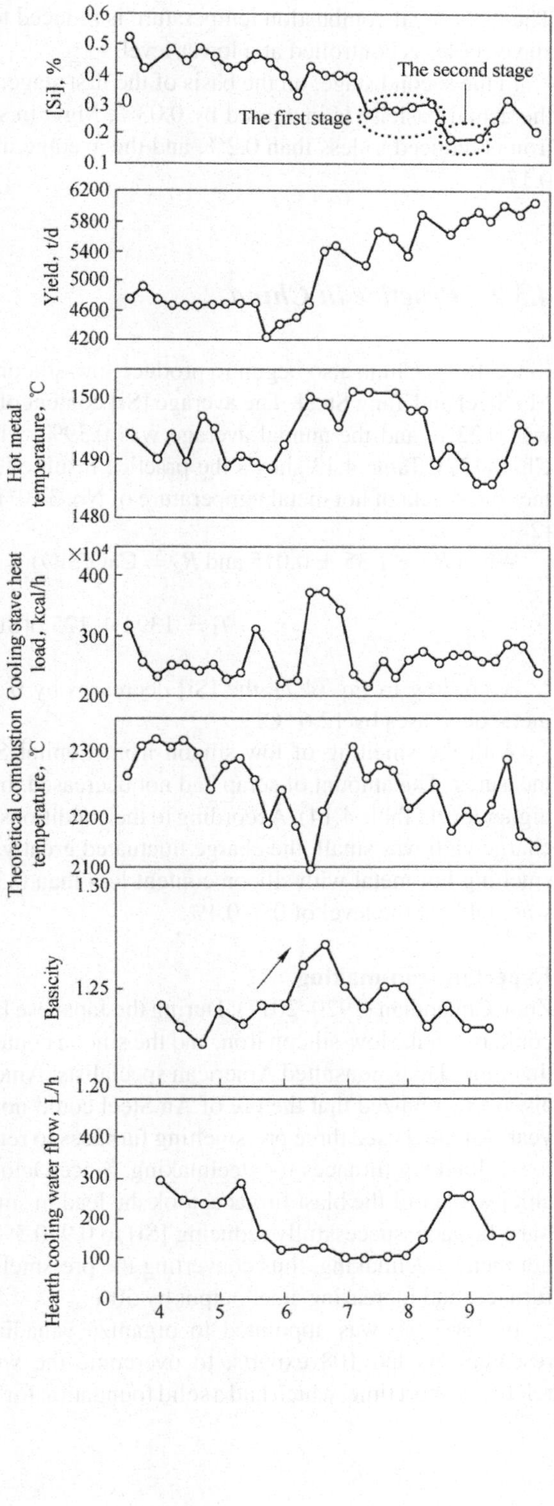

Fig. 4.14 Changes in parameters

The theoretical combustion temperature is reduced to 2200 °C, and the heat load of stave cooler is controlled at a lower level.

In the second stage, on the basis of the first stage, the basicity of slag is adjusted, the actual basicity is increased by 0.03%, MgO in slag is increased by 2%, [Si] in iron is reduced to less than 0.2%, and the average monthly in August is reduced to 0.17%.

4.5.2 *Practice in China*

Since 1980, China also begun to produce low-silicon iron effectively in Hang Steel, Ma Steel and Tang Steel. The average [Si] content of Hang Steel in September 1983 was 0.22%, and the annual average was 0.39%, which was the first-class level in China [27]. Table 4.13 shows the practice result of No. 3 BF of Capital Steel. The measured data of hot metal temperature of No. 3 BF in January 1983 were sorted out [28].

When $R_3 = 1.35 \pm 0.015$ and $R_2 = \mathrm{CaO/SiO_2} = 1.1 \pm 0.10$,

$$T_{\mathrm{I}} = 1394 + 125.6[\mathrm{Si}] \tag{4.7}$$

According to Eq. (4.7), the [Si] decreases by 0.1% and the temperature of hot metal decreases by 12.6 °C.

With the smelting of low-silicon iron, Capital Steel had taken many effective measures. The amount of scrap had not decreased, but increased, and stabilized at a higher level (Table 4.14). According to the conditions of Capital Steel at that time, the charge yard was small, the charge fluctuated greatly, and there was no qualification smelting hot metal with silicon content less than 0.3%. During the better period, it was stable at the level of 0.3–0.4%.

Expert in Iron-making

Zhou Chuandian (1920–2017): During the Japanese Invasion puppet period, An Steel could not make low-silicon iron, and the silicon content of hot metal was 1.5–2%. At that time, Japan consulted American specialists. American and Japanese experts had always recognized that the ore of An Steel could not make low-silicon iron. In that year, An Steel used three pre-smelting furnaces to remove silicon and then sent them to steelmaking furnaces for steelmaking. Since October 1950, Zhou Chuandian, the chief worker of the blast furnace, took the lead in smelting low-silicon iron in No. 2 blast furnace, successfully reducing [Si] to 0.9–0.5%, creating conditions for direct hot metal steelmaking, thus converting the pre-smelting furnace into a steelmaking furnace, and increasing steel output by 50%.

In 1965, he was appointed to organize vanadium and titanium ore smelting research. He led 108 experts, to overcome the world's smelting problems in a relatively short time, which laid a solid foundation for Pan Steel to put into production.

Table 4.13 Changes of indicators after silicon reduction in No. 3 blast furnace of Capital Steel

Time	[Si],%	Standardized air speed, m/s	Actual air speed, m/s	Kinetic energy, kg·m/s	Air volume, m^3/min	Air volume, MPa	Penetrability index	Comprehensive smelting intensity, t/(m^3·d)	Average daily output, t	Coke ratio, kg/t	Comprehensive coke ratio, kg/t
August, 1982	0.48	112	183	4845	1953	0.199	1588	1.036	2135	392	499
October, 1982	0.47	114	183	5031	2047	0.208	1551	1.150	2333	384	507
November, 1982	0.47	117	188	5480	2071	0.205	1593	1.184	2435	388	502
January, 1983	0.43	115	180	4900	2069	0.214	1488	1.17	2424	386	499
February, 1983	0.43	120	185	5550	2262	0.220	1407	1.221	2419	394	508
March,1983	0.42	128	190	6104	2280	0.222	1539	1.210	2439	402	509
April,1983	0.40	130	197	6623	2335	0.220	1624	1.225	2548	389	490
May, 1983	0.36	190	6629	2390	0.216	1695	1.225	2605	396	496	

Table 4.14 Actual scrap consumption of Capital Steel in that year

Year	1978	1979	1980	1981	1982	1983	1984	1985	1986
Steelmaking hot metal [Si], %	0.86	0.50	0.53	0.57	0.44	0.40	0.40	0.35	0.40
Scrap steel consumption in steelmaking plant, kg/t-steel	67.77	76.41	83.05	114.41	123.34	133.03	132.91	132.41	134.88

Baiyun'ebo Iron Mine of Baotou Steel has a high content of fluorine and alkali metals, and its high and long-term production is not smooth after its commissioning. From 1973 to 1977, as team leader, he organized relevant colleges and universities and the company to tackle key problems together to make the production of Baotou Iron and Steel Co. turn to normal.

He is the author of "Zhou Chuandian Collection" (4 volumes) [29] and editor-in-chief of "Technical Manual of Blast Furnace Iron-making Production," "Formation and Development of Iron-making Technology in An Steel" and several other essays [30].

4.6 Chilled Hearth and Treatment

Chilled hearth refers to the fact that hot metal cannot tap out of the tap hole due to the coldness of the furnace, the stickiness or condensation of the slag. Modern blast furnace chilled hearth seldom occurs. In the past years, chilled hearth happened occasionally because of poor equipment and insufficient measuring instruments.

4.6.1 Reasons and Types of Chilled Hearth

In "One Hundred Cases of Blast Furnace Accident Handling" edited by Xu Juliang and Liu Qi, two experts of the Ministry of Metallurgy, in 1984, the chilled hearth was listed as the first chapter, a total of 19 cases [31]. These 19 cases occurred in famous iron and steel companies in China at that time and can be classified into four categories:

Category :	Operating failure	Equipment fault	Water leakage	Long – term blowing off
Number :	4	5	4	6

In fact, these four categories are interrelated.

The freezing caused by operation errors mainly refers to the freezing caused by serious furnace cooling caused by operation reasons, such as the chilling of hearth caused by continuous channel, deep collapse or weighing errors.

When equipment failure happened to force accidentally blowing off, production was forced to interrupt. Some factories are forced to shut down for days or tens of days due to fan failure; or feeding equipment failure, long-term shutdown, inadequate sealing of blast furnace, or the cooling intensity of cooling system failing to reduce in time and too much heat dissipation of blast furnace all result in chilled hearth.

"Watering" into the furnace consumes a lot of heat for whatever reason. Leakage not only happens in the water tank or the stave. There was a 576 m^3 blast furnace. Because the temperature at the top of the furnace was too high, it was forced to water the top. After the shift, the foreman did not explain to the next shift, only recorded on the shift record: “Because the top temperature is too high, the top is watering.” The successor saw the record and thought that the water was instantly watering, which did not pay attention. He did not check the top of the furnace and cut the water. The water was continuously poured for 7 h until iron could not be discharged from the tap. When there were partial slag gushing and slag filling at the tuyere, it realized that the problem was serious, but it was too late.

When there is long-term blowing off or long-term banking of the furnace, freezing also occurs: Although the blowing off or banking with preparation is different from long-term blast shutoff caused by equipment failure, the plan is not in line with reality or taking longer time than expected or inappropriate corresponding operation results in chilled hearth.

The freezing types listed in Table 4.15 are actually the main reasons for chilled hearth.

4.6.2 Hearth Freezing Caused by Weighing Errors

In the initial stage of self-construction of large blast furnaces in China, the equipment level, especially the automatic testing facilities, including measuring elements and management software, has gone through a period of exploration and accumulation. An Steel No. 11 blast furnace (2025 m^3) is the first large blast furnace with the high degree of automation designed and constructed independently in China. On March 3, 1986, 17:58–23:22, No. 11 BF of An Steel was loaded with 27 batches of 2106 t sinters for 5 h 24 min without coke, which was caused by mid-shift operation errors. After the shift leader took over, the iron was discharged at 16:30, and then, the pressure difference increased and the permeability index decreased. The shift leader thought that the furnace was not in good condition, reducing blast volume and coal, and carried out the wrong procedure (this feeding procedure was defective, only adding ores without coke). It was only around 23:00 that the wrong feeding procedure was found. At this time, estimably, the coke-free burden had reached to the lower part of the furnace body (Fig. 4.15).

If 2106 t coke-free cold burden fell into hearth, the whole furnace would be frozen. How to deal with it? There were different opinions, after discussion and decision at the meeting:

The first step was to blast with a small number of tuyeres to remove slag–iron and replace cold burden.
The second step was to blow out the cold burden from the selected tuyere.
The third step was to cut the furnace jacket and remove the cold burden.

4.6.2.1 Blowing on and Burden Exchanging

As shown in Table 4.16, starting from 3:45 on the March 4th, No. 19, 20 and 21 tuyere were used to supply air, and the rest were blended. Because the furnace was

Table 4.15 Chilled hearth and types

Factory name	Furnace number	Volume, m^3	Freezing date	Processing time	Cause of freezing	Freezing classification
Ben Steel	5	2000	January 22, 1978	5	Channel + collapse	1
An Steel	3	831	January 12, 1971	4	Furnace cooling + collapse	1
Baotou Steel	3	1800	June 26, 1972	10	Furnace cooling + equipment failure + continuous blowing off	1 + 2
Ji Steel	2	100	January 5, 1981	5	Blast stoppage, furnace cooling + operation failure in iron tapping accident	1 + 2
Shui Steel	2	1200	January 30, 1979	9	Poor coke strength, furnace cooling + operation failure	1
An Steel	1	568	March 12, 1975	7	Stave cooler leakage	3
An Steel	11	2085	March 12, 1975	7	32 h of blowing off overhaul + leakage, 16 h later found out	4
Baotou Steel	2	1513	March 12, 1975	10	Overhaul 37 h + leakage	3 + 4
Tai Steel	3	1200	March 12, 1975	8	Water leakage	3

(continued)

Table 4.15 (continued)

Factory name	Furnace number	Volume, m^3	Freezing date	Processing time	Cause of freezing	Freezing classification
An Steel	6	936	March 12, 1975	8	Due to the earthquake, no equipment was used to seal the furnace for 25 days, the sealing was not strict, and the feeding line dropped by 7 m	4
Anyang Steel	2	255	March 12, 1975	51	19 days unplanned blowing off due to equipment failure	2
An Steel	10	1513	March 12, 1975	7	Feeding equipment failure + frequent blackout	2
Wu Steel	2	1436	March 12, 1975		Closed for 71 days	4
Capital Steel	4	1200	March 12, 1975	10	19-day maintenance plus continuous blowing off	4 + 1
Wu Steel	4	2516	March 12, 1975	10	After 8 months of medium repair, in the furnace steel there is 700–800 t residual iron, 100–200 t residual slag and 20 t flux	4
Shui Steel	1	568	March 12, 1975	10	Banking for 22 days	4

(continued)

Table 4.15 (continued)

Factory name	Furnace number	Volume, m^3	Freezing date	Processing time	Cause of freezing	Freezing classification
Ji Steel	2	255	March 12, 1975	12	Leakage + blast break	3
Ma Steel	11	300	March 12, 1975	5	Equipment failure, blowing off for 8 h	2
Xiang Steel	2	750	March 12, 1975	8	Equipment failure + furnace cooling	2 + 1

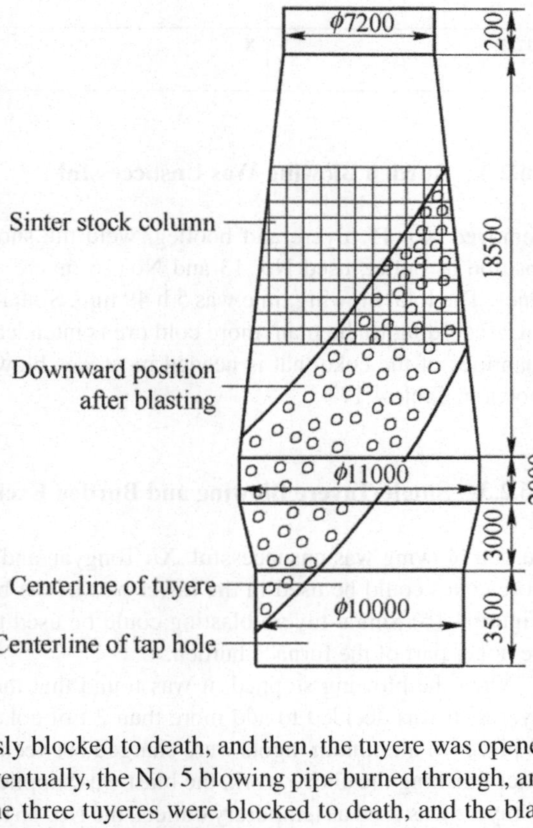

Fig. 4.15 Schematic diagram of burden descending position after mischarging and blasting

cold, the tuyere was continuously blocked to death, and then, the tuyere was opened again until 12:20 on the 4th. Eventually, the No. 5 blowing pipe burned through, and the blast was forced to stop, the three tuyeres were blocked to death, and the blast and burden exchanging were finished. A total of 17 batches and 166 t of net coke were added from the discovery of wrong burden to the burnout of blowpipe.

Table 4.16 Working state of tuyere

Time	Tuyere number of blast furnace									
	18	19	20	21	22	1	2	3	4	5
3:45	X	O	O	O	X	X	X	X	X	X
4:00	X	O	O	O	X	X	O	X	X	X
						(Automatic blowing)				
4:45	X	O	O	O	O	X	O	X	X	X
	(Automatic blowing)									
6:30	X	⊠	⊠	⊠	⊠	X	O	X	X	X
	(Tuyere of blast furnace automatic blooding of)									
6:40	X					X	O	O	X	X
						(Manual opening)				
11:00	X					X	O	O	O	X
						(Manual opening)				
12:26	X					X	⊠	⊠	⊠	⊠
						(Automatic blooding of tuyere) (Burn through)				

4.6.2.2 Burden Blowing Was Unsuccessful

Removed No. 15 tuyere and bootleg, weld the short section, used No. 15 tuyere position as outlet, used No. 13 and No. 16 tuyere to blast intermittently for eight times. The total blowing time was 5 h 49 min. Sinter 9.8 t and coke 28 t were blown out. According to the plan, more cold ore is intended to blow out, but instead, large quantities of the coke that is needed most was blown out. The blowing results are shown in Table 4.17.

4.6.2.3 Single-Tuyere Blowing and Burden Exchange

Burden blowing was unsuccessful. Xu Tongyan and others proposed [32] that coke (core coke) could be used in the lower part of the blast furnace (hearth deadman). (Figure 4.16) Single-tuyere blasting could be used to burn the coke, heat and melt the upper part of the furnace burden.

When the blowing stopped, it was found that there was room in front of some tuyeres. It was decided to add more than 2 t of coke before the No. 3 tuyeres, and 13 pieces of aluminum ingots were charged between the No. 13 and No. 14 tuyeres. After finishing the above work, the blast was supplied at 15:30 on March 6, the No. 15 tuyere was used as temporary outlets, the No. 16 tuyere was used for blowing, the blast pressure was 0.02 MPa, the blast temperature was 500 °C, and the blast volume was not shown, but the activeness before the tuyeres was observed, and the No. 13 and No. 17 tuyeres were blown out, and the plugging was made at 22:40 and No. 16 as outlet was kept for blasting to maintain the initial formation of the channel. On

Table 4.17 Blowing conditions for every time

Number	Time	Number of blowing tuyere	Number of charge blowing tuyere	Highest blast pressure, MPa	Blast volume, m^3/min	Blast temperature, °C	Blow out materials volume
1	March 4th, 23:30–23:55	13, 16	15	0.08		Cold wind	Blow out a small amount of coke and sinter
2	March 5th, 1:10–2:01	13, 16	15	0.12	800	600	Coke 15t sinter 5t
3	March 3th, 16:09–16:15	13, 16	15	0.14		500	Coke 10t sinter 4t
4	March 5th, 20:30–21:00	13, 16	Diameter 160–180 mm	0.20	0	600	Blow out a small amount of coke
5	March 5th, 22:04–23:15	13, 10	Removed tuyere	0.165	1100	730	Blow out 1t coke
6	March 5th, 23:53–24:00	13, 16	15	0.23	1200	720	a small amount of coke
7	March 6th, 1:09–1:26	13, 16	15	0.23	1200	700	Coke 1t Sinter 0.2t
8	March 6th, 2:11–4:13	13, 16	15	0.20	1000 ~ 1150	670	Coke 1t Sinter 0.3t

Fig. 4.16 Diagram of coke heap at the lower part of blast furnace

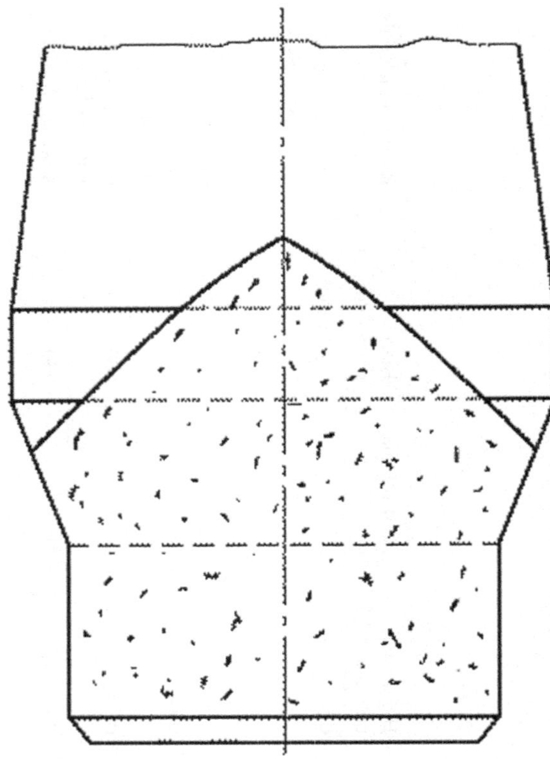

the 7th night shift, the blast pressure was still 0.02 MPa. It was observed that there were large slides in front of the tuyere and no slag gushing. At 7:40, No. 14 tuyere blowed open, the blast pressure remained 0.02 MPa, and the blast volume raised to 529 m^3/min and opened the temporary tap at 13:00 and tap 30 t iron. At 17:10, open No. 17 tuyere. The tuyere worked normally. The blast pressure was 0.04 MPa, and the blast volume was 650 m^3/min. The slag notch was opened many times without slag flowing out.

A tuyere was opened at the prescribed interval of every 4–8 h to "flush the charge stock column with weak gas flow, so as to make it peeled off layer by layer, keeping the formed passage, and preventing sudden collapse and blockage" [33].

A total of 105 t of iron were discharged from the temporary tap hole on March 7th, once every 1–2 h during the period. 134 t coke was added all day, and there was no slag tapping from the slag notch.

On March 8th, three tuyeres, No. 13, No. 16 and No. 17, worked with blast pressure of 0.05 MPa and blast temperature of 840–850 °C and blowing off at 9:30–11:00, by passing No. 18 and No. 19 tuyeres and plugging mud (with residual iron, 0.8–1 m); at 17:00, No. 18 was opened, a total of four tuyeres worked. There was no slag flowing out from the temporary tap hole, and 30 t slag flowing out when the west slag notch was opened at 20:35. At 23:20, No. 19 tuyere was opened. By this time,

532 t coke had been added, which was equivalent to the amount of coke loaded with in the wrong burden charging. Beginning at 17:25 on March 8th, the first truck of ore was added. After that, the charging began according to the method of charging, equivalent to a new start-up, as detailed in Table 4.18.

On March 10th, blowing off from 8:00 to 12:00, the temporary tap hole of No. 15 tuyere was handled as slag notch was made; then, the temporary tap hole would be moved down to the slag notch plane. On March 13th, it resumed normal. The whole process lasted 12 days and products lost 23,000 t. It was the cost of the development of China's iron and steel industry.

4.6.3 Chilled Hearth Treatment Method

The above examples, including almost all the chilled hearth treatment methods, have been successful in exploring and handling such a serious and unprecedented freezing accident. Examples include:

(1) Recovery as quickly as possible. Therefore, we should try our best to reduce tuyere slag blooding and burnout in order to reduce blowing off time, heat loss and production loss.
(2) Make every effort to maintain less or single tuyere to blast, so as to make the stock line down, add enough coke to replace the cold burden to compensate the serious insufficient heat in the furnace.
(3) Remove chilled iron and slag as far as possible.

When the furnace is cold, it is difficult to avoid slag blooding, while the low-temperature slag often contains hot metal. Once the slag is blooding, it often burns off the blowpipe, forcing blowing off suddenly, which results in all tuyere blooding. This kind of slag is treated for a long time, which not only loses time, but also heat. In order to reduce slag blooding, the number of working tuyeres should be reduced as much as possible and the number of blowpipes with slag blooding should be reduced too.

The best way to speed up the replacement of cold burden by coke is to use a single tuyere to blow. In the front of the tuyere, a combustion zone is formed, which can make the stock column drop locally and make the following coke reach the tuyere with the minimum volume. Therefore, it is also necessary to treat chilled hearth with a single tuyere for the purpose of burden exchange.

Blast furnace tap hole is quite far away from tuyere. Blast furnace with slag notch can use it as a temporary tap hole to reduce the distance between cold slag, cold iron discharging port and tuyere. This is an effective method to deal with chilled hearth. Remove the three sets of slag notches and the slag notches, remove the nearby slag–iron condensate with oxygen or other tools at the front end, fill the coke inside and burn the adjacent tuyeres to form a path. At the slag notch outside the newly filled coke, fill it with gun mud, dig out the space for installing the temporary tap hole with a small shovel and install the temporary tap hole.

Table 4.18 Coke load change process

Time	Sinter batch weight, t	Manganese ores batch weight, t	Silica batch weight, t	Coke batch weight, t	Coke rate, t/t	Charging system	Charging line
March 8th 19:00	3.9	0.30		9.57	4.57	2A + B A = JJJ,B = JJJK	2.5
March 8th 22:10	7.837	0.20		8.507	1.888	2A + B A = JJJ,B = JJKK	2.0
March 9th	9.400	0.20		8.294	1.538	3A + 2B A = JJJ,B = JJKK	1.5
	11.50	0.25	0.125	7.957	1.219	A + B A = JJJ,B = JJKK	1.5
March 10th	14.100	0.30	0.125	7.656	0.990	2A + 3B A = JJJ,B = JJKK	1.5
March 11th	23.200	0.50	0.3	10.620	0.802	All A inverted unit charging JJJ 1.5 m KK 0.5 m	1.5
	23.300	0.50	0.2	9.45	0.711	All A inverted unit charging JJJ 1.5 m KK 0.5 m	1.5
	25.400	0.50	0.1	8.76	0.633	All A inverted unit charging JJJ 1.5 m KK 0.5 m	1.5
	25.400	0.50	0.1	9.54	0.605	All A inverted unit charging JJJ 1.5 m KK 0.5 m	1.5

(continued)

Table 4.18 (continued)

Time	Sinter batch weight, t	Manganese ores batch weight, t	Silica batch weight, t	Coke batch weight, t	Coke rate, t/t	Charging system	Charging line
March 12th	29.350	0.55	0.1	9.54	0.571	All A inverted unit charging JJJ 1.5 m KK 0.5 m	1.5
	29.45	0.50		9.54	0.569	All A inverted unit charging JJJ 1.5 m KK 0.5 m	1.5
	32.40	0.60		9.54	0.518	All A inverted unit charging JJJ 1.5 m KK 0.5 m	1.5

The temporary tap hole is made of carbon bricks or fire clay bricks with the size of three sets of slag notch. After installation, according to tap hole requirements, brick sleeve and mud sleeve are built (Fig. 4.17). External tap hole is connected with slag runner. According to the requirements of main tapping runner, it is used as temporary slag skimmer and iron runner after baking.

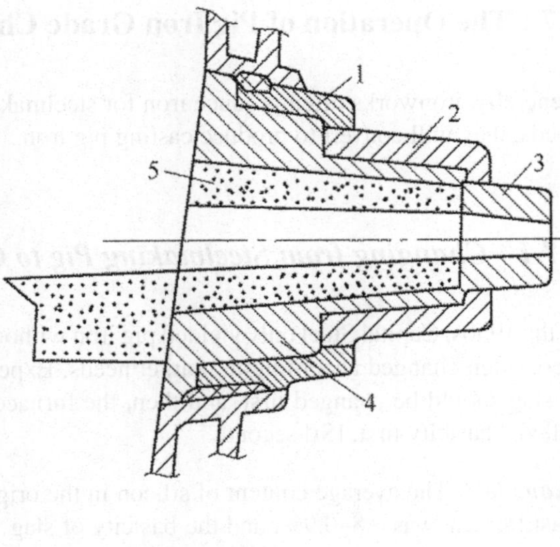

Fig. 4.17 Schematic diagram of temporary tap hole. 1—slag notch large set; 2—slag notch two sets; 3—temporary tap hole of carbon brick or fire clay brick; 4—ring brick; 5—temporary mud sleeve for tap hole

More severe chilling, as described in the previous example, is the discharge of iron from the tuyere plane that might be an unavoidable choice for modern large blast furnaces without slag notches.

Of course, modern blast furnaces have little chance of chilling. The method of choosing tuyere tapping is of high risk and that of changing tuyeres into temporary tap holes is similar with slag notches, but the distance between tuyere and tap hole is 4–5 m, and the distance between temporary tap holes and furnace front working plane is very long. It needs special platform to install tapping device. Pull down the tuyere and make a temporary tap hole with the same shape as the tuyere according to the size of the tuyere with carbon bricks or fire clay bricks. The inner diameter is 80–90 mm and is supported by a welded bracket. Two sets and large sets of fire clay bricks are built and baked with gun mud. Steel frame, iron runner, bricklaying, mud filling and sand laying were installed outside the steel frame. They are striving to connect with the runner of ladles in front of the furnace. Gooseneck tube is used to weld blind plate.

4.6.4 Quantity of Coke Blank

How much coke should be added depends on the chilling conditions. Xia Zhongqing, Engineer of An Steel, gave a reference quantity with a high reference value.

Table 4.19 shows that when the coke volume is larger than the hearth volume, the treatment would be smooth, but when it is less than the hearth volume and only half of the hearth volume, the treatment might fail. This is a very important experience and lesson.

4.7 The Operation of Pig Iron Grade Changing

Generally, ironworks often produce iron for steelmaking, but sometimes for market needs, they will convert to produce casting pig iron.

4.7.1 Changing from Steelmaking Pig to Casting Pig Iron

In the 1950s, Capital Steel only yielded pig iron without steel, and the pig iron grades were often changed according to market needs. Experience shows that the basicity of slag should be changed first, and then, the furnace temperature changed that so called, "basicity first, [Si] second."

Example 1 The average content of silicon in the original steelmaking pig of No. 1 blast furnace was 0.8–0.9%, and the basicity of slag was about 1.05. Several times

Table 4.19 Coke added quantity and recovery of furnace conditions

Furnace number	Add coke		Furnace volume, m^3	Coke volume/furnace volume	Change of the furnace conditions
	t	m^3			
An Steel No. 1 BF	88.8	197.3	72.5	2.722	Temperature of furnace rose rapidly and recovered smoothly
An Steel No. 3 BF	21.8	48.4	101.3	0.476	Fail
An Steel No. 3 BF	65.4	145.3	101.3	1.435	Temperature of furnace rose and recovered smoothly
An Steel No. 3 BF	88.0	195.6	101.3	1.931	Temperature of furnace rose and recovered smoothly
An Steel No. 6 BF	28.3	62.9	126.9	0.496	Fail
An Steel No. 11 BF	160	335.6	274.7	1.294	Temperature of furnace rose and recovered smoothly
Mei Steel No. 1 BF	67.5	145.6	134	1.086	Temperature of furnace rose and recovered smoothly

changed to casting pig iron containing 2.0% silicon, and the basicity requirement was 0.9. The ingredients were as follows:

Match ratio	Batch weight, kg		Ingredient rate, kg				Load	The yield of pig per batch, t
	Coke	Ore	Sinter	Hainan ores	Manganese ores	Limestone		
Original burdening rate	2760	9100	10,500	1400	0	500	3.18	5.15
New burdening rate	?	10,500	8700	1800	180	0	?	5.63

(1) Calculating according to the basicity of 0.9, sinter should be reduced to 8700 kg/batch and raw ores increased to 800 kg/batch in the new proportion.

According to the requirement of [Mn], 180 kg/batch of manganese ores should be added without limestone.

(2) The [Si] content in iron increased from 0.9 to 2.0%. According to the table of quantitative adjustment parameter coefficient of Capital Steel, the coke rate should be increased: $(2 - 0.9) \times 60 = 72$ kg/t.
(3) The [Mn] in iron increased by about 0.7% and the coke rate increased by $0.7 \times 14 = 9.8 \approx 10$ kg/t.
(4) Remove 500 kg/batch of limestone and save coke: $500 \times 0.3 = 150$ kg.
(5) The coke rate of the original batch rate is 2760/5.45 = 506 kg/t (5.45 is tapping quantity, t/batch); the coke rate of the new batch rate should be: 506 + 72 + 10 = 588 kg/t.
(6) The new blending rate was $588 \times 5.63 - 150 = 3160$ kg for each batch of coke, and the load was 10,500/3160 = 3.32.

The order of feeding is as follows:

13:30, 96 batches: added 20 batches of raw ingredients (without limestone) to reduce basicity;
16:10, 113 batches: added 3 batches of coke blank and raised furnace temperature;
16:40, 116 batches: added the burden for iron grade charging, change basicity and load.

The smelting results as follows:

Date	May5th				May 6th			
Time	13:15	16:40	19:12	22:05	0:40	3:00	5:39	11:50
[Si], %	0.86	0.83	1.25	0.96	1.07	1.90	2.08	1.88
CaO/SiO_2	1.053	1.058	1.08	1.06	1.005	0.95	0.985	0.885

Throughout the iron grade transformation, slag basicity declined after 11–14 h to reach the required level, and furnace temperature reached the required level after 11 h from the coke blank charging.

Example 2 The pig iron grade was transformed at No. 3 BF in October 1966.

(1) Ingredients ratio are as follows:

Time	Coke, kg	Sinter, kg	Pangjiabao ores, kg	Hainan ores, kg	Manganese ores, kg	Load	Remarks
Original ingredient rate	2400	9300	700	0	200	4.14	5A + 5B
23:20	3000						Add 3 cars of net coke

(continued)

(continued)

Time	Coke, kg	Sinter, kg	Pangjiabao ores, kg	Hainan ores, kg	Manganese ores, kg	Load	Remarks
23:25	3000	8750	0	1250	200	3.33	The 20 batches of acid burden
23:55	3000	9300	0	700	150	3.33	7A + 13B

From 325 to 329 batches (23:25–23:55), one batch of coke was added to each batch, totaling 5 batches of coke.

(2) The composition of the burden is as follows (%):

Mine name	Fe	SiO_2	CaO	S
Sinter	52.8	10.06	9.4	0.035
Hainan ores	58.8	13.05		
Pangjiabao ores	51	15.6		

(3) The smelting results as follows:

Date	Time	[Si], %	[S], %	[Mn], %	CaO/SiO_2
24th	22:26	0.68	0.017	0.68	0.93
25th	0:45	0.57	0.019	0.74	0.918
	3:30	0.73	0.017	0.90	0.954
	6:20	0.83	0.021	0.76	0.878
	9:13	1.39	0.019	0.78	0.90
	11:50	2.51	0.011	0.80	0.95
	14:30	3.01	0.011	0.80	0.925
	17:12	2.44	0.010	0.72	0.917
	19:40	1.94	0.019	0.61	0.78
	22:15	2.1	0.016	0.72	0.802

4.7.2 From Casing Iron to Steelmaking Iron

The conditions for converting casting iron to steelmaking iron were as follows:

[Si], %	[Mn], %	CaO/SiO_2	Ore batch, t	Coke batch, t	Load
2.4	0.58	0.9	10	2.64	3.79

The iron grad changing operation was as follows:

(1) The basicity of slag increased from 0.9 to 1.06 by removing raw ore in 134 batches at 16:10.
(2) 21:40, 183 batches: increased load, coke batch weight reduced from 2640 to 2300 kg, load increased from 3.79 to 4.34, the next day due to high furnace temperature, load increased to 4.50, coke batch weight decreased from 2300 to 2150 kg.

The smelting results as follows:

Date	16th			17th			
Time	16:32	19:10	21:35	2:25	6:00	8:40	11:00
[Si], %	2.58	3.32	3.22	2.77	2.02	1.17	1.20
CaO/SiO_2	0.91	0.873	0.915	0.975	0.94	1.03	1.075

The basicity of slag increased from 0.91 to 1.03, 16 h after feeding the basicity of burden, and the furnace temperature decreased from 3.22 to 1.2% after light burden charging in 11 h. According to Table 4.6 "Table of quantitative adjustment parameter coefficient," the [Si] decreased from 2.5 to 0.8% (Si decreased 1.7%) and the actual load should be 20–25%, only 14% increased. Therefore, the furnace temperature was higher and the load should be increased by 7% at 15:20 the next day, so that the furnace temperature could meet the requirement, squandering 30 t coke.

4.7.3 Summary of Iron Grade Transformation

(1) Change the basicity of slag 4 h in advance.
(2) Change the load 4 h after changing the basicity. Generally, from casting iron containing 2.5% of [Si] to steelmaking (steelmaking grade number P08, containing about 0.8% of [Si]), the load should be changed by 20–25%. The number of changes should be determined by the basic furnace condition.
(3) In order to reduce the amount of transitional hot metal, steelmaking iron converted to casting iron, and coke should be added for 3–5 batches before burden changing.

The dynamic process of iron grade transformation is shown in Fig. 4.18.

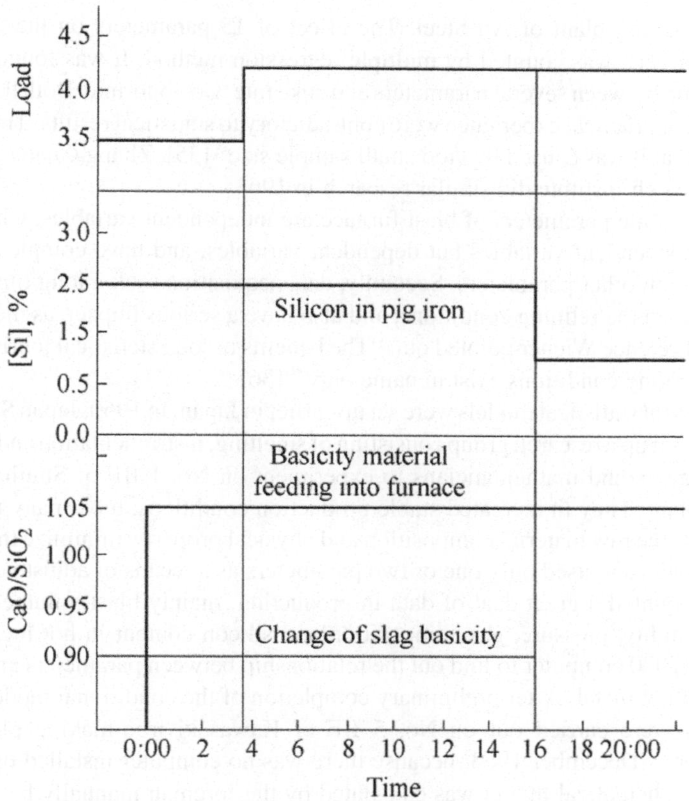

Fig. 4.18 Dynamic curve of cast iron converted to steel

4.8 Expert System for Furnace Temperature Adjusting

4.8.1 The Development and Research of Automation Control for Blast Furnace Process

The development of blast furnace mathematical model began in the 1960s. At the first International Conference on Iron and Steel Automation, there were four process papers (B1–4) and three practical control papers (B5–7) on sintering, iron-making and steelmaking [34]. The conference was held in Amsterdam, the Netherlands, March 29–31, 1965.

The earliest mathematical model of blast furnace was dealing with the change of silicon content in hot metal by analyzing the thermal state of blast furnace. The earliest statistical method to study the coke rate change of blast furnace in China was used in the cooperation between the Teaching and Research Group of Applied Mathematics of Northeast Institute of Technology (Northeast University now) and the iron-making plant of An Steel. The effect of 15 parameters on the coke rate of blast furnace was counted by multiple regression method. It was found that the relationship between several parameters and coke rate was abnormal. Both theoretical analysis and practical experience were contradictory to statistical results. The authors believed that it was caused by "too small sample size" [35]. Zhang Guangan of Ben Steel Research Institute did similar research in 1963.

In fact, some parameters of blast furnace are independent variables, while some are not independent variables but dependent variables, and have complex internal relations with other parameters. Secondly, data acquisition took a long time, which have changed the refining conditions, and also have a serious impact, as the founder of control science Wiener pointed out: "The benefits of long statistical journey under many changing conditions exist in name only." [36]

Successful statistical models were set up earlier in Japan. In 1960, Japan Steel Tube Company set up a research group consisting of smelting, instrumentation and automation engineers and mathematicians to experiment on No. 1 BF of Shuijiang iron-making plant. They first created stable production conditions, took many measures to stabilize the raw material composition and physical property, minimize the change of parameters and used only one or two parameters as a means of adjustment.

They counted a great deal of data in production, mainly blast volume, temperature, humidity, pressure, gas composition and silicon content in hot metal. They used IBM7070 computer to find out the relationship between parameters and silicon content in hot metal. After preliminary completion of the conditional model, industrial tests were carried out on No. 5 BF of Kawasaki iron-making plant from November to December 1963, because there was no computer installed on the BF, and the mathematical model was calculated by the foreman manually between the two casts.

In November 1962, a set of HOC-300 computer was installed on Shuijiang No. 1 BF, which was reopened to record the operation information of BF. The computer was modified in 1964 with the function of a control computer.

Figure 4.19 illustrates the above statistical model.

The blast furnace reaction zone was divided into two parts; the input of lower part was blast and heavy oil, and the output was iron and slag. The input of upper part was burden, and output was gas. ①–⑥ in Fig. 4.19 represents the smelting reaction in the furnace. The development process and experimental results were reported in the literature [37, 38].

The work was in the preliminary stage, based only on statistics, with great limitations. If the using range was beyond the statistical region, the original orderliness would not be available. In addition, the production conditions of blast furnaces often change, and it might be difficult to get the results only from the original data.

The significance of this exploration was pioneering the practice of using computers in blast furnaces, the first one in Japan, serving as the pioneering test of blast furnaces in the early stage of the world.

Statistical model only dealt with the judgment of furnace temperature, but producers would like to concern about the productivity of blast furnace under certain conditions. In the practice of blast furnace temperature determination, the biggest

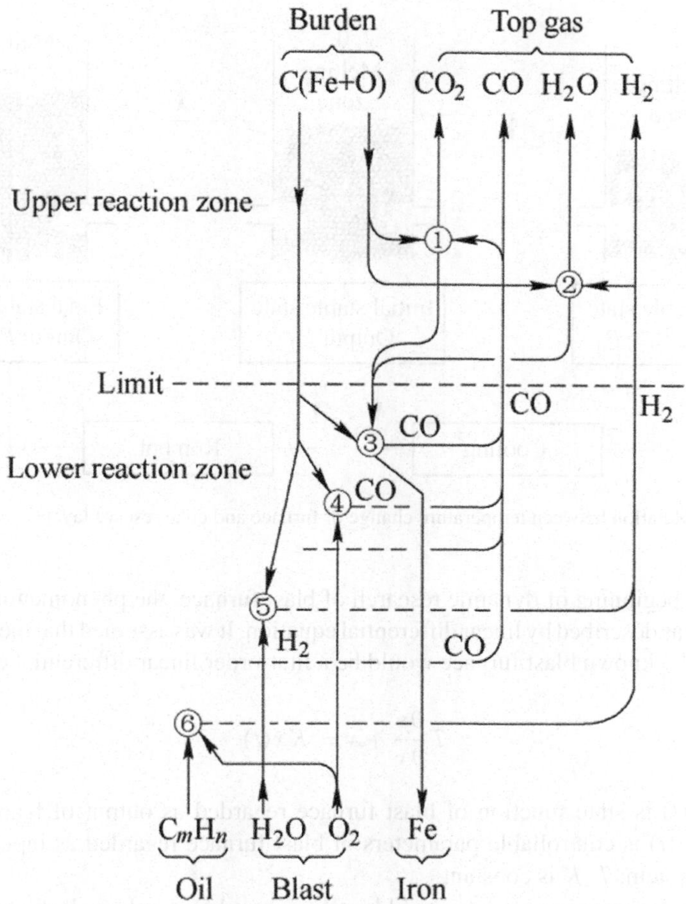

Fig. 4.19 Reaction diagram of blast furnace smelting process

problem was the operating time after the blast furnace parameters change with totally different under different smelting conditions. Blast furnace workers so called this phenomenon "inertia." The dynamic control model of blast furnace developed by the scholars of IRSID solved the problem of inertia to a certain extent. C. Staib, et al. assumed that when the burden removed from the top of the furnace to the melting zone for 6 h, the amount of carbon produced from the melting zone was corrected by the change of the average carbon solution loss (direct reduction) in one hour. In t time, the change of carbon content in the coke reserve layer can be calculated by formula. Figure 4.20 shows the change of coke content in the reserve layer and explains the principle of the temperature change of the blast furnace. The carbon content in the coke reserve layer decreased and the furnace temperature became cooler, while it became hotter when the carbon content increased [39].

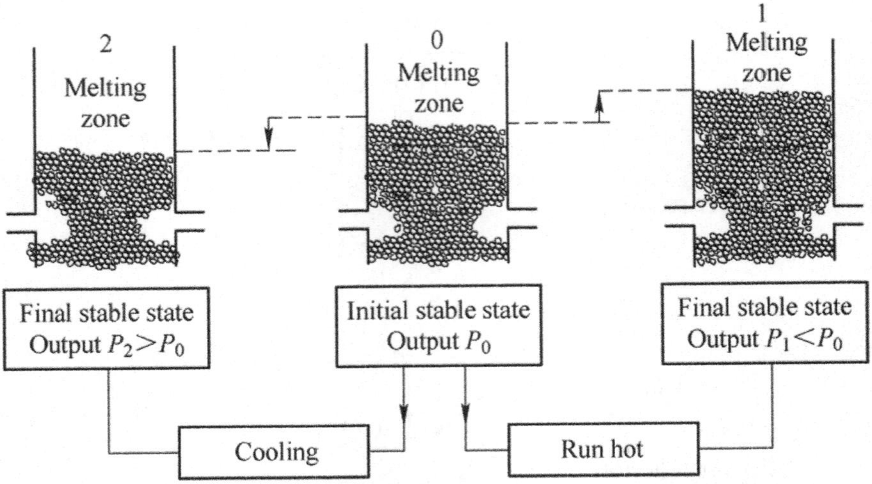

Fig. 4.20 Relation between temperature change of furnace and coke reserve layer

At the beginning of dynamic research of blast furnace, the phenomenon of blast furnace was described by linear differential equation. It was assumed that the equation of state of a known blast furnace would be a first-order linear differential equation:

$$T\frac{\partial y}{\partial x} + y = Kx(t) \tag{4.8}$$

where y (t) is state function of blast furnace regarded as output of blast furnace system; y (t) is controllable parameters of blast furnace regarded as input of blast furnace system; T, K is constant.

The above equation of state could be transformed by Laplace. Before and after transformation, x (t) and x (s) are the input of the system, and y (t) and $y(s)$ are the output of the system. Both of them can also describe the state change of blast furnace. The ratio of state function and input parameter is defined as "transfer function." The physical meaning is as follows:

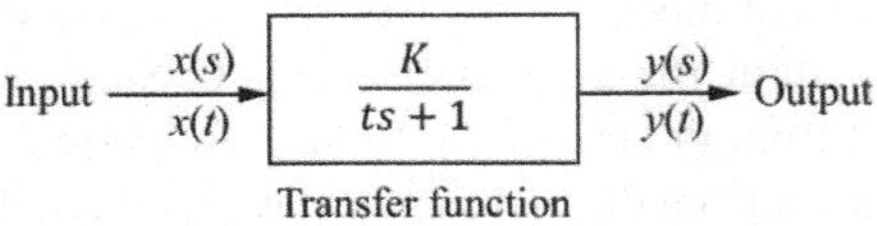

C. Staib, et al. had researched quantitatively on furnace temperature adjustment parameters commonly used in blast furnaces. The dynamic response curves of Sect. 4.3, Figs. 4.4 and 4.5 represent a part of their research. *B.* Wood sorted out the transfer functions of some parameters, which are listed in Table 4.20.

In 1963, the French model was tested in No. 4 furnace of Homecourt Plant of Sidelor Company, in Monteville Plant of Normandy Company in 1964, and then at No. 5 blast furnace of Longbas Plant successively. Good results were obtained.

Table 4.20 Transfer functions

Wood	$\frac{-0.058}{1+28s}$	$\frac{0.005}{1+6s}$	$\frac{0.014e^{-8s}}{1+8s}$	$\frac{0.028e^{-2s}}{1+11s}$	$\frac{0.05}{1+3.5s}$
Vidol	$\frac{-0.021}{1+s}+\frac{0.007e^{-8s}}{1+1.5s}$	$\frac{0.004}{1+6s}$	$\frac{0.015e^{-12s}}{(1+4.3s)^2}$		
Barlier	$\frac{-0.053}{1+3.4s}$	$\frac{0.0025}{1+6.2s}$	$\frac{0.015e^{-3s}}{1+16s}$		
Staib	$\frac{-0.023}{1+2s}$	$\frac{0.002}{1+14s}$	Augmentation = 0.016	$\frac{0.02e^{-14s}}{1+14s}$	
Rebeko			$\frac{0.016e^{-4.5s}}{1+7.5s}$		$\frac{0.03}{1+2.25s}$
Botler		$\frac{0.003e^{-4s}}{1+9s}$		$\frac{0.017e^{-4s}}{1+16s}$	

Details could be seen in the literature [38]. The static and dynamic models of France had opened up a new height of BF process control, especially the introduction of transfer function to carry out dynamic research, which provided a great space for development and broadened the way for future development.

4.8.2 BF Expert System

In the 1970s, Ben Steel, An Steel and Capital Steel began to develop blast furnace models in cooperation with relevant departments. The prediction of silicon content in hot metal was the basic content. The model of Bao Steel No. 1 BF was introduced from Japan. Since 1989, Wu Steel and Capital Steel cooperated with University of Science and Technology, Beijing to develop an artificial intelligence expert system for blast furnace smelting. At that time, the main parameters for judging the thermal state of blast furnace and their influence on the thermal state are shown in Tables 4.21 and 4.22 [40, 41]. The Automation Research Institute of Capital Steel continued to carry out development and research. In November 2007, the No. 2 BF of Capital Steel contrasted the Finnish model operating in parallel. Because the development was on the basis of the Finnish model, the actual expert system of Capital Steel has the dominance.

Table 4.21 Main parameters for judging the thermal state of blast furnace

Mark	Meaning	Mark	Meaning
R_{F_e}	Variation of hot metal formation rate	Q_1	Oxygen enrichment
Q_W	Blast volume	T_W	Blast temperature
D_P	Pressure drop	H_T	Comprehensive load
T_T	Top temperature	CO	Throat CO quantity
V_B	Burden velocity	T_1	Throat temperature
Q_c	Permeability index	T_2	Furnace temperature

Table 4.22 Effects of various parameters on thermal state

Number	State / Parameter	Over heat	Run hot	Run cool	Over cooling (Freezing)
1	Variation of hot metal formation rate	↑	↑	↓↓	↓
2	Blast volume	↓↓	↓	↑	↑↑
3	Pressure drop	↑	↑	↓	↓
4	Top temperature	↑	↑	↓	↓
5	Burden velocity	↓↓	↓	↑	↓↓
6	Permeability index	↓	↓	↑	↑
7	Oxygen enrichment	↑	↑	↓	↓
8	Blast temperature	↑	↑	↓	↓
9	Water tank temperature	↑	↑	↓	↓↓
10	Water tank temperature difference	↑	↑	↓	↓
11	Front end temperature of tuyere	↑	↑	↓	↓
12	Hearth temperature	↑↑	↑	↓	↓↓

Note ↑↑ is a sharp rise; ↑ is a rise; ↓ is a decline; ↓↓ is a sharp decline

References

1. J. Chen, *Data Manual of Charts and Charts for Steelmaking*, 2nd edn. (Metallurgical Industry Press, Beijing, 2010), p. 99
2. X. Huang, *Principles of Iron and Steel Metallurgy*, 4th edn. (Metallurgical Industry Press, Beijing, 2013), pp. 452–453
3. E. Wigman et al., *Iron-Making* (Metallurgical Industry Press, Beijing, 1993), pp. 193–194
4. G. Zhang et al., Study on pig iron properties of Ben Steel. Iron Steel **3**, 1–8 (1983)
5. H.H. Barekini et al., Temperature and composition of damaged iron in BF bottom. Steel **3**, 198–200 (1961)
6. V.W. Lyboz, H. Weber, Metall. Trans. **2**, 14–18 (1956)
7. Capital Steel Iron-making Plant, Timely Regulations for Iron-making Process (2006) (internal data)
8. C. Staib, in *Iron-making Proceedings* (1976), pp. 66–84
9. B. Wood, in *Developments in Iron-making Practice* (1972), 146–151
10. T. Yang, D. Cang, Y. Ding, *Oxygen-Enriched Pulverized Coal Injection In Blast Furnace* (Metallurgical Industry Press, Beijing, 1996), pp. 1–10
11. Shijingshan Iron and Steel Company, BF coal blending test. Iron Steel **2**, 1–6 (1966)
12. Y. Liu, in *Capital Steel Science and Technology (Trial)* (1979), 79–97
13. *Iron-making Proceedings* (1975), pp. 307–328
14. Iron and Steel Engineer **10**, 30–34 (1977); **8**, 51–52 (1976)
15. Stahl und Eisen, **23**, 1243–1244 (1978)
16. Steel, **12**, 11–14 (1972); **7**, 587–589 (1974); **9**, 788–792 (1976); **11**, 979–982 (1976)
17. Metallurgist, **12**, 11–14 (1974)
18. Ferrous Metallurgy, **6**, 35–39 (1976)
19. Metallurgy and Mining Industry, **4**, 7–8 (1974)

20. Soviet Union. Application of pulverized coal in iron-making (1974) (internal data)
21. Y. Liu, Practice of high pulverized coal injection rate in blast furnace. Iron Steel **6**, 7–11 (1981)
22. Y. Liu, Seeing the strengthening of blast furnace operation from the practice of blast furnace treatment in Capital Steel. Iron Steel **1**, 25–28 (1979)
23. Editor-in-chief of Zhejiang Institute of Metallurgy. Compilation of foreign patent documents on low silicon ferrous smelting in blast furnace (1987) (internal data)
24. X. Huang, Fukuyama No. 2 BF set a new record of 0.15% silicon content in hot metal. Iron-Making **5**, 30 (1985)
25. C. Che, A brief comment on the smelting of low silicon pig iron. Iron-Making **3**, 46–51 (1988)
26. U. Toshio et al., Control technology of silicon content in molten iron. Foreign Steel **10**, 9–16 (1987)
27. J. Song, *Theory and Practice of Blast Furnace Smelting* (Metallurgical Industry Press, Beijing, 2005), p. 257
28. S. Su, Some problems in smelting low-silicon pig iron with high carbon quality. Iron-Making **4**, 7–11 (1989)
29. C. Zhou, *Collected Works of Zhou Chuandian*, vol. 1–4 (Metallurgical Industry Press, Beijing, 2001)
30. C. Zhou, *Suggestions and Chronicles (Sequel)* (Metallurgical Industry Press, Beijing, 2006)
31. J. Xu, Q. Liu, *100 Cases of Blast Furnace Accident Treatment* (Metallurgical Industry Press, Beijing, 1986), pp. 1–90
32. T. Xu, Analysis of coke migration in hearth of blast furnace. An Steel Technol. **2**, 19–25 (1988)
33. Z. Xia, *BF Operation and Practice* (Liaoning People's Publishing House, Shenyang, 1988), pp. 256–269
34. Intemational Conference on Iron and Steelmating, Automation (1965)
35. An Steel Iron-making Plant, et al. Statistical analysis of coke ratio influenced by smelting intensity and other factors, Anshan Metal Society, 1964
36. N. Wiener, in *Cybernetics* (Science Press, Beijing, 1962), p. 25. (in Chinese)
37. Y. Fujii, et al., in *Iron-making Proceedings* (1967), pp. 57–65
38. K. Katsura, et al., in *Interactional Conferenee on Iron and Steelmaking (I)* (1965), 207–216
39. C. Staib, et al., J. Metal **1**, 33–39 (1965)
40. T. Yang, Y. Liu, et al., Development and application of expert system for No. 2 blast furnace smelting in Capital Steel. Iron-Making **6**, 1–10 (1995)
41. T. Yang, J. Xu et al., Development of artificial intelligence expert system for blast furnace smelting. Capital Steel Sci. Technol. **1**, 15–23 (1995)

Chapter 5
Control of Slag Composition

Abstract In the process of iron-making, the slag and gas with dust are accompanied by the production of hot metal. The slag comes from burden and the ash of the fuel. Slag affects blast furnace production very much, especially the quality controlling of hot metal. The viscosity and stability of slag play significant role in the smooth operation of blast furnace. The impact of slag on the campaign life of blast furnace is not only reflected in the corrosion of the lining, but also in the repair ability of the corroded lining.

The composition of slag is very complex. The slag prepared for special purposes has special functions. The main functions of slag are as follows:

1. Desulfurization;
2. To clean up scaffolding;
3. To repair lining;
4. To remove alkali metals;
5. To flush hearth accumulation;
6. To enrich useful compositions in burden for the extraction or exclusion from the slag, such as manganese enrichment by two-step smelting for ferromanganese, which is equivalent to the enrichment of manganese ores.

Iron-making mainly depends on practical experiences in the early years. People gradually and profoundly studied the composition of slag and eventually reached the present level through long-term practice and accumulation with constantly using the relevant science. Table 5.1 shows the slag composition of some blast furnaces in 1928, indicating the actual situation of advanced blast furnaces in the world in early period.

The hot metal and slag data of fifteen blast furnaces in eleven countries are listed in Table 5.1. The slag composition varies greatly due to the different origin of the burdens and different kinds of fuel. In those years, the method of controlling sulfur in hot metal by changing slag basicity had been mastered. Through using the knowledge of ternary phase diagram to analyze the composition and viscosity of the slag, some experts and professors have measured the viscosity of the slag and implemented the

Y. Liu, *The Operation of Contemporary Blast Furnaces*,
https://doi.org/10.1007/978-981-15-7074-2_5

Table 5.1 Slag in some blast furnaces in 1928 and corresponding hot metal composition [1]

Country	Austria	Belgium	Czechoslovakia	France	Germany	Romania	Japan	Australia	India	Canada	The USA				
											1	2	3	4	5
Hot metal composition, %															
Si	0.1	0.5	0.5	0.7	0.58	0.7	1.57	1	1	1.05	0.91	1.2	1.9	1.5	2
S	0. 05	0.07–0.1	0.02–0.1	0.09–0.12	0.035	0.015	0.04	0.034	0.035	0.3	0.037	0.3	0.035	≥0.035	0.08
Mn	2	1	0.7	1.1	2.16	3.3	1.13	2.07	1.6–1.9	1.63	2.36	1.64	0.32	1.5	1
Slag composition, %															
SiO_2	34.4	28	28.6	31.5	33.05	37	33.52	35.38	28.27	37.2	36.7	33	35.5	28.9	29.6
CaO	29.9	46	44.8	44	45.5	37.6	47.9	40.25	32.13	42.1	46.2	48	49.02	49	41.24
MgO	12.1	2.75	5.8	3	4.89	6.15	1.78	1.92	14.72	2.74	2.55	5.2	0.56	3	3.78
Al_2O_3	12.1	18	15.7	13	8.16	11.6	14.21	18.45	21.15	14.3	11.5	11.8	12.64	10	16.14
MnO	4.2	1	1.4	1.2	2.82	6.4	0.54	–	1.04	–	2	0.8	–	–	0.78
(CaO + MgO)/SiO_2	1.22	1.74	1.77	1.5	1.52	1.18	1.48	1.19	1.66	1.2	1.33	1.61	1.4	1.8	1.66
Blast temperature, °C	300	800	810	679	560	627	625	635	607	593	616	621	728	593	760

synthesis of the slag. Extensive and in-depth studies were carried out. Since then, these experiences and research fruits were used in production till now.

The method of controlling sulfur in hot metal by changing slag basicity had been mastered, while the measurement and research of the phase diagram had been developed. Molecular theory and ion theory had been developed successively to provide basis for the synthesis of complicated slag. The reasonable slag composition cannot be determined directly by formulas and models so far, but might be realized soon.

5.1 Hot Metal Quality Control

5.1.1 The Harm of Sulfur in Hot Metal

The sulfur in hot metal exists in the form of [FeS] with melting point of 1190 °C. [FeS] is not a compound, but a solid solution of iron and may completely dissolve in hot metal. In solid iron, the solubility of [FeS] is as low as 0.013% at the temperature of 998 °C (see Fig. 5.1).

In the process of solidification and cooling of hot metal, the sulfur-rich liquid precipitates from the γ-Fe crystals and distributes as network structure. Once heating solid iron or steel to 900–1100 °C, "the sulfur-rich eutectic crystal becomes liquid

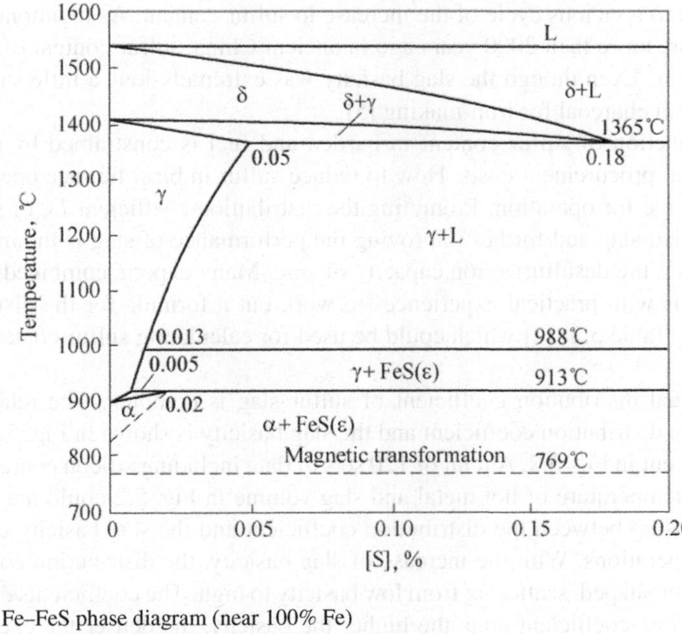

Fig. 5.1 Fe–FeS phase diagram (near 100% Fe)

again while the compression of the metal ingot or steel could cause the sulfur-rich liquid to slide along the grain boundary to break the metal ingot or steel that is so-called hot brittleness. This is why sulfur considered as a harmful element" [2].

5.1.2 Sulfur Distribution and Calculating Formula

Sulfur is derived from burdens and fuel. Based on the study of the material balance, the similar equations [3, 4] calculating the amount of sulfur in iron can be acquired in many iron-making textbooks:

$$[S]_{\%} = \frac{0.1\left((S)_{\text{burden}} - (S)_{\text{gas}}\right)}{1 + 0.001 L_s Q_{\text{slag}}} \tag{5.1}$$

where $[S]_{\%}$ is the content of sulfur in hot metal (mass percent); $(S)_{\text{burden}}$ is sulfur load, kg/t; $(S)_{\text{gas}}$ is The amount of sulfur escaped with furnace gas, kg/t; L_S is the distribution coefficient of sulfur in slag vs iron, $L_S = \frac{(S)}{[S]}$; Q_{slag} is the slag volume per ton iron, kg/t.

Equation (5.1) shows that sulfur content in burden and fuel should be reduced and the slag volume should be increased to "release" more sulfur. But increasing the amount of slag, definitely resulting in a decrease in the iron content of burden, would not conducive to blast furnace operation or the reduction of fuel consumption. Actual sulfur in burdens mainly comes from fuel. Therefore, the increase in slag volume would lead to a vicious cycle of the increase in sulfur content. As mentioned in the introduction, more than 2000 years ago in ancient China, sulfur content of pig iron was very low. Even though the slag basicity was extremely low, a little sulfur was brought from charcoal for iron-making [5].

The reduction of sulfur content in burden and fuel is constrained by resource and material procurement costs. How to reduce sulfur in blast furnace operation is a central topic for operation. Promoting the distribution coefficient L_S of sulfur in hot metal and slag and further improving the performance of slag is the main way of enhancing the desulfurization capacity of slag. Many experts combined theoretical analysis with practical experiences to work out a formula for the distribution coefficient (Table 5.2) [2] which could be used for calculating sulfur content in hot metal.

The actual distribution coefficient of sulfur slag is complex. The relationship between the distribution coefficient and the slag basicity is shown in Fig. 5.2. But it is not apparent in Fig. 5.2. A total of 120 sets of data including silicon content in hot metal, the temperature of hot metal and slag volume in Fig. 5.2 could not indicate the relationship between the distribution coefficient and the slag basicity closed to the daily operations. With the increase of slag basicity, the distribution coefficient appeared fan-shaped, scattering from low basicity to high. The coefficients distribute less in the low coefficient area, the higher the basicity, the denser the coefficients distribution, which shows that improving the slag basicity would be beneficial to the

Table 5.2 Formula of slag distribution coefficient

Formula	Remarks
$\frac{(S)}{[S]}=\frac{(S)_{Ca}}{[S]}+\frac{(S)_{Mn}}{[S]}=\frac{(Cao)}{K_1(FeO)}+\frac{[Mn]}{K_2}$ $1gK_1=\frac{5700}{T}-3.72+0.05(SiO_2)$ $1gK_2=-\frac{3840}{T}+1.17$	By mass percent
$\frac{(S)}{[S]}=2+10[\%Mn]$	Slag: 33–44% CaO, 16–20% SiO_2 Metal: 0–0.6% Mn
$\frac{(S)}{[S]}=n_S\left[2.6+1.6\frac{(CaO)_f}{(FeO)}+11.0[\%Mn]\right]$ n_s is the molecules of various components for 100 g slag $(CaO)_f$ is the molecules of free CaO for 100 g slag, and $(CaO)_f=[(CaO)-4(P_2O_5)](L-2)$ $L=\frac{(CaO)-4(P_2O_5)}{(SiO_2)}$	Calculated by molecules for 100 g slag components, which is suitable for open hearth melting period
$\frac{(S)}{[S]}=1.4+16[(CaO)+(MnO)-2(SiO_2)-4(P_2O_5)-$ $(Fe_2O_3)-2(Al_2O_3)]$	Calculated by molecules for 100 g slag FeO 3–70%, Mn 13–39%
$\frac{(S)}{[S]}=[0.5+2.25(FeO)]\times\left[\frac{4(CaO)'+2.5(MnO)}{(FeO)}+1\right]$ $(CaO)'=(CaO)-2(SiO_2)-3(P_2O_5)-(Al_2O_3)$	Suitable for refining period of open hearth and oxidation period of electric furnace
$\frac{(S)}{[S]}=[0.5+2.25(FeO)']\times\left[\frac{2.5(MnO)}{(FeO)'}+1\right]$ $(FeO)'=\sum(FeO)+(CaO)+(MgO)+(MnO)$ $-2(SiO_2)-3(P_2O_5)-(Al_2O_3)$	Suitable for melting process of open hearth. Calculated by molecules for 100 g slag components
$\frac{(S)}{[S]}=\left[1.23\frac{(CaO)}{(SiO_2)}-0.92\right](1+5[Mn])$	By mass percent
$\frac{(S)}{[S]}=2+9.2\frac{(CaO)'}{(FeO)}$ $(CaO)'=(CaO)-(CaO)_{2CaO-SiO_2}-(CaO)_{4CaO-P_2O_5}$ $=(CaO)-1.86(SiO_2)-1.57(P_2O_5)$ $\frac{(S)}{[S]}=2+13.2(CaO)'\frac{[Mn]}{(Mn)}$	By mass percent, when $(CaO)'\frac{[Mn]}{(Mn)}<0.8$

process of desulfurization. This figure also reminds of the fact that the slag basicity is not the only decisive factor for sulfur content of hot metal that is totally different from the feeling of daily operations.

The fan-shaped area was divided into five parts with five and half rays. But no apparent regulations emerged after comparing carefully the relationship between

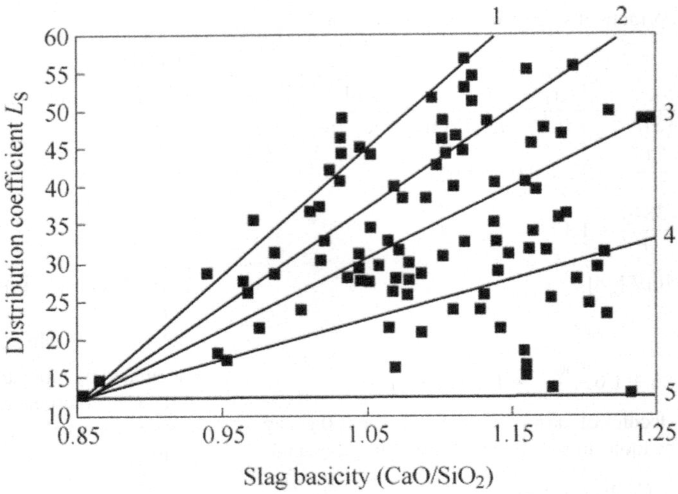

Fig. 5.2 Relationship between the distribution coefficient and the slag basicity

the distribution coefficient and the slag basicity, the amount of slag, [Si] and the temperature of hot metal. Because under certain conditions, various characteristics of slag have more influences.

Table 5.3 shows the annual (or monthly) average of 26 large-sized blast furnaces in China in 2014. It truly reflected the real level.

5.1.3 The Slag Basicity

Blast furnace slag is mainly made up of four oxides including CaO, MgO, SiO_2 and Al_2O_3. In addition, there are small amounts of other oxides including MnO, FeO and TiO_2. Basicity is generally expressed in three ways: $R_2 = CaO/SiO_2$, $R_3 = (CaO + MgO)/SiO_2$, $R_4 = (CaO + MgO)/(SiO_2 + Al_2O_3)$. Among them, R_2 and R_3 are commonly used.

About desulfurization in blast furnaces, many experts have given the calculation formulas mainly resulting from thermodynamics. In addition, some textbooks have given useful conclusions, such as references [3, 4]. In daily BF operations, the sulfur is controlled mainly by changing the slag basicity. When [*S*] is high, the slag basicity should be increased. Figure 5.3 shows the actual results of the slag basicity (R_3) and sulfur content in hot metal of some large-sized blast furnaces in China in 2014.

The relationship of the slag basicity (R_3) and sulfur content in iron is shown in Fig. 5.3 and Table 5.3. Sulfur content in hot metal is also affected by many factors including temperature, silicon content in hot metal, slag volume and so on, so the basicity is not the only decisive factor. The higher the temperature, the lower the sulfur content in hot metal. The temperature of hot metal and slag affects it in a

Table 5.3 Slag composition and sulfur distribution coefficient (In early 2014)

Category	Furnace number	Ma Steel A	Ma Steel B	Sha Steel	Tai Steel No. 5	Tai Steel No. 6	Qian Steel No. 3	Jingtang Steel No. 1	Jingtang Steel No. 2	Ba Steel No. 1	Ba Steel No. 2	Bao Steel No. 7	Anyang Steel No. 3	Mei Steel No. 5
	Volume, m^3	4000	4000	5800	4350	4350	4000	5707	5707	4038	4038	4150		4070
Hot metal	Si, %	0.42	0.44	0.27	0.57	0.56	0.49	0.26	0.25	0.50	0.47	0.54	0.41	0.42
	Temperature of hot metal, °C	1511	1514	1500	1513	1517	1516	1495	1500	1498	1487	1500	1504	
Slag	R_2	1.14	1.13	1.21	1.21	1.20	1.19	1.16	1.16	1.21	1.20	1.12	1.20	1.24
	R_3	1.37	1.37	1.45	1.39	1.37	1.44	1.40	1.40	1.41	1.43	1.51	1.38	1.46
	Slag volume, kg/t	307	303	299	304	307	318	290	290	361	341	348	377	282
	L_s	41.00	39.26	34.27	41.45	42.87	53.19	16.47	16.40	44.50	38.70	51.50	24.63	44.64

Category	Furnace number	Wu Steel No. 5	Wu Steel No. 6	Wu Steel No. 7	Xing Steel	Tang Steel No. 3	Tang Steel No. 4	Ji Steel No. 4	An Steel No. 1	An Steel No. 2	An Steel No. 3	Han Steel No. 7	Jiu Steel
	Volume, m^3	3200			3200	3200		3200				3200	
Hot metal	Si, %	0.51	0.56	0.47	0.93	0.44	0.47	0.43	0.59	0.45	0.47	0.62	0.63
	Temperature of hot metal, °C	1489	1496	1498	1481	1515	1521	1528	1495	1507		1484	1504
Slag	R_2	1.17	1.16	1.17	1.07	1.17	1.18	1.19	1.16	1.19	1.19	1.06	1.14
	R_3	–	–	–	1.31	1.42	1.43	1.45	1.36	1.39	1.38	1.29	1.31
	Slag volume, kg/t	354.7	367.5	364.5	315	332	330	399	376	334	328	450.1	387.2

(continued)

Table 5.3 (continued)

Category	Furnace number	Wu Steel No. 5	Wu Steel No. 6	Wu Steel No. 7	Xing Steel	Tang Steel No. 3	Tang Steel No. 4	Ji Steel No. 4	An Steel No. 1	An Steel No. 2	An Steel No. 3	Han Steel No. 7	Jiu Steel
	Volume, m^3	3200			3200	3200		3200				3200	
	L_s	36.92	33.93	36.15	24.17	47.8	44.76	27.67	50.45	43.2	43.6	32.27	30.19

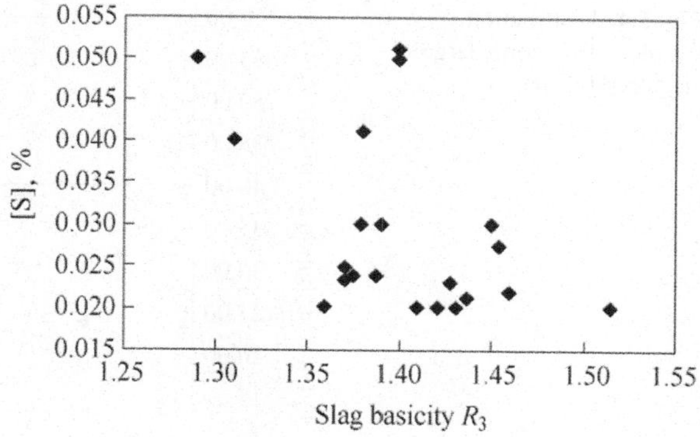

Fig. 5.3 Diagram of slag basicity (R_3) and sulfur content in hot metal of some large sized blast furnaces in China in 2014

larger range. The high temperature will inevitably lead to the increase in heat and fuel consumption, so the decision should be made comprehensively. Basicity adjustment is the major way in daily operation.

It is worthy for attention that the distribution coefficients of two large-sized blast furnaces were less than 20 (Table 5.3). Although there were external desulfurization facilities, the desulfurization efficiency of blast furnace is too low. High sulfur content (>0.05%) in hot metal would lead to the erosion of furnace line to reduce hearth life and increase the external desulfurization load. Whether it was economically reasonable or not, the judgment should be made from the whole process of production. It is very easy to improve the desulfurization efficiency for such kind of blast furnaces:

1. To rise [Si] content in hot metal from <0.3 to >0.3%, controlling between 0.3 and 0.4%. The temperature of hot metal would increase by about 10 °C.
2. To rise slag basicity to 1.2 or more, the distribution coefficient and the temperature of hot metal would both increase.

By above measures, it would be possible to eliminate the external desulfurization for general steel and reduce the desulfurization load for advanced steel. Whether the cost is reasonable or not, it should be studied and verified. Reducing the influence of alkali metals with lower basicity, at about $R_2 = 1.15$, has little effect on alkali discharging with slag, which can be regarded as spiritual "support" without any practical effects.

5.1.4 The Comparison Among Sulfur Distribution Coefficients

Figure 5.4 shows the schematic representation of the distribution coefficient (Table 5.3). The distribution coefficients of large-sized blast furnaces in China are boundary with 40, above 40 and below 40. Above 40, the slag with high desulfurization efficiency is close to that of Former USSR (in 1987), as shown in Table 5.4. Below 40, the slag is close to the type of slag in Europe in 1998, as shown in Table 5.5.

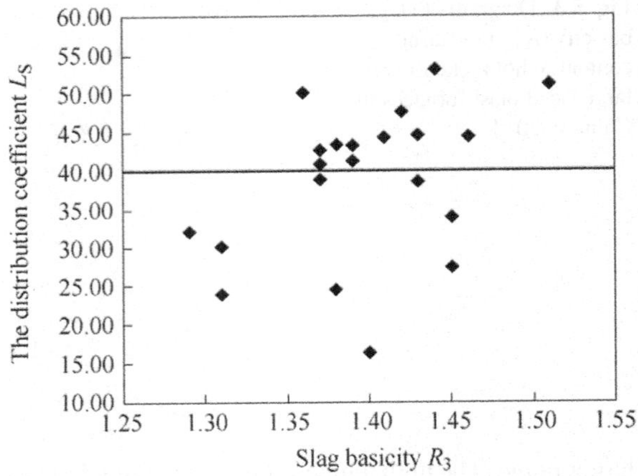

Fig. 5.4 Distribution coefficients of some large sized blast furnaces

In those years, Former USSR was similar to China, but in contrast to Europe, it emphasized the desulfurization in blast furnace process for general types of steel instead of external desulfurization. Table 5.6 shows the distribution coefficients of some blast furnaces in three countries and regions, reflecting the operating differences in that year. It is obvious that the desulfurization of China's current blast furnaces still has great potential. Some factories emphasized furnace productivity, putting the output of blast furnace in the first place. Actual high coefficient is not always the way of production for less cost. The production of low consumption, long campaign and low cost may be obtained by appropriately lowering the output. Improving desulfurization capacity of slag may be an effective measure for low cost.

5.2 Stable Slag and Viscosity

5.2.1 Stable Slag

The stability of slag mainly depends on slag composition and furnace temperature. It is difficult to avoid a slight change in composition. Even if the burden and fuel composition were stable, it would be very difficult to avoid the effects of temperature changes caused by operation. The metallurgical property, especially the viscosity of slag, must be stable to eliminate the adverse impacts of composition fluctuation. When a little fluctuation of temperature happened with a little change of viscosity of slag, it would be necessary for slag in the temperature range of normal operation and for blast furnace production.

Slag has been studied for a long time. Up to now, a lot of existing research results can be used to guide production. The method of determining the best slag composition completely by formula or algorithm is coming soon.

Table 5.4 Compositions and distribution coefficient of iron and slag of some blast furnaces in Former USSR in 1987

Company	Furnace number	Furnace volume, m^3	Hot metal composition, %		Slag composition, %						Slag basicity			L_S
			Si	S	SiO_2	Al_2O_3	CaO	MgO	FeO	S	R_2	R_3	R_3	
MMK	10	2014	0.54	0.017	36.94	11.07	41.05	7.83	0.18	0.84	1.11	1.32	1.02	49.41
NTMK	4	1513	0.64	0.019	38	13.5	40.3	6.9	0.39	0.66	1.06	1.24	0.92	34.74
	6	2711	0.61	0.019	39	12.1	43.4	5.6	0.39	0.66	1.11	1.26	0.96	34.74
CheMZ	1	1719	0.82	0.028	40.09	10.81	35.74	11.59	0.26	0.71	0.86	1.18	0.93	25.36
OHMK	4	2002	0.85	0.015	38.4	9.3	45.8	4.6	0.58	1.04	1.19	1.31	1.06	69.33
KarMK	3	2700	0.7	0.018	36.57	15.54	39.43	7.5	0.55	0.89	1.08	1.1	0.77	49.44
	4	3200	0.83	0.02	36.6	15.8	39	7.8	0.56	0.87	1.07	1.09	0.76	43.5
KMK	5	1719	0.58	0.016	37.5	13.61	36.47	10.66	0.47	0.53	0.97	1.26	0.92	33.13
SMK	1	3000	0.66	0.02	37.67	15.58	34.16	12.01	0.49	0.49	0.91	1.23	0.87	24.5
UNMK	3	2002	0.96	0.019	39.89	7.72	43.18	7.13	0.29	1.01	1.08	1.26	1.06	53.16
CoMK	1	3000	0.88	0.02	38.1	6.6	48.1	5.3	0.42	1.87	1.26	1.4	1.19	93.5
AST	3	1719	0.84	0.033	37.9	8.08	43.4	4.47	0.78	1.74	1.15	1.26	1.04	52.73
	6	1719	0.88	0.032	38.14	7.02	46.9	5.06	0.85	2.06	1.23	1.36	1.15	64.38
EMZ	1	1386	0.81	0.024	38.96	6.4	48.4	3.45	0.36	1.8	1.24	1.33	1.14	75
ZMK	5	2300	0.85	0.022	39.52	5362	48.2	5.65	0.3	1.6	1.22	1.36	0.01	72.73
NLMK	4	2002	0.68	0.02	38.1	8.3	42.2	10	0.38	1.14	1.11	1.37	1.13	57
	5	3200	0.7	0.021	38.9	8.3	41.8	9.52	0.4	0.84	1.07	1.32	1.09	40
	6	3200	0.72	0.016	38.1	9.44	39.5	9.6	0.52	0.56	1.04	1.29	1.03	35
CheMK	1	1007	0.7	0.016	41.1	9.2	40.5	10.4	0.44	0.83	0.99	1.24	1.01	51.88
	4	2700	0.69	0.017	40.4	9.8	40.7	10.3	0.41	0.8	1.01	1.26	1.02	47.06

(continued)

Table 5.4 (continued)

Company	Furnace number	Furnace volume, m^3	Hot metal composition, %		Slag composition, %						Slag basicity			L_S
			Si	S	SiO_2	Al_2O_3	CaO	MgO	FeO	S	R_2	R_3	R_3	
	5	5580	0.67	0.016	39.7	9.4	40.8	10.8	0.47	0.82	1.03	1.3	1.05	51.25
KST	7	2002	0.94	0.027	38.52	8.2	46.7	5.9	0.38	2.03	1.21	1.37	1.13	75.19
	8	2700	0.85	0.029	38.7	7.9	47.6	4.82	0.37	1.96	1.23	1.35	1.12	67.59
	9	5000	0.8	0.022	37.4	7	46.9	6.2	0.3	1.79	1.25	1.42	1.2	81.36
ZST	4	1513	0.63	0.027	38.94	5.94	47.72	5.73	0.24	1.48	1.23	1.37	1.19	54.81

Note This set of data is from Professor Wang Xiaoliu who used to carry out such investigation in Former USSR. I would like to extend my thanks to him!

Table 5.5 Compositions and distribution coefficients of iron and slag of some blast furnaces in Europe in 1998

Company and furnace number		Furnace volume, m^3	Slag volume, kg/t	Hot metal composition, %		Hot metal temperature, °C	Slag composition, %						Slag basicity		L_S
				Si	S		SiO_2	Al_2O_3	CaO	MgO	S	$K_2O + Na_2O$	R_2	R_3	
Linz	A	2772	293	0.57	0.051	1465	39.07	8.64	38.06	9.37	1.1	1.8	0.974	1.214	21.57
	5	1268	272	0.97	0.066	1443	39.54	9.07	37.4	9.07	1.18	1.7	0.946	1.175	17.88
	6	1261	269	1.01	0.064	1440	39.34	9.3	37.49	9.05	1.12	1.65	0.953	1.183	17.50
Donawiz	1	1250	340	0.91	0.064	1424	41.78	12.53	35.66	9.98	0.8	1.94	0.854	1.092	12.50
	4	1273	340	0.74	0.059	1432	38.02	12.03	32.84	9.14	0.84	2.1	0.864	1.104	14.24
Rautaruukki	1	1196	203	0.42	0.082	1476	37.2	8.8	39.77	10.9	1.34	0.64	1.069	1.362	16.34
	2	1255	203	0.48	0.056	1456	37.27	8.8	39.82	11.2	1.48	0.67	1.068	1.369	26.43
	2	5513	272	0.33	0.03	1513	36.4	11.42	42.52	7.76	1.18		1.168	1.381	39.33
Svelgen	1	4419	273	0.36	0.029	1511	36.35	11.49	42.25	7.94	1.18		1.162	1.381	40.69
	9	2132	282	0.41	0.04	1508	36.4	10.78	41.49	7.7	1.15		1.140	1.351	28.75
Phoenix	3	1858	239	0.52	0.044	1504	36.6	11.31	41.99	6.51	1.38	0.59	1.147	1.325	31.36
Westphalia	4	1812	246	0.62	0.055	1498	37.06	12.39	41.78	5.32	1.31	0.76	1.127	1.271	23.82
	7	2120	254	0.45	0.038	1499		12.09	42.32	5.55	1.23	0.74			32.37
Huckingen	A	2785	253	0.51	0.049	1499	37	1.3	39.8	9.8	1.27	0.71	1.076	1.341	25.92
	B	2581	248	0.49	0.047	1486	37.3	9.2	40.5	9.8	1.34	0.73	1.086	1.349	28.51
Seitz kit	A	2771	232	0.57	0.053	1486	37.4	9.5	36	12	1.48	1.08	0.963	1.283	27.92
	B	3023	234	0.58	0.055	1475	37	10.3	35.7	11.7	1.45	1.05	0.965	1.281	26.36
Dillingen	4	2065	256	0.54	0.044	1473	37.9	11	40.5	7.8	1.23	0.98	1.069	1.274	27.95
	5	3067	262	0.5	0.039	1483	37.8	11	40.7	7.7	1.19	0.9	1.077	1.280	30.51

(continued)

Table 5.5 (continued)

Company and furnace number		Furnace volume, m^3	Slag volume, kg/t	Hot metal composition, %		Hot metal temperature, °C	Slag composition, %						Slag basicity		L_S
				Si	S		SiO_2	Al_2O_3	CaO	MgO	S	$K_2O + Na_2O$	R_2	R_3	
Bremen	2	3198	214	0.57	0.042	1445	36.2	10.8	40.4	8.8	1.37	0.91	1.116	1.359	32.62
	3	1623	184	0.49	0.047	1448	36	9.9	38.3	11.1	1.55	0.98	1.064	1.372	32.98
Eisenstaedt	3	814	280	0.61	0.045	1438	35.5	9.6	39.1	10.5	1.38	0.86	1.01	1.397	30.67
Rosenberg	3	805	251	1.07	0.056	1454	36.1	9.3	38.7	11.3	1.77	1.01	1.072	1.385	31.61
Hoogovens	6	2678	205	0.43	0.036	1517	32.1	16.8	36.5	10.9	1.27	(K_2O) 0.38	1.137	1.477	35.28
	7	4450	206	0.45	0.032	1511	32.1	16.8	36.5	10.9	1.3	(K_2O) 0.38	1.137	1.477	40.63

Note This set of data was provided by the European side at the China–Europe exchange meeting for directors of iron-making plants in 1989

Table 5.6 Distribution coefficients of some blast furnaces in three countries or regions

Country (region)	L_S					
	<20	21–30	31–40	41–50	51–60	>60
China	7.7	15.4	30.8	34.6	11.5	0
Europe	20	48	32	0	0	0
Former USSR	0	7.7	19.2	15.4	23.1	34.6

Table 5.7 Dolomitized limestone

Production area	Chemical constituents, %					
	CaO	MgO	SiO_2	Al_2O_3	P	S
First layer of Longquanwu village	50.7	3.5	1.5	0.4	0.008	0.016
Second and third layer of Longquanwu village	39.35	13.7	2.4	0.4	0.006	0.032

In early years, Capital Steel used dolomitized limestone, so slag contained higher MgO. When we review the slag of Capital Steel over the years today, we can find some slag, with bad but stable composition, is able to adapt to the fluctuations of raw material composition and fuel composition. Or when furnace temperature was not stable, slag would still flow normally.

In the 1950s to 1960s, the limestone production base of Capital Steel was located in Longquanwu Village in the western part of Beijing. The ore was divided into three layers including limestone as the first layer and dolomitized limestone as the second and third layer (Tables 5.7). Therefore, early slag of Capital Steel had high MgO content. It was realized from the production that when binary basicity is the same as ternary basicity (R_3), the desulfurization ability of slag contained MgO was lower than that of slag contained equal amounts of CaO, but its fluidity and stability were better.

In those years, Capital Steel did not establish steelmaking plants. Except a little part of iron produced for pipe casting most of them were used for steel-making of side-blown converters and open hearths. Therefore, there were contained high manganes for converters and high silicon for casting in hot metal. The slag basicity of casting iron was low, and MnO_2 contained in slag of side-blown converters was often more than 2% with excellent fluidity. Table 5.8 shows the main compositions of hot metal and slag of different iron types including some products of Capital Steel in that year.

5.2.2 *Effect of MgO on the Viscosity of Slag*

In 1931, R. S. McCaffery gave a special report named "Effect of MgO on the viscosity of slag" at a conference in New York, systematically introducing the research results [6, 7] by his team. The studies on MgO in slag, which is so classical that had been

Table 5.8 Compositions of hot metal and slag of different iron types

Types	Number	Composition of hot metal, %				Range of slag composition, %								Slag bacisity	
		[Si]		[Mn]		SiO_2		Al_2O_3		CaO		MgO		CaO/SiO_2	
		Max	Min	Max	Min	Max	Min	Max	Min	Max	Min	Max	Min	Max	Min
For steel making	261	0.96	0.27	1.58	0.08	41.78	31.03	17.48	5.36	48.4	30	16.24	3.45	1.262	0.811
For casting	22	2.77	1.44	0.99	0.24	38.85	34.67	16.54	10.77	38.1	29.75	13.86	8.02	1.02	0.766
Contained high manganese	28	1.56	0.71	2.26	1.74	39.46	35.6	14.6	12.2	34.3	28.88	15.21	8.03	1.07	0.732

widely used for many years, also quoted in China's first iron-making textbook used in university [8], and later classic textbooks [9, 10], "Iron-making Manuals" [11]. Figure 5.5 shows the effect of MgO on the viscosity of slag.

In 1955, Zhuang Zhen'e, Tang Naiwu researched on the smelting of high MgO slag. "When the content of MgO in slag is 18%, the furnace runs smoothly, and it is no different from general slag as the content is 10%." "The characteristics of high MgO slag which are beneficial to smooth operation can be explained by slag particularity: (1) Slag fluidity is relatively good. (2) The viscosity of slag changes relatively less when the temperature and the slag basicity change. In other words, viscosity of slag is relatively stable." "These can be explained by McCaffery's datum" [12]. Figure 5.5 shows that other compositions in slag exert a little effect on viscosity of slag with good runnability when MgO > 15% at a temperature of 1400 °C.

Figure 5.5 also shows that when MgO = 15%, slag compositions calculated in accordance with the proportion of Al_2O_3/SiO_2 as shown in Table 5.9.

Fig. 5.5 Influence of MgO on viscosity of slag

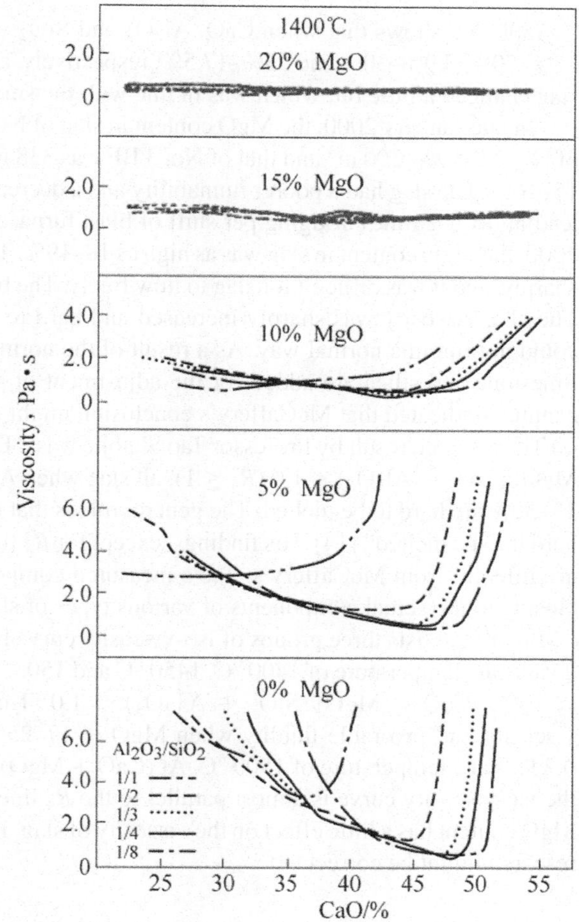

Table 5.9 Slag composition when MgO = 15% (%)

Slag composition		MgO	15					
		CaO	25	30	35	40	45	50
Al_2O_3/SiO_2	1:1	Al_2O_3	30	27.5	25	22.5	20	17.5
	1:2		20	18.3	16.7	15	13.3	11.7
	1:3		15	13.8	12.5	11.3	10	8.8
	4:4		12	11	10	9	8	7
	1:8		6.7	6.1	5.6	5	4.4	3.9
	1:1	SiO_2	30	27.5	25	22.5	20	17.5
	1:2		40	36.7	33.3	30	26.7	23.3
	1:3		45	41.2	37.5	33.7	30	26.2
	4:4		48	44	40	36	63	28
	1:8		53.3	48.9	44.4	40	35.6	31.1

Table 5.9 shows that when CaO, Al_2O_3 and SiO_2 were controlled in a range of 25%–50%, 3.9%–30% and 53%–17.5%, respectively, and MgO $\geq$ 15%, viscosity of slag changed a little bit, which was in line with the smelting practice in AISC [12].

"In late January 2000, the MgO content in slag of No. 2 blast furnace (the volume of No. 2 BF was 200 m^3 and that of No. 3 BF was 338 m^3) in Weiyuan Steel reached 15–16%. The slag had a poorer runnability and a decreased desulfurization capacity, leading to 1–2 times bridging per shift of blast furnaces. In the middle of February 2000, the MgO content in slag was as high as 18–19%. The viscosity of slag increases sharply, and it was difficult for slag to flow freely. The blast furnace was out of order. The slag viscosity was sharply increased and hard to flow. Therefore, the furnace could not run in a normal way. As a result of the normal production of the furnace, limestone and silica are added for the adjustment of slag composition" [13]. This example indicated that McCaffery's conclusion might not be comprehensive.

The research result by Professor Tao Shaojie was: "Except for few slag of (CaO + MgO)/(SiO_2 + Al_2O_3) $\leq$ 1.0 ($R_4 \leq 1$), all slag when Al_2O_3 = 15–25% and MgO = 15–30% are hard to be melted. The general rule is that the higher Al_2O_3 content, the harder to be melted" [14]. His findings, except for R_4 (quaternary basicity) $\leq$1, were all different from McCaffery's. Tao's measured components were relatively dense, closing to the actual components of various types of slag.

Figure 5.6 lists three groups of iso-viscosity curve in which Al_2O_3 = 15%, 20%, 25% at the temperature of 1400 °C, 1450 °C and 1500 °C, respectively. When Al_2O_3 = 15%, (CaO + MgO)/(SiO_2 + Al_2O_3) = 1.0. This kind of slag has the least viscosity and favorable fluidity when MgO = 17–25%. The viscosity is less than 0.2 Pa·s at a temperature of 1500 °C. At (CaO + MgO):(SiO_2 +Al_2O_3) = 1.0 or so, the iso-viscosity curve is almost parallel to the R_4 line, and the change in CaO and MgO content has a little effect on the viscosity of slag. In this condition, McCaffery's research might be correct.

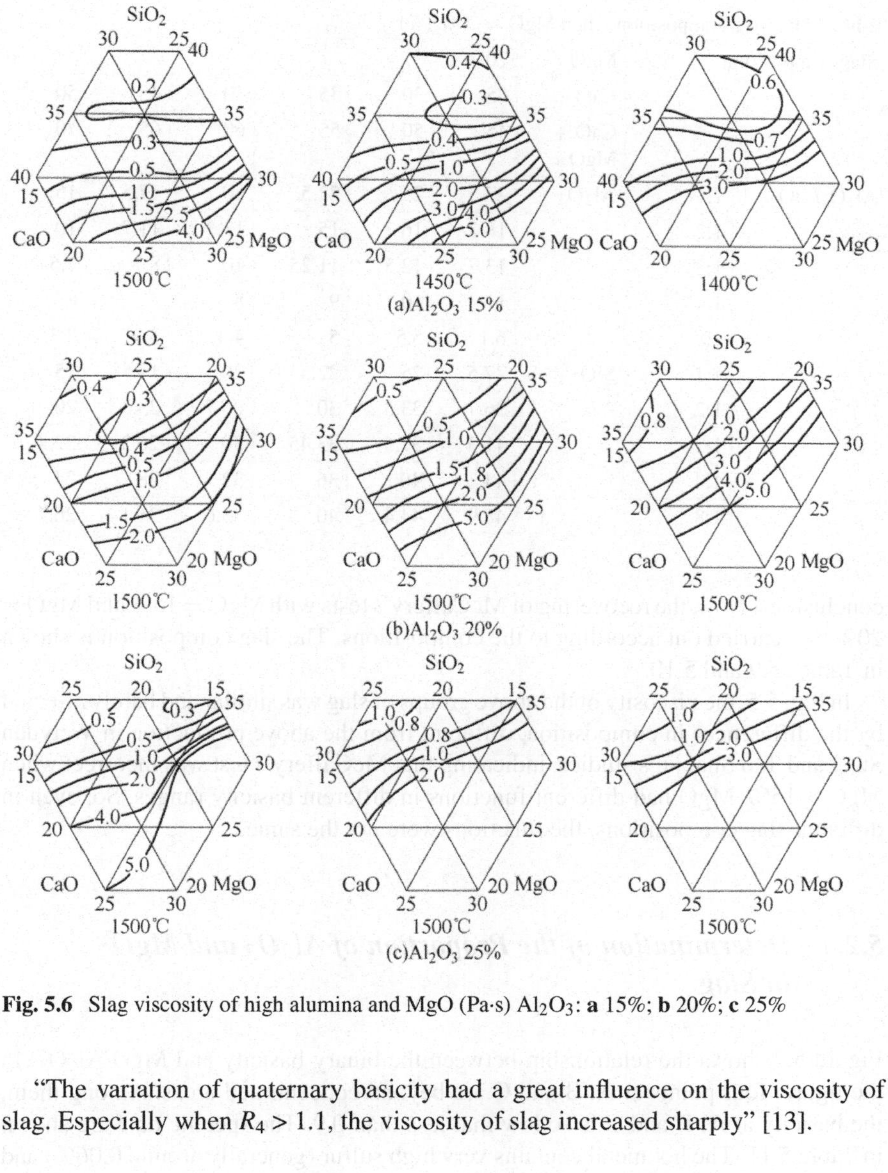

Fig. 5.6 Slag viscosity of high alumina and MgO (Pa·s) Al_2O_3: **a** 15%; **b** 20%; **c** 25%

"The variation of quaternary basicity had a great influence on the viscosity of slag. Especially when $R_4 > 1.1$, the viscosity of slag increased sharply" [13].

5.2.3 Analysis of McCaffery Viscosity Diagram

The results of McCaffery's test shown in Fig. 5.5 were sometimes very different from the results of later studies. For example, Tao Shaojie's result that R_4 (quaternary basicity) ≥ 1 was different from McCaffery's results when the content of MgO was high. Other production practices indicated that there are problems in McCaffery's

Table 5.10 Slag composition when MgO = 20% (%)

Slag composition								
		MgO	20					
		CaO	25	30	35	40	45	50
		CaO + MgO	45	50	55	60	65	70
Al_2O_3/SiO_2	1:1	Al_2O_3	27.5	25	22.5	20	17.5	15
	1:2		18.3	16.7	15	13.3	11.7	10
	1:3		13.7	12.5	11.25	10	8.7	7.5
	1:4		11	10	9	8	7	6
	1:8		6.1	5.5	5	4.4	3.9	3.3
	1:1	SiO_2	27.5	25	22.5	20	17.5	15
	1:2		36.7	33.3	30	26.7	23.3	20
	1:3		41.3	37.5	33.45	30	26.3	22.4
	1:4		44	40	36	32	28	24
	1:8		48.9	44.4	40	35.6	31.1	26.7

conclusions. Now, the recovering of McCaffery's tests with MgO = 15% and MgO = 20% was carried out according to the compositions. The slag composition is shown in Tables 5.9 and 5.10.

In Fig. 5.5, the viscosity of the above groups of slag was similar and barely affected by the differences in composition, different from the above production in Weiyuan Steel and Tao Shaojie's studies, indicating that McCaffery's test was incorrect when MgO > 15%. MgO had different functions in different basicity ranges. So, even in different slag compositions, the functions were not the same.

5.2.4 *Determination of the Proportion of Al_2O_3 and MgO in Slag*

Figure 5.7 shows the relationship between the binary basicity and MgO/Al_2O_3. In the figure, four points at A, B and C are beyond concentrated areas. Among them, the basicity at A, B and C is so low that less than 0.9. The specific data are shown in Table 5.11. The hot metal contains very high sulfur (generally around 0.06%) and relatively low temperature (about 1424–1450 °C) due to low basicity, which leads to the need for external desulfurization. Among these three cases, two of them are low phosphorus iron produced by Donawtz's No. 1 and No. 4 blast furnaces (the volumes of them are 1250 and 1273 m^3) in Austria, and another one was low phosphorous hot metal produced by Germany with high-magnesium low-phosphate ores. This is the lowest slag basicity for iron for steelmaking ever seen. However, after 10 years, the lowest slag basicity 0.811 was no longer used, because the basicity of the slag was promoted to 0.922 (Table 5.12). Due to the use of magnesium-rich ore, magnesium

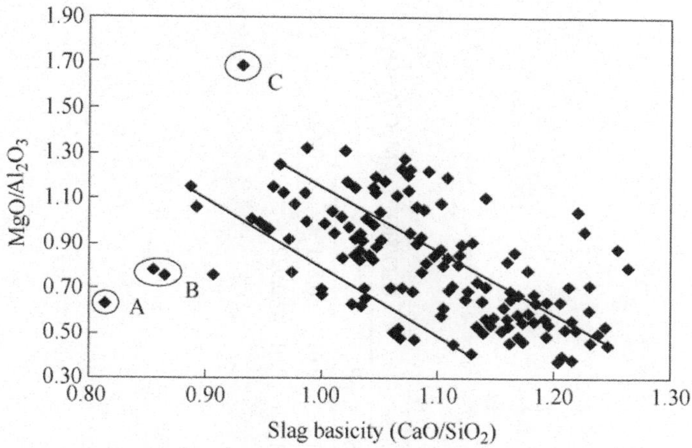

Fig. 5.7 Relationship between binary basicity and alumina, MgO

Table 5.11 Low basicity slag

Point position	Year	Slag composition, %						Slag basicity	Slag volume, kg/t
		SiO_2	Al_2O_3	CaO	MgO	MnO	P	CaO/SiO_2	
A	~1980	37	17	30	11	1.1	0.01	0.811	325
B	1998	41.78	12.53	35.66	9.98	1.61	0.005	0.854	340
B	1998	38.02	12.03	32.84	9.14	1.59	0.005	0.864	340

oxide in slag was as high as 13.5% and MgO/Al_2O_3 in slag as high as 1.668. It was based on resource characteristics instead of smelting demand. The slag performed well.

Table 5.13 shows the appropriate interval (dense area) and the allowable lower limit for MgO/Al_2O_3 ratio in slag. The actual data in Fig. 5.6 show that if the ratio was lower than the limit, the slag would be too viscous to contribute to the smooth operation of blast furnaces.

According to above data, there were three regulations to determine the appropriate composition of slag:

1. The relationship between binary basicity of slag and MgO/Al_2O_3 generally presented as a straight line. The higher the basicity, the lower the ratio.
2. The appropriate ratio range of slag basicity and MgO/Al_2O_3 is shown in Table 5.13.
3. MgO/Al_2O_3 data in dense area are more reliable than that in lower limit area, so it can be used as reference for slag composition adjustment. Lower limit ratio would be also a feasible option, but less be used in the past.

It is an attempt to standardize slag composition with empirical data, which needs to be verified by practices or laboratory tests in the future. According to the results of recent studies, that:

Table 5.12 Different compositions controlled low phosphorus ore with magnesium enrichment [15, 16]

Point position	Hot metal temperature, °C	Slag composition, %								CaO/SiO_2	MgO/Al_2O_3
		SiO_2	Al_2O_3	CaO	MgO	MnO	FeO	S	P		
A	1450	37	17	30	11	1.1	0.9	1.2	0.01	0.811	0.647
C	1500	38.5	8	35.5	13.5	0.26	0.26	1.5	0.01	0.922	1.688

Table 5.13 Appropriate area and lower limit of MgO/Al_2O_3 in slag

Slag basicity (CaO/SiO_2)	0.9	1	1.1	1.2	1.25
Appropriate ratio (dense area)	1.30	1.06	0.82	0.58	0.46
Lower limit	1.06	0.84	0.62	0.40	0.29

1. "The appropriate slag composition of the original slag of TISCO was consisted of MgO (about 11%) and Al_2O_3 (14–15%). The basicity R_2 of slag (1.20) [17] and Al_2O_3/MgO in slag (14/11 = 1.27 and 15/11 = 1.36) was both higher than the appropriate range.
2. "Under the existing conditions of No. 2 iron-making plant of Tangshan Steel, when the basicity is controlled within the range of 1.05–1.15 and the content of Al_2O_3 is up to 14%, the slag fluidity and melting temperature could meet the production requirements with the stable operation and adequate heat in the hearth." "MgO was added mainly to reduce the slag viscosity and improve the fluidity. It might be feasible to adding about 11% MgO in the slag" [18]. Calculated with data above, 14/11 = 1.27 was also higher than the appropriate range. Whether the above cases could reduce MgO/Al_2O_3 to about 0.5 or not, still to be studied.

5.3 Slag with Special Functions

There are various requirements for blast furnace operation based on slag characteristics, mainly including:

1. Cleaning up scaffolding;
2. Relining;
3. Discharging alkali metals;
4. Dealing with hearth accumulation;
5. Concentrating specific compositions in the burden.

Relining and dealing with hearth accumulation would be discussed in detail in Chaps. 4 and 8, no repeating here.

5.3.1 Cleaning up Scaffolding

Cleaning up scaffolding on the furnace line with slag of low melting point and high fluidity is called "flushing furnace." The low melting compounds such as FeO, CaF_2 and MnO_2 are commonly used as flushing agents. In the 1960s to 1970s, continuous casting was still underdeveloped. Ingot should be heated in the soaking furnaces before rolling to produce a large amount of "soaking pit slag." It was also commonly used for flushing agents. The general composition is shown in Table 5.14.

Table 5.14 Soaking pit slag (iron phosphate) composition (%)

Position	Fe	FeO	P	S	SiO_2	Al_2O_3	CaO	MgO
Capital steel in 1989	60.98	44.52	0.015	0.035	8.83	3.14	1.8	1.45
Baotou Steel in 1989	66.2	44.62			3.14	4.73	0.8	0.72

With the development of continuous casting, steel ingot had been taken place by continuous casting slab without soaking pit slag any more. Using soaking pit slag as flushing agents had become a history.

The fluorite is usually used for flushing furnace. Slag containing CaF_2 has been studied intensively due to the use of fluoride-containing iron ores. In the former Soviet Union, a specific manual was published to record the research results of fluoride-containing slag [19]. In Baotou Steel, Chinese iron and steel company, it was deeply and extensively studied for the smelting characteristics of fluoride-containing slag due to the use of fluoride-containing iron ores of Baiyun'ebo Iron Mine with guiding significance to flushing furnace [20]. Figures 5.8 and 5.9 show part of the research results.

Figure 5.8 shows that the lower the slag temperature, the greater CaF_2 effect. The more the slag contains CaF_2, the lower the slag viscosity. When CaF_2 content exceeds 10%, the effect on viscosity decreases. In fact, the content required for flushing blast furnace is far less than 10%. As can be seen from Fig. 5.9, when CaF_2 is 1.14%, the viscosity decreases a lot, but CaF_2 increased from 1.14 to 2.2%, and the viscosity reducing becomes smaller than the former. When the fluorine content increases to 4.58%, although the fluorine content increases more than doubled, the

Fig. 5.8 Effect of CaF_2 on the viscosity at slag basicity 1.0

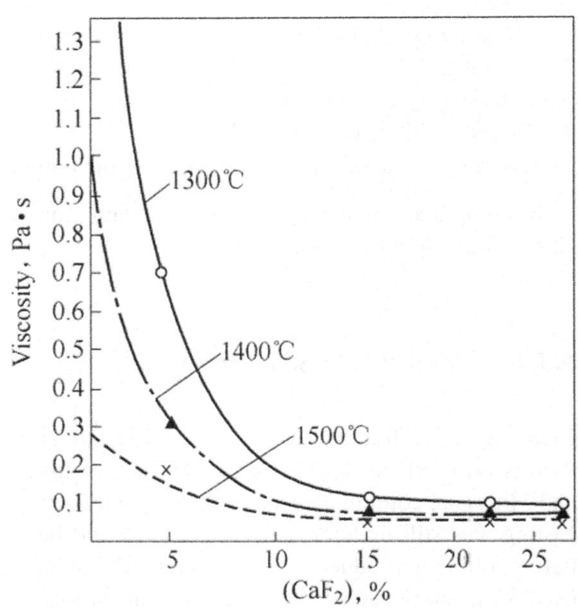

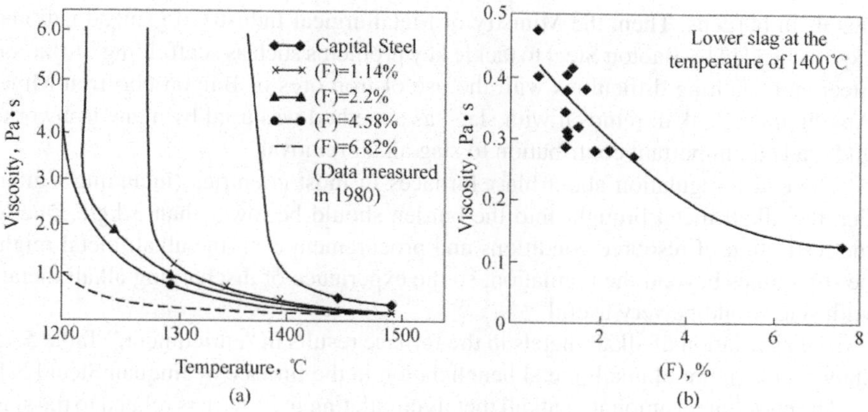

Fig. 5.9 Effect of fluorine on slag viscosity **a** viscosity measured at different temperatures; **b** viscosity measured under different fluorine contents

Fig. 5.10 Slag viscosity curve under different CaF_2 contents

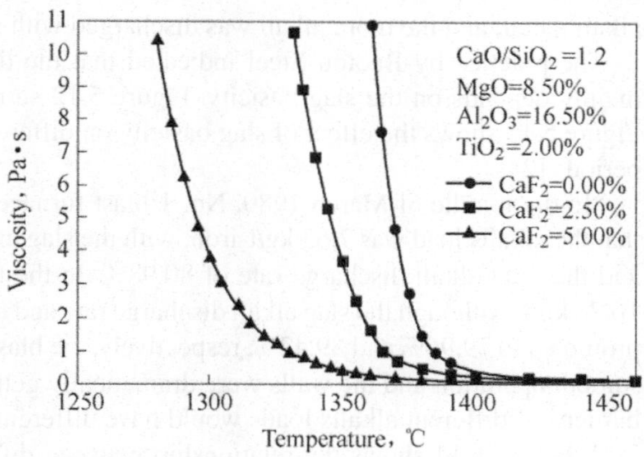

viscosity reducing is still limited. In other words, the slag viscosity would be similar at the fluorine content between 1 and 6% and at the temperature 1400 °C. Therefore, the fluorine content is unnecessary to be very high in the slag composition of blast furnaces for flushing furnace. The slag composition of Capital Steel for flushing furnace generally contains 1.5% F, equivalent to about 3% CaF_2 (Fig. 5.10) [21].

5.3.2 Discharging Alkali Metals with Slag

Alkali metals are harmful to blast furnace production. It not only erodes brick lining, but also leads to the scaffolding in blast furnace. In the 1960s and 1970s, the blast furnaces of Baotou Steel were getting thick, mainly due to the high content of alkali metals in burdens. Then, the Ministry of Metallurgical Industry organized national experts headed by Baotou Steel to tackle key problems such as scaffolding of Baotou Steel and smelting difficulties with the use of iron ores of Baiyun'ebo Iron Mine. Among them, "alkali removal with slag" as a method was used by many ironworks and made an important contribution to slag alkali removal.

There is a regulation about blast furnaces in most countries (including China) that the alkali metal brought into the burden should be lower than 3 kg/t. Due to the restriction of resource conditions and procurement cost, the alkali metal might be sometimes beyond the regulation, so the experience of discharging alkali metals with slag would be very useful.

The circulation of alkali metals in the furnace results in "enrichment." Table 5.15 shows the amount of alkali metal beneficiation in the furnace of Jiuquan Steel [22].

The enrichment amount of alkali metals circulating in furnace is related to the slag basicity and temperature. Studies by Zhou Shizhuo, Xu Ruxiong and other experts showed that "Reducing the reduction, volatilization and cycle accumulation of alkali metals in blast furnace could discharge more alkalis along with slag. The experiment (see Fig. 5.11) proved that the lower the slag basicity, the less volatilization rate the alkali metal, and the more alkali was discharged with slag" [23].

The practice by Baotou Steel indicated that the discharge of alkalis with slag mainly depends on the slag basicity. Figure 5.12 shows the result of the practice. Figure 5.13 shows the effect of slag basicity on different alkali contents in different periods [24].

"In the middle of March 1980, No. 1 blast furnace was in no smooth operation that the alkalis load was 7.63 kg/t-iron, with the slag alkali discharge rate of 74.06% and the total alkali discharge rate of 80.98%. In the first of May, alkalis load was 10.74 kg/t. Although the slag alkali discharge rate and total alkali discharge rate were promoted to 79.96% and 89.42%, respectively, the blast furnace was in obviously no smooth operation and the walls were dramatically getting thick, which showed that burdens of different alkalis loads would have different effects on the blast furnace" [24]. Figure 5.14 shows the relationship between different contents of [Si] in hot metal and slag basicity in Baotou Steel. It coincided with many later research results: Reducing the basicity of the slag would be an effective way to improve the ability of slag to discharge alkali. Table 5.16 shows the practice results [25] of No. 6 blast furnace of Shaoguan Steel.

1. $(K_2O + Na_2O) = 0.1431e^{0.0714/R'}$, [Si] = 0.6–0.8%, 5–13 kg$(K_2O + Na_2O)$/t (Alkalis load);
2. $(K_2O + Na_2O) = 0.2313e^{0.0100/R'}$, [Si] = 0.6–0.8%, 7–9 kg$(K_2O + Na_2O)$/t;
3. $(K_2O + Na_2O) = 0.1720e^{0.0434/R'}$, [Si] = 1.3–1.2%, 9–7 kg$(K_2O + Na_2O)$/t

Table 5.15 Alkali metal balance of No. 6 blast furnace in Jiuquan steel

Project		Unit consumption, kg/t	K_2O		Na_2O		$K_2O + Ka_2O$	
			Proportion, %	Mass, kg/t	Proportion, %	Mass, kg/t	Proportion, %	Mass, kg/t
Input	Sinter	1172.90	0.288	3.378	0.105	1.232	0.393	4.610
	Imported pellets	153.2	0.111	0.170	0.172	0.264	0.283	0.434
	Local pellets	445.60	0.208	0.927	0.108	0.481	0.316	1.408
	Lamp ore	70.70	1.190	0.841	0.053	0.037	1.243	0.879
	Local pulverized coal	421.90	0.068	0.287	0.084	0.354	0.152	0.641
	Coarse coke	146.00	0.062	0.091	0.129	0.188	0.191	0.279
	Total input			5.694		2.556		8.251
Output	Slag	474.10	1.150	5.452	0.478	2.266	1.578	7.030
	Gravitational dust collection	11.65	0.180	0.021	0.323	0.038	0.503	0.059
	Bag dust collection	8.29	0.747	0.062	0.710	0.059	1.457	0.121
	Environmental dust collection	0.96	0.447	0.004	0.192	0.002	0.639	0.006
	Total output			5.539		2.365		7.216
Beneficiation				0.155		0.192		1.035

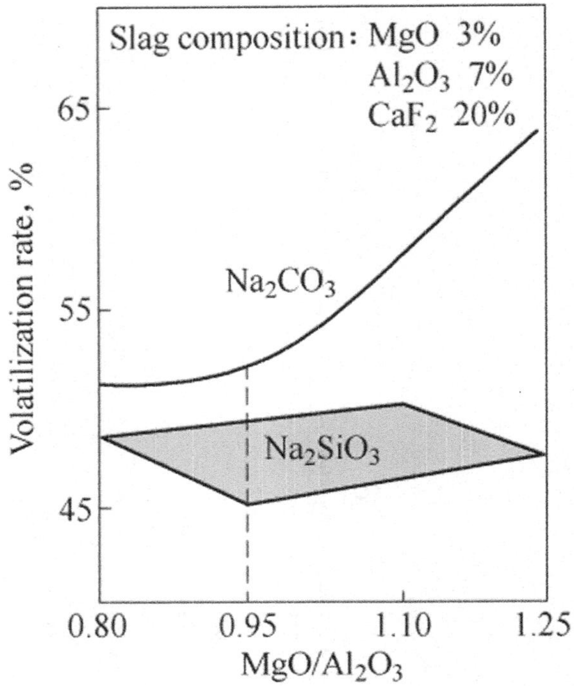

Fig. 5.11 Effect of slag basicity on the volatilization of alkali metals

5.3.3 To Make Use of Lean Ores with Slag

In the smelting process of blast furnace, the proportion of metal manganese distribution in the hot metal and in slag is called distribution coefficient of manganese. To control and make more metal manganese accessing into the slag under corresponding smelting conditions, for the purpose of the enrichment of poor manganese ores and the improvement of grades, is a special process usually for making high manganese ferromanganese with poor manganese ores which is so-called two-step process.

In the smelting process of blast furnace, all of the phosphorous and more than 99% iron in burden enter into hot metal. The proportion of manganese entering into hot metal is related to many factors but mainly depends on the basicity of the slag. Figure 5.15 shows the distribution coefficient of manganese when making high manganese ferromanganese with "two-step process" at Capital Steel in 1962. The result of practice at iron plants of Zijiang, Hunan Province, in 1958 was used as reference in this figure. It was found that the slag basicity was closely related to the distribution coefficient. When the alkalinity was lower than 0.8, the MnO in slag increased sharply. When the alkalinity was lower than 0.6, the distribution coefficient reached the maximum. Equation (5.2) [26] can be obtained by the analysis of 69 groups of data on hot metal and slag composition.

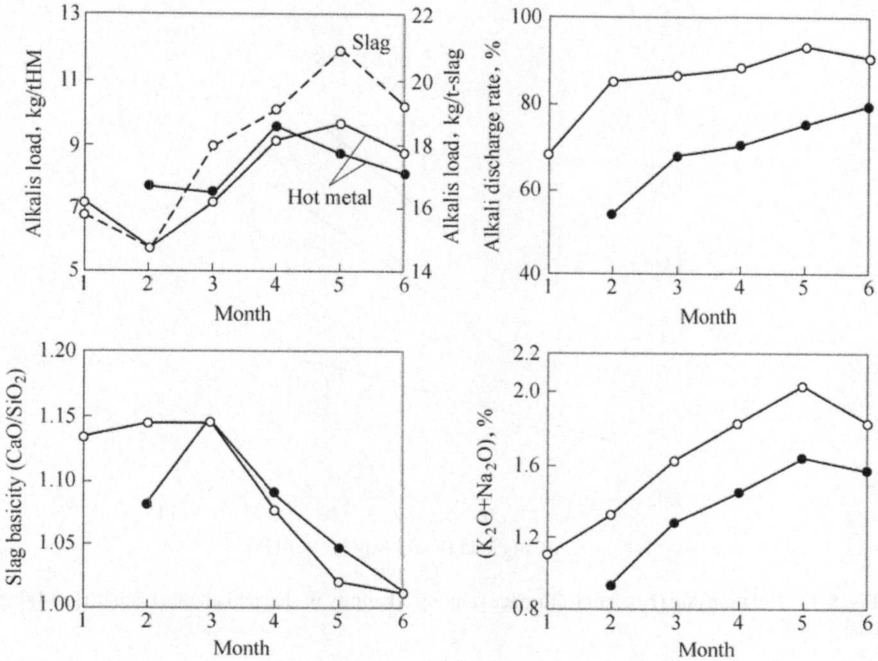

Fig. 5.12 Alkali monthly discharged from slag in No. 1 and No. 2 blast furnace of Baotou Steel in the first half of 1980

Fig. 5.13 Regression relationship between alkalinity in slag and the basicity. Average daily data of no. 2 blast furnace from December 24, 1979–July 7, 1980

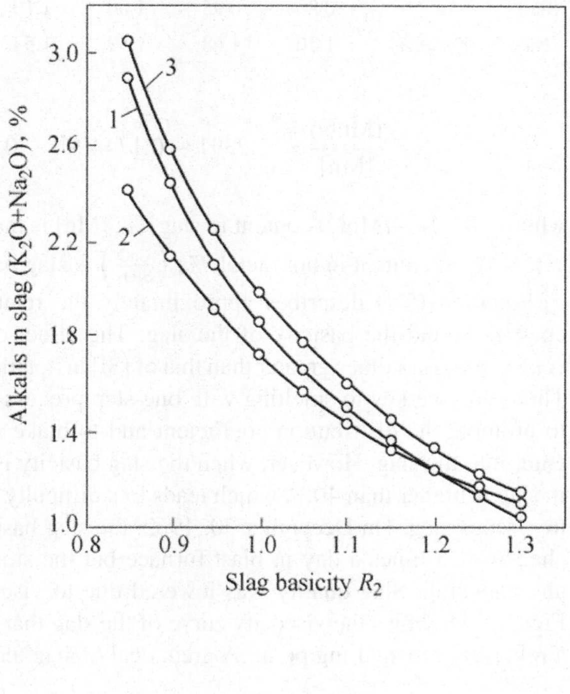

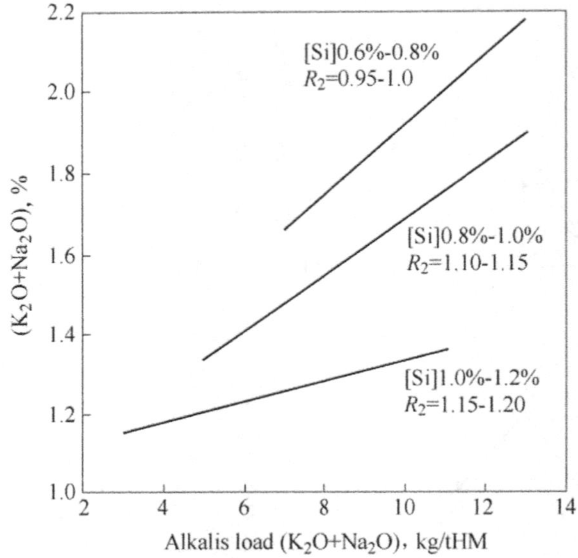

Fig. 5.14 Relationship (No. 2 blast furnace) among alkalinity in slag and alkalis load, [Si] and slag basicity (*R*)

Table 5.16 Alkali discharge capacity with slag in No. 6 blast furnace of Shao Steel under different alkalinities [25]

Slag *R*	0.80	0.95	1.00	1.10	1.20	1.30	1.40
$(Na_2O + K_2O)$, %	1.70	1.68	1.59	1.54	1.53	1.50	1.46

$$\frac{(\text{MnO})}{[\text{Mn}]} = 2.2391 - 0.1744[\text{Si}] - 0.998\left(\frac{\text{CaO}}{\text{SiO}_2}\right) \tag{5.2}$$

where (MnO) is (MnO) content in slag, %; [Mn] is manganese content in hot metal, %; [Si] is Si content in hot metal, %; $\left(\frac{\text{CaO}}{\text{SiO}_2}\right)$ is slag basicity.

Equation (5.2) describes approximately the relationship between distribution coefficient and the basicity of the slag. The effect of the basicity on distribution coefficient is six times greater than that of [Si] in hot metal on distribution coefficient. Therefore, the key to smelting with one-step process is to control the slag basicity, to promote the distribution coefficient and to make a large amount of manganese enter into the slag. However, when the slag basicity is lower than 0.4, SiO_2 content in slag is higher than 40%, which leads to a difficulty in blast furnace operation due to viscous slag. On December 30, 1962, the slag basicity dropped to less than 0.3, checking 25 times a day in blast furnace but the stock almost no moving without any checking. Slag quality was lowered due to viscous slag with too much iron. Figure 5.16 shows the viscosity curve of the slag that the slag with low basicity had a relatively low melting point. A great deal of slag temperature indicated that when

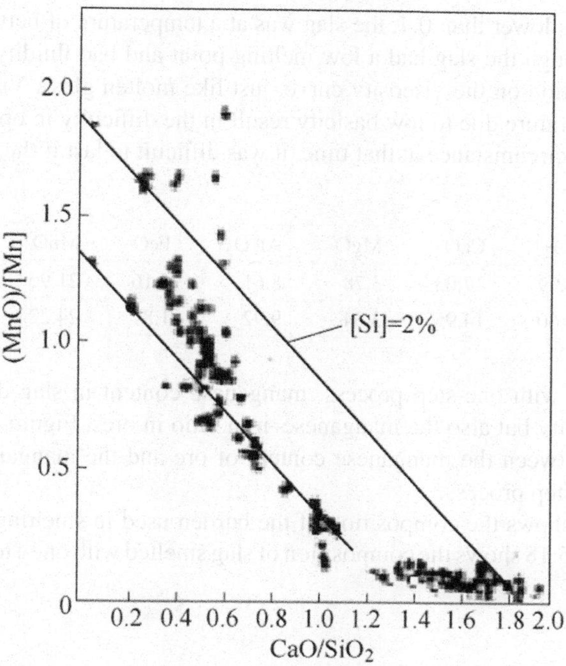

Fig. 5.15 Relationship between distribution coefficient of manganese in slag and basicity

Fig. 5.16 Slag viscosity curve

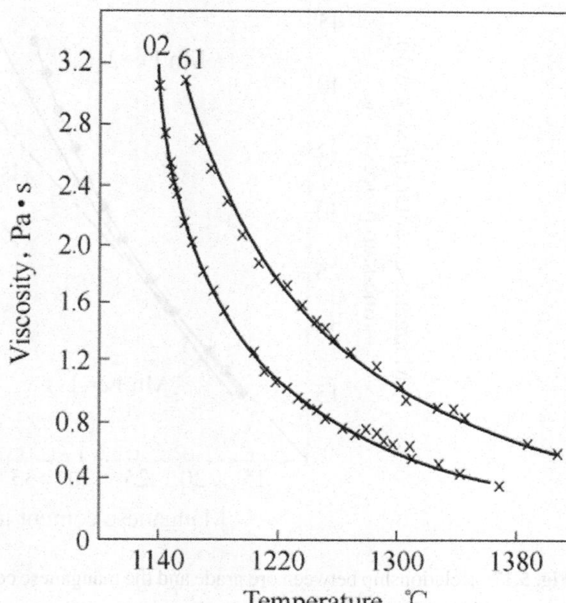

the basicity was lower than 0.4, the slag was at a temperature of between 1240 and 1340 °C. Although the slag had a low melting point and bad fluidity, there was no apparent inflection on the viscosity curve, just like molten glass. Viscous slag and low slag temperature due to low basicity result in the difficulty in operation. Under the production circumstance at that time, it was difficult to last if the basicity is less than 0.4.

Number	SiO_2	CaO	MgO	Al_2O_3	FeO	MnO	CaO/SiO_2
02	36.99	17.03	3.78	8.61	2.16	21.96	0.46
61	39.60	14.95	3.04	9.92	1.87	24.22	0.38

In smelting with one-step process, manganese content in slag depends on not only slag basicity but also the manganese–iron ratio in ores. Figure 5.17 shows the relationship between the manganese content of ore and the manganese content in slag with one-step process.

Table 5.17 shows the composition of the burden used in smelting with one-step process. Table 5.18 shows the composition of slag smelted with one-step and two-step process.

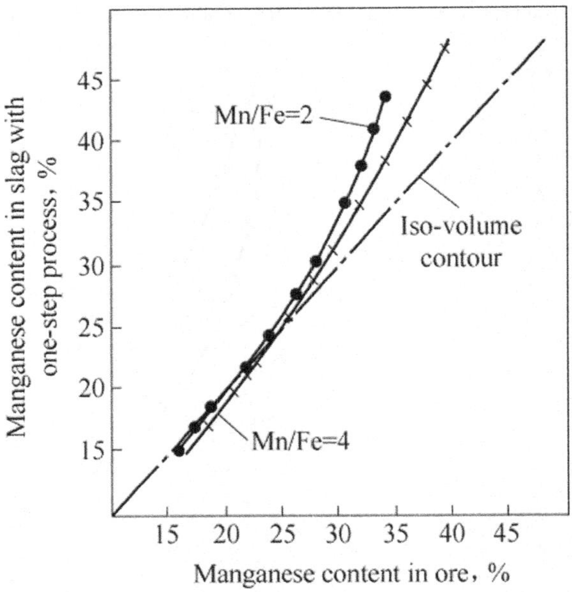

Fig. 5.17 Relationship between ore grade and the manganese content in slag with one-step process

Table 5.17 Composition of the burden smelted with one-step process (%)

Type	TFe	TMn	SiO_2	Al_2O_3	CaO	MgO	S	P	C
Mugui manganese deposit	10.74	23.30	36.69	4.4	0.38	0.26	0.034	0.35	–
Limestone	0.2	–	1.73	0.74	42.3	10.40	0.004	0.04	–
Coke	0.79	–	0.63	5.68	0.81	0.16	0.65	0.048	83.93

In Table 5.18, the ferromanganese produced with one-step process had a high phosphorous content of 1.61%. The slag produced with one-step process contains 28.7% MnO and little Fe and P. The manganese ore fed into the furnace produced with one-step process had a Mn content of 23.3% (Fig. 5.17). The manganese-rich slag produced with one-step process contains 28.71% MnO (for a period of time). The ferromanganese produced with two-step process contains 75% Mn. The production target is shown in Table 5.19.

All ferromanganese smelted with two-step process with a manganese content of 75% on average in a short time were obviously affected by products in the transition period. However, it is difficult to smelt ferromanganese containing more than 80% Mn with two-step process. It can be inferred from Fig. 5.18 the relationship between Mn/Fe in ore and manganese content in ferromanganese.

Table 5.18 Composition of the slag smelted with one-step and two-step process

Composition	Ferro manganese, %					Slag, %							Note
	Si	Mn	S	P	C	SiO_2	CaO	MgO	Al_2O_3	MnO	S	P	
One-step method	3.01	27.9	0.016	1.61	5.88	39.17	16.78	4.64	8.68	28.71	0.402	0.002	$CaO/SiO_2 = 0.46$
Two-step method	1.29	75.0	0.012	0.48	5.25	32.7	40.6	10.12	9.74	6.37			$(CaO + MgO)/SiO_2 = 1.50$

Table 5.19 Smelting production index with one-step and two-step process

Index	Energy consumption of per ton iron			Smelting intensity, t/(m^3 d)	Blast temperature, °C	Top gas temperature, °C	Mn content in ore, %	Check (total number/daily number)	Manganese recovery, %
	Coke	Ore	Lime						
One-step	2.503	6.758	2.136	0.70	530	166	22.89	76/6.9	Input 14.3
Two-step	7.275	6.741 (slag-rich)	7.076	0.785	674	330	21.43 (slag-rich)	71/5.1	Input 85.7

Fig. 5.18 Relationship between the manganese content in manganese ore and that in hot metal

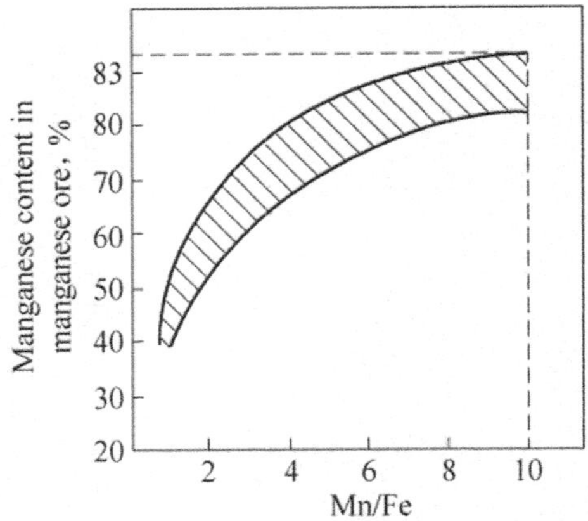

References

1. F. Clements, *Blast Furnace Practice* (Richard Clay and Sons Limited, 1929), I Appendix
2. W. Shoukun, *Thermodynamics of Metallurgical Processes* (Science Press, Beijing, 2010), pp. 187–190
3. W. Xiaoliu, *Metallurgy of Iron and Steel (Iron-making Sector)*, 3rd edn. (Metallurgical Industry Press, Beijing, 2013), pp. 144–145
4. N. Shuren, *Iron-smelting Technology* (Metallurgical Industry Press, Beijing, 2014), pp. 122–126
5. H. Rubin, K. Jun, *Mining and Metallurgy Sector of History of Science and Technology in China* (Science Press, Beijing, 2007), pp. 344–389
6. R.S. McCaffery, J.F. Oesterle, O.O. Fritsche, Effect of magesia on slag vicosity. Trans. AIME. 120–140 (1932)
7. R.S. McCaffery, J.F. Oesterle, O.O. Fritsche, Determination of viscosity of iron blast furnace slags. Trans. AIME. 86–120 (1932)
8. L. Zongcai, Z. Quding, *Iron-making* (The Commercial Press, Beijing, 1952), pp. 235–242
9. Iron-making Teaching and Research Group in University of Science and Technology Beijing. *Iron-Making*, vol 2 (Metallurgical Industry Press, Beijing, 1960), pp. 261–262
10. Iron-making Teaching and Research Group in Northeastern University. *Modern Iron-Making*, vol 1 (Metallurgical Industry Press, Beijing, 1959), pp. 259–260
11. Z. Chuandian, *Technical Manuals of Blast Furnace Iron-making Production* (Metallurgical Industry Press, Beijing, 2002), p. 125
12. Z. Zhuang, N. Tang, D. Li, et al., Problems of smelting high magnesia in blast furnace. Acta Metallurgica Sinica **3**, 19–235 (1957)
13. L. Zhongyu, H. Lin, D. Nengyun et al., Study on physical properties of slag in blast furnace of Weiyuan steel. J. Rare Earth **9**, 87–89 (2002)
14. T. Shaojie, A study on properties of high Al_2O_3 slag and high MgO slag. Iron Steel **1**, 11–17 (2007)
15. Verein Deutscher Eisenheuttenleute (Translated by W. Jian, et al.) Slag Atlas (Metallurgical Industry Press, Beijing, 1989), p. 138
16. VDEh, Slag Atlas, 2nd edn, p. 216 (1995)
17. G. Hao, Z. Jianliang, Z. Hua et al., Slag composition suitable for 4350 m^3 BF of Tai steel. J. Iron Steel Res. **9**, 58–60 (2008)
18. G. Mingshan, Z. Yuzhu, X. Li et al., Effect of MgO content and basicity on BF slag viscosity. Henan Metall **3**, 7–8 (2005)

19. A.A. Aberdin et al., *CaO–SiO_2–Al_2O_3–MgO–CaF_2 Manual of Physical Properties of Slag Smelting System* (Metallurgy, Moscow, 1987)
20. L. Donglu, L. Chunguang, W. Hulin, *Key Research and Technological Progress of Mining and Smelting Technology in Baiyun'ebo Special Ore* (Metallurgical Industry Press, Beijing, 2007), pp. 332–339
21. Z. Heshun, M. Hongbin, C. Zun, Some opinions on the slag system of blast furnace in Capital steel. Tech. Commun. Iron-making **4**, 12–14 (2011)
22. X. Jun, W. Qingxue, Analysis and analysis and control measures of alkali metals in blast furnace of Jiuquan steel. Iron-making **6**, 32–39 (2014)
23. S. Zhou, R. Xu, M. Yu et al., Experimental study on blast furnace scabs and slag alkali discharge in Baotou steel. Iron Steel **9**, 1–8 (1982)
24. Z. Shimin, Balance of alkali metals in blast furnace. J. Iron Steel Res. **1**, 1–6 (1982)
25. S. Wu, S. Liu, X. Xianming, Control of alkali metal hazard in No. 6 blast furnace of SISG. Iron-Making **1**, 39–41 (2009)
26. Y. Liu, Z. Fan, A method to deal with low-grade manganese ore. Capital Steel Sci. Technol. **1**, 60–68 (1981)

Chapter 6
Iron and Slag Tapping

Abstract In the process of smelting in blast furnace, part of burden descends into the hearth in the form of slag and hot metal; the other part rises in gasification and discharge from the top of the furnace. The slag and hot metal gradually accumulate in the hearth. Figure 6.1 shows an example of the process of iron and slag accumulation in the hearth of No. 3 blast furnace of Capital Steel within interval 2.5 h between two tapping.

Let us suppose that the hearth space is fully filled with hot metal and slag. From the former iron plugging time (7:25) until the next iron plugging at 9:55, the thickness of hot metal and slag in the hearth during this period were 917.4 mm and 1127.1 mm, respectively. The total thickness of hot metal and slag was 2044.5 mm. The dotted lines in the figure represented the iron and slag accumulation depth, respectively. The tapping of slag began at 8:15. The intersection of the time coordinate and the slag accumulation line represented the relative height of slag interface, until the slag notch plugging as iron tapping, the thickness of 749.2 mm upper slag was discharged. The solid line in the middle of Fig. 6.1 represented the change of the slag interface.

Iron tapping began at 9:14 and ended at 9:55. The height of hot metal discharged from the tap hole and lower slag was equal to 1372.4 mm in the hearth. From 7:25 to 9:55, the amount of slag and hot metal entering into the furnace was 2044.5 mm, and the total amount of tapping was 2121.6 mm. The input was less than the output, so the liquid level of the hearth was 77.1 mm lower than the previous one. It is common that input is not equivalent to output, resulting in the fluctuation of the hearth liquid level. Figure 6.2 shows ten successive changes of the hearth liquid level of No. 3 blast furnace of Capital Steel. Figure 6.2 also shows that the system of iron and slag tapping in 1987. The constant variation of liquid level of the hearth and the rise and descent of the deadman in the furnace resulted in the variation of up-thrust, all of which played an important role in the blast furnace operation.

Y. Liu, *The Operation of Contemporary Blast Furnaces*,
https://doi.org/10.1007/978-981-15-7074-2_6

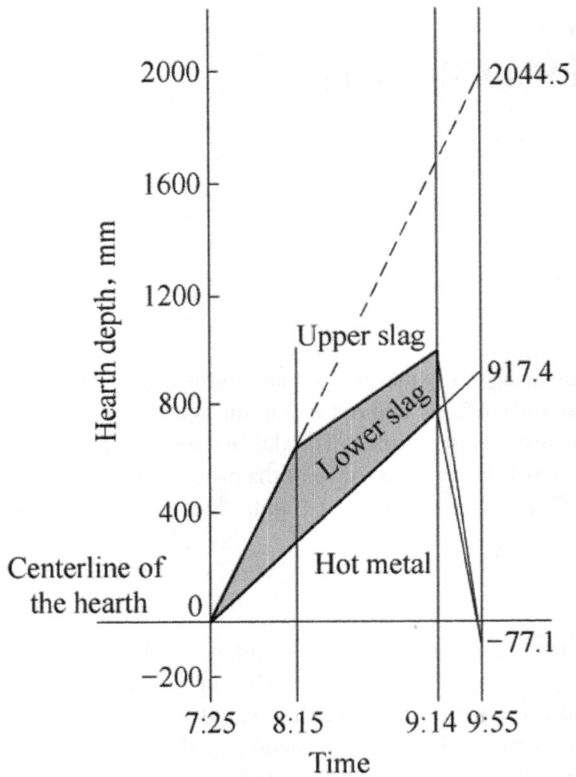

Fig. 6.1 Variation of slag and hot metal depth in hearth

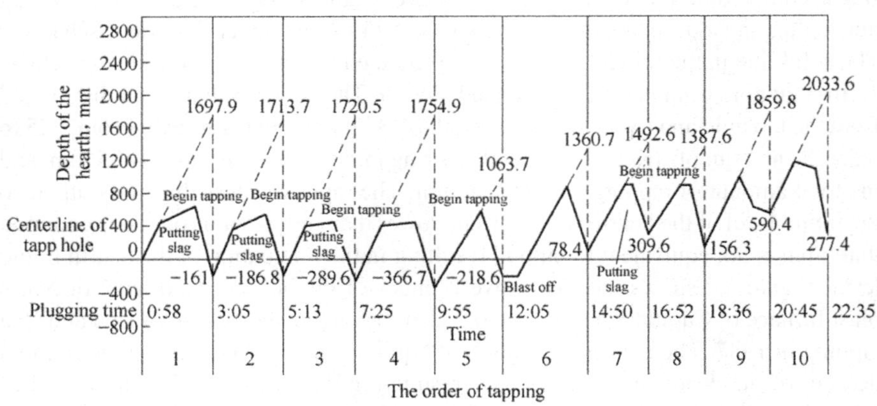

Fig. 6.2 Liquid level changes of the hearth

6.1 Up-Thrust in Blast Furnace

Professor Yang Yongyi pointed out the action of furnace core coke (deadman) in his later work. "Deadman is not absolutely stationary. It sinks when tapping and then rises slowly as the accumulation of slag and hot metal in the hearth" [1]. Professor Yang's clear discussion on the movement of deadman actually proposed a new force, the upward pressure of the rise of deadman on the stock column in the furnace. This force was named as "up-thrust [2]" in this book.

6.1.1 The Action of up-Thrust

When the liquid level rises, the stock column is compressed due to the up-thrust action, resulting in the decrease of stock column porosity and increase more resistance for gas. The meter record shows that the blast pressure has risen and the blast volume has reduced. In this case, the blast furnace operators usually take measures including tapping hot metal and slag to reduce the liquid level of the hearth and reduce the compression of the up-thrust on the stock column.

Sometimes the belated discharge of slag and hot metal may lead to the increase of up-thrust. Besides, the stock column permeability deterioration, the blast furnace may be not running as smoothly as before, even causing the blast furnace accident."On January 24, 1992, slag and hot metal of No. 2 blast furnace could not be discharged in time. Then the blast pressure increased from 0.31 to 0.328 MPa, and the blast volume was reduced by 100 m^3/min. At the moment, the permeability won't be improved only by reducing the blast pressure to 0.315 MPa (higher than original pressure). However, slag and hot metal still could not be discharged normally in 39 min. The blast pressure was no longer reduced, leading to bridging and twice blowing off and checking, because of slag entering into the tuyere and the redness appearing at blowpipes. After emptying slag and hot metal and turning off the blast, the hanging phenomenon still existed. The burden was not descended until all tuyere caps were opened to reduce the pressure in furnace from 0.11 to 0.03 MPa with hanging in 2 h and 41 min [3]." It can be seen that tapping iron and slag in time is very important.

6.1.2 Stabilizing the up-Thrust of Blast Furnace

In the blast furnace operation, up-thrust fluctuations occur frequently, although these fluctuations do not affect the uniformity of process as timely tapping iron and slag. To basically stabilize up-thrust, the amount of slag and hot metal produced from burdens into the furnace should be kept the roughly same with the amount of discharged slag and hot metal, and the liquid level of slag in the hearth should be maintained stability. Burdens are put into the furnace constantly, so the only way to stabilize up-thrust

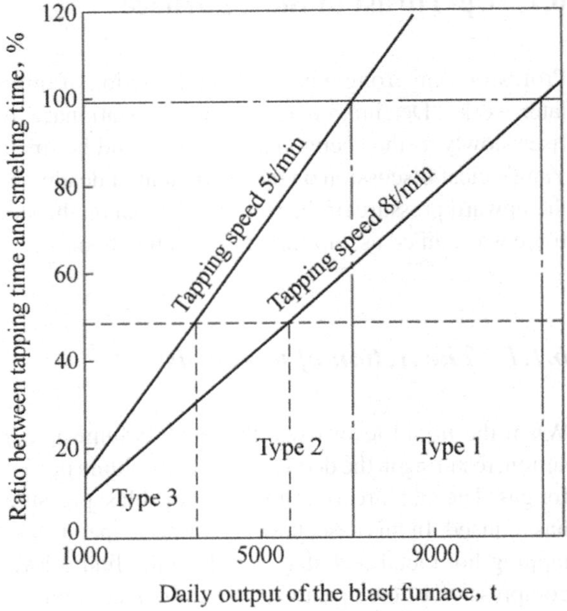

Fig. 6.3 Relationship between the daily output of the blast furnace and tapping time

would be continuous tapping, which is an effective way to maintain the stability of up-thrust according to analyzing from the perspective of BF smooth operation.

Continuous tapping and the control of slag flow make a balance between the input of slag into the hearth and the output of slag discharged from the blast furnace and maintain the basic stability of liquid level of the hearth. Therefore, the up-thrust fluctuations are reduced. The tapping speed of the large blast furnace is generally 5–8 t/min. Some large blast furnaces have their tapping speeds of more than 9 t/min.

In Fig. 6.3, the tapping system of blast furnace is divided into three types according to the tapping speed of 5–8 t/min.

Type 1, the blast furnace with a daily output of more than 7200 t, the continuous tapping should be adopted.

Type 3, the blast furnace with a daily output of less than 3600 t. This type of blast furnace is hard to tap continuously. In order to reduce the large fluctuation of up-thrust, it could be reduced with upper slag discharging or longer tapping period. Without upper slag discharge, tapping must be very frequent; frequently tapping might increase labor for hot metal transportation and cast house operation. For this type of blast furnace, the best way is to discharge slag in an appropriate amount and control the upper slag ratio not higher than 80% to avoid burnout of the slag notch. Besides, while maintaining a sufficient amount of hot mental each time, increasing tapping times to reduce the up-thrust fluctuations is a good way to stabilize the blast furnace. In order to increasing tapping times, it is necessary to set two tap holes on the blast furnace. The double-tap hole not only improves the working conditions of blast furnace settler, but also contributes to maintaining the required depth and ensuring sufficient calcinations time of tap holes. The damage of the hearth is generally intensified from the bed tap hole. Maintaining the depth of tap holes is an important measure to reduce the damage of the hearth. The double-tap hole is conducive to the promotion of blast furnace. Inappropriately, part of 1000 m^3 blast furnaces in China only has one tap hole, which may hinder the promotion of blast furnaces. It is necessary to take advantage of the overhaul opportunity to change them into double-tap holes.

Type 2, the blast furnace with a daily output of 3600–7200 t, with output more than Type 3 and less than Type 1. The tap system is also between other two types. Blast furnaces with the output closing to the upper limit can be used for continuous tapping without discharging upper slag. Blast furnaces with the output closing to the lower limit are impossible to adopt continuous tapping in general conditions, which is similar to Type 3. Therefore, upper slag is discharged in some blast furnaces.

There is a large allowance range of tap speed for BF. The smaller the blast furnace volume, the slower the tap speed. The division shown in Fig. 6.3 is rough. Conclusively, blast furnaces can be divided into three categories from the perspective of cast house operation, including continuous tapping, no discharging upper slag and discharging upper slag.

6.2 Hot Metal Tapping

6.2.1 The Movement of Hot Metal in Hearth

Hearth working diagram is shown in Figs. 2.2 and 3.3. Figures indicate that hot metal in the hearth is accumulated in "core coke zone" and "coke free zone." Hot metal in coke free zone is stationary without tapping. In the 1950s, the isotope was used in Germany to determine that the hot metal in the hearth is stationary [4] (see Chap. 4). Today, the BF volume has been already enlarged. Blast furnaces with the volume of more than 2000 m^3 are relatively universal. In some countries (such as Japan), the volume of every blast furnace is more than 5000 m^3. Due to the continuous tapping, the state of hot metal in the core coke zone and coke free zone of hearth is quietly different. Owing to the continuous tapping, the movement of hot metal and slag continues, only the hot metal in lower part of coke free zone could keep stability without disturb. Figure 6.4 shows the results of model tests and calculations.

Figure 6.4a shows the deadman contacts the hearth bottom and (b) the deadman is floating and slag flows to iron notch through deadman. "Coke free gutter (zone)" is formed under deadman. The resistance in this area is very small. So researchers named it "coke free zone." Professor Du Hegui translated it as "Channel flow" in Chinese. The iron-flow flows along the small-resistance path and may change its path in coke free zone, as shown in Fig. 6.4b. If the deadman layer is shallow and close to the bottom of the furnace, the hot metal must flow near the furnace wall. Over time, the furnace wall is eroded into a garlic-like shape (like an elephant foot). In order to

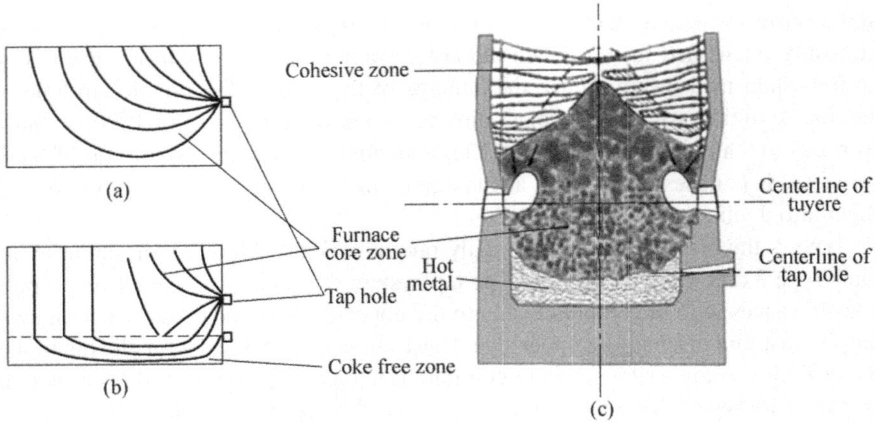

Fig. 6.4 Flow of hot metal in the hearth (left) and the hearth diagram in the model [5]. **a** The deadman contacts the hearth bottom; **b** the deadman is "floating"; **c** hearth working diagram

reduce the erosion of the furnace wall, it is necessary to reduce the flow of hot metal near the furnace wall.

The hot metal flowing along the furnace wall, also called "circulation flow," is the main cause of elephant foot-shaped erosion. Reducing tapping rate and deepening deadman layer are the main measures to reduce the circulation flow.

6.2.2 Tapping Speed

The tapping speed of blast furnace is related to its volume. Figure 6.5 shows the average tapping speed of some blast furnaces [6].

Modern large blast furnaces all have very deep deadman layer, some of which have the layer depth of 3 m with very fast tapping speeds, average speed of 8 m/min, and even above 8 m/min in a period of time. The way to control tapping speed is mainly to control the diameter of the tap hole. The variation of the tap hole diameter of a 5700 m^3 blast furnace in Jintang, Capital Steel is shown in Fig. 6.6. The diameter

Fig. 6.5 Tapping speed of blast furnace

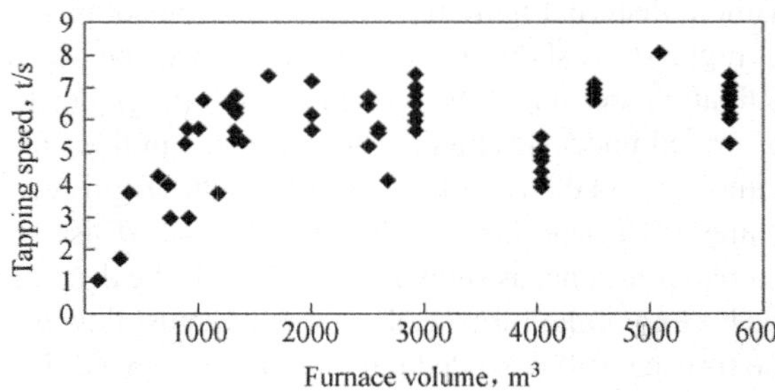

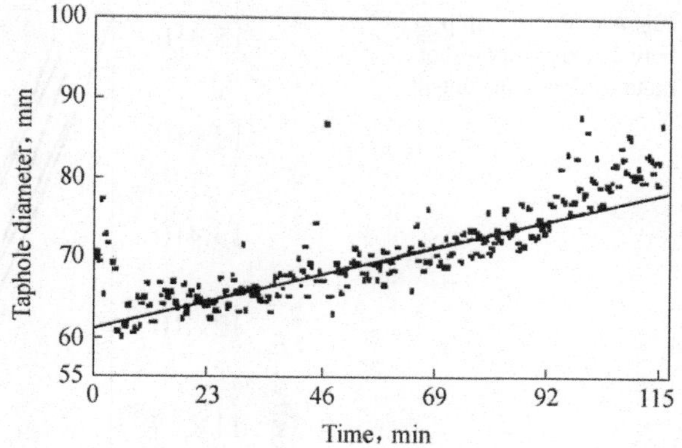

Fig. 6.6 Changes in taphole diameter in the tapping process of Jingtang blast furnace in Capital Steel

of the bit used at tap hole drill was 60 mm in initial stage and then was expanded to 80 mm as output rising [7].

6.2.3 *The Viscosity of Hot Metal*

For most modern blast furnaces, the hot metal temperature is around 1500 °C with good fluidity. Figure 6.7 shows viscosity curves of pure iron, pig iron and steel [8]. The viscosity of hot metal is related to the carbon content. Figure 6.8 shows the relationship between carbon content and viscosity [9]. The carbon content of hot

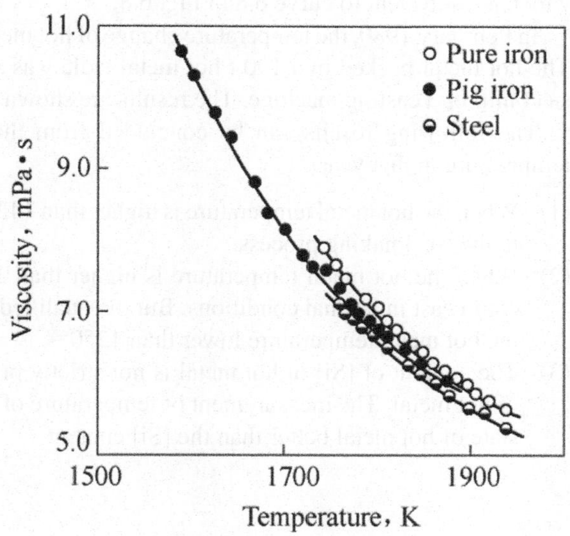

Fig. 6.7 Viscosity of pure iron, pig iron and steel

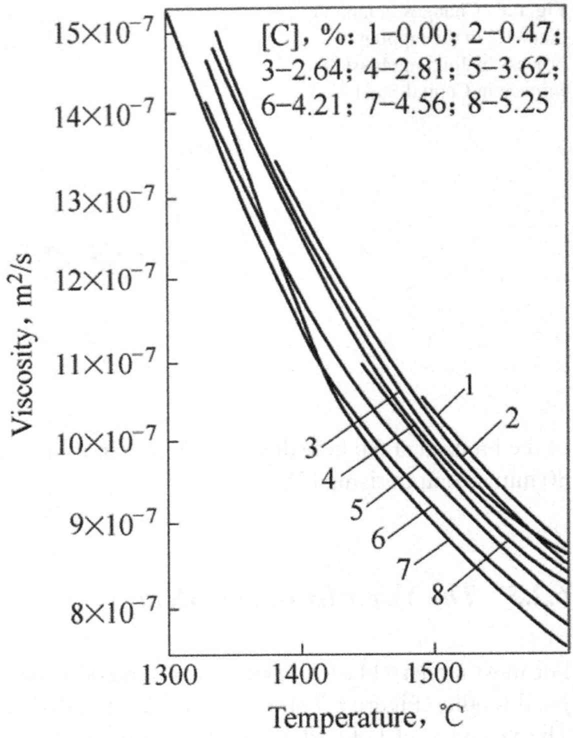

Fig. 6.8 Relationship between viscosity of hot metal and carbon content

metal in blast furnace is generally 4–5%. Only a small part is beyond this range, which is equivalent to curve 6.8 in Fig. 6.8.

In February 1980, the temperature change of hot metal was tested in Capital Steel. The hot metal packed in a 100 t hot metal ladle was sent from a blast furnace to a steel mill or a casting machine. The results are shown in Table 6.1 [10].

The following results can be concluded from the measurement of hot metal temperature in that year:

(1) When the hot metal temperature is higher than 1420 °C, scrap steel can be added in the steelmaking process.
(2) When the hot metal temperature is higher than 1250 °C, the casting machine could cast in normal conditions. But the qualified ingots could not be casted, if the hot metal temperature lower than 1250 °C.
(3) The content of [Si] in hot metal is not strictly proportional to the temperature of hot metal. The measurement of temperature of hot metal reflects the thermal state of hot metal better than the [Si] content.

Table 6.1 Cooling of hot metal in transportation

Furnace number	Measurement times	Blast furnace volume, m^3	Cast house hot metal [Si], %			Temperature of cast house hot metal, °C			Cooling rate, °C/min
			Max	Min	Average	Max	Min	Average	
1	28	576	0.62	0.2	0.44	1470	1340	1417	1.6
4	21	1200	0.82	0.27	0.49	1448	1385	1408	1.66

6.2.4 Continuous Tapping

As the volume of the blast furnace expands, continuous tapping becomes a reality. Table 6.2 is a set of data for the evolution of tapping in Japan [11].

In Table 6.2, filling coefficient of hearth "is a parameter, which indicates the volume of the hearth occupied by iron and slag accumulation during the time between the end of the last iron tapping and the upper slag are discharged or start tapping." Table 6.2 is an indication that the expansion of BF volume and the promotion of daily output turn continuous tapping into a reality. Continuous tapping is a leap in blast furnace operation, creating conditions for controlling the up-thrust stability. With the accumulation of continuous tapping experience and the improvement of the quality of the taphole clay, the number of tapping began to reduce. Table 6.3 shows the evolution of tapping operation in No. 6 blast furnace (5153 m^3) of Chiba Works, Kawasaki Steel. The tapping times per day reduced from 13.9 to 10.9 times [11].

Table 6.2 Comparison of tapping data among blast furnaces

Blast furnace	*A*	*B*	*C*	*D*	*E*
Year	1968	1958	1965	1973	1980
Hearth diameter, m	4.9	6.55	7.2	11.1	14.1
Hearth volume (from tap hole to slag notch), m^3	18.9	80.9	107.9	343.5	663.6
Slag notch number	1	1	1	2	4
Pig iron output, t	760	890	1700	6200	10,500
Slag volume, kg/t	350	450	320	315	310
Daily tapping times	9	7	8	13	11
Storage period, min	70	95	50	40	Continuous tapping
Volume production, m^3/min	0.150	0.198	0.322	1.16	1.96
Filling coefficient of hearth h_f	0.556	0.232	0.149	0.135	–

Table 6.3 Evolution of tapping operation in blast furnaces

Date	March, 1979	September, 1979	March, 1980	September, 1980
Cumulative tapping time per day, min	1630	1540	1610	1500
Tapping times per day	13.9	11.8	10.7	10.9

6.3 Tapping Slag

6.3.1 The Flow of Slag in Furnace Core Zone

The viscosity of slag is more tens of times than hot metal. Slag flows in furnace core zone and is affected by furnace core coke. Fukutake, et al. found that the residual ratio of slag in furnace reached 0.36 [11]. The slag left in the furnace at the tap hole side is approximately at the same level with the tap hole, but the slag level away from the tap hole is higher. According to the actual results of blast furnace dissection in Japan, the distribution diagram of slag in the furnace is shown in Fig. 6.9 [12].

When tap holes are located on the same side, the distribution of the slag in the blast furnace with multiple tap holes must be uneven. The research of Japanese experts indicated that the distance of slag interface is related to the hearth diameter. According to the formula, as the furnace diameter is 11.5 m, the distance of liquid level is 1.5 m. When the diameter is 14.7 m, the distance reaches 2.5 m [12]. Uneven liquid level lasting for a long period has less influence in general conditions. However, when the furnace is in an abnormal condition, especially in chilled hearth, due to high slag viscosity, the hot metal fell from the dropping zone through different slag layers with different speeds, inevitably affecting the evenly fall of burdens and the uniform distribution of gas flow, which will cause the changes of flow to local orientation. Gas flow or "channeling," as blast furnace workers so-called, is a "killer" for lining. Of course, the effect of channeling on the blast furnace is not only for lining, which is well known and unnecessary to discuss here. Therefore, two or more tap holes should be installed and evenly distributed in circle and the tap holes should be used alternately or symmetrically, so as to flatten the liquid level in furnace core zone, which is an effective measure for eliminating "channeling."

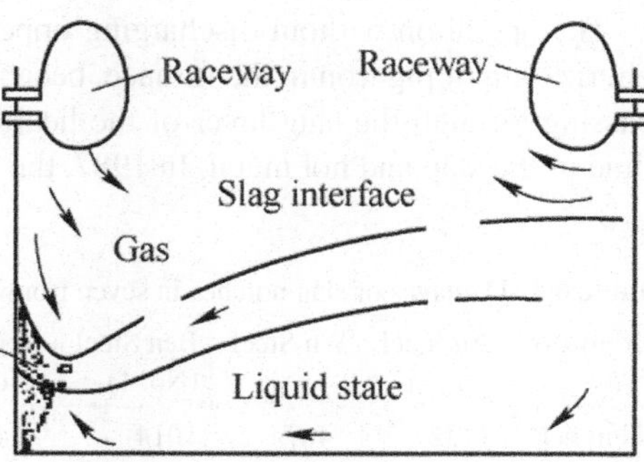

Fig. 6.9 Residual liquid level of slag in furnace

6.3.2 Without Discharging Upper Slag

For many years, slag is discharged from the slag notch. Once the hearth is out of order or the slag notch is seriously damaged, it may not only affect production, but sometimes cause malignant accidents such as slag notch explosion. As the volume of the blast furnace expanding and the grade of burden improving, the time for tapping becomes longer and longer, and the amount of slag becomes less and less. The furnace top pressure is also getting higher along with furnace volume increase which makes slag tapping difficult through the slag notch. Damages often occur to the slag notch, causing a lot of loss to production. Table 6.4 shows the damages of the slag notches of some ironworks [2].

The slag discharged from tap holes is the fundamental method to solve the problems of the damage of slag notches. If the blast furnace is without upper slag tapping and does not install with slag notches and also other slag notch devices including slag notch coolers, blacking slag machine and upper slag runner, the blast furnace platform would be flatter and easier to operate in the casting house. Early in 1971, No. 4 blast furnace in Cherebawitz, the Soviet Union did not discharge upper slag. Until 1973, large blast furnaces in MMK, NTMK and Lipesk did not discharge upper slag [13]. The first BF that did not discharge upper slag is No. 4 blast furnace in Wuhan Steel which stopped discharging upper slag since November 1987. In early years, 170 slag notches in Wuhan Steel were damaged due to the discharge of upper slag, accounting for 43% of the whole ironworks [14].

The advantage without discharging upper slag also attracted small blast furnaces: "We (Yili Steel) put forward the idea without discharging upper slag when the two 20 m^3 blast furnaces opened in early 1988 with actively and boldly practice. In the past few years, except that the No. 2 blast furnace opened the slag notch as dealing with the chilling of the hearth, these two blast furnaces did not discharge upper slag during three years of the fourth generation of the furnace" [15].

The operation without discharging upper slag would be beneficial to the desulphurization of pig iron in the furnace, because it increases the thickness of hot metal passing through the slag layer of the hearth, prolongs the desulfurization reaction time of the slag and hot metal. In 1987, the first-class and second-class high-quality

Table 6.4 Damages of slag notches in seven ironworks in 1988

Ironwork	An Steel	Wu Steel	Ben Steel (No. 1)	Ben Steel (No. 2)	Capital Steel	Pan Steel	Baotou Steel	Total
Number of damaged slag notch	1132	467	1014	143	255	169	164	3344
Life of each slag notch, t	6273	9046	2767		12,331	12,284	11,973	6493

foundry pigs accounted for only 79.11% of all pig iron, but in 1992 it had risen to 96.18%, which greatly improved the quality of products and improved the economic benefits of enterprises.

No. 5 blast furnace (420 m^3) in Zhangye Steel opened in November 2001. Fourteen tuyeres including one tap hole and one slag notch were equipped. From the opening of furnaces to May 2006, No. 5 blast furnace had discharged upper slag all the time with manual slag notch plugging. With the improvement of the smelting strength of the blast furnace, the slag notch was damaged frequently, which affected the normal production order of the blast furnace with many accidents that the slag notch could not be plugged. Since June 2006, the discharging upper slag had been canceled with good results.

The diameter of the hearth of No. 5 blast furnace was 5.4 m. The distance between the centerline of the tap hole to the centerline of the slag notch was 1.3 m. The safe volume of iron was 130.8 t. According to the highest daily output of 1320 t during January to May in 2006, the average tapping amount was 82.5 t as tapping 16 times a day, which was nearly 50 t less than the safe volume of iron. Therefore, it would be feasible to cancel the discharge of upper slag from the perspective of the safe volume of iron [16].

6.3.3 To Stabilize the up-Thrust

Continuous tapping of blast furnaces creates conditions for stabilizing the up-thrust. "Continuous tapping and controlling the flow of iron and slag may balance (equalize) the amount of slag and hot metal falling into the hearth and the amount of slag and hot metal discharged, and keep the liquid level in the hearth basically stable." The stability of coke free zone results in the stability of up-thrust. The slag and hot metal interface in the hearth is near the inner side of the tap hole. As shown in Fig. 6.4c, the hot metal and slag flow out at the same time without interruption, thereby concluding the stability of the slag and hot metal interface and the up-thrust. Figure 6.10 shows a tapping example of a large blast furnace [7]. The figure shows that the slag and hot metal tap at the same time, although both flow volume fluctuate.

After the use of the slag separator (so-called small pit), it's observed or measured that slag and hot metal flow continuously, and that the interface of slag and hot metal is basically stable. One of them is interval, indicating a change of the interface. Within an interval proportion of time, it is an indicator of stable up-thrust. It is completely continuous tapping with a 100% stability rate. If completely interval tapping, the hot metal and slag are, respectively, left alone, there is hot metal without any slag or slag without any hot metal with a stability rate of zero.

Interval tapping as hot metal flowing may result in the fall of deadman and the reduction of coke free zone, which can be seen in Fig. 6.11a. While the height of slag and hot metal interface is near tap hole (b), slag and hot metal flow out at the same time.

S_F is assumed as the volatility of up-thrust (%), S_F:

$$S_F = \frac{t_2}{t\Sigma} \tag{6.1}$$

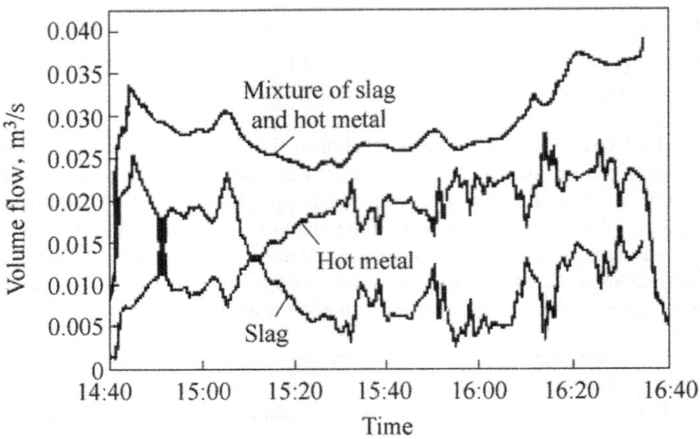

Fig. 6.10 Volume flow changes of hot metal, slag and their mixture in the tapping process of blast furnace

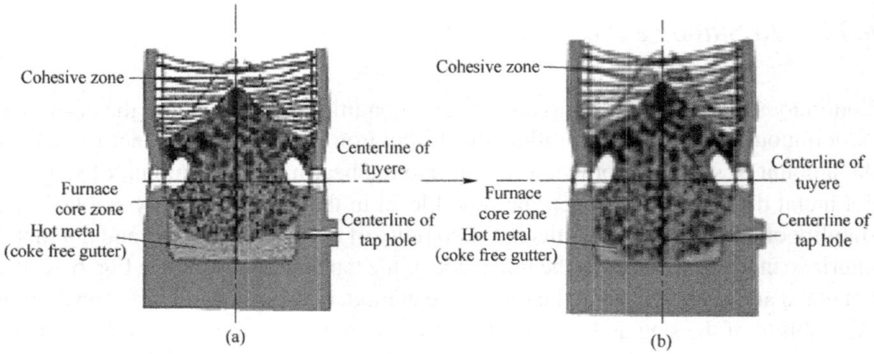

Fig. 6.11 Changes in coke free gutter as intermittent tapping

where t_2 is the co-flowing time of slag and hot metal, min; t_{Σ} is overall tapping time, min.

When operating, determine a value and follow this value regulating the tapping speed to stabilize the interface.

Automatic detection is very convenient. A temperature measuring device may be separately installed at the slag–iron separator. According to the temperature difference between the slag and iron with or without flow volume, the cutoff time of slag and hot metal would be determined by Eq. (6.1).

6.4 Gas Emitting in Tapping Process

The tapping process is accompanied with the gas emitting from hot metal. Less attention had been paid because it is hard to be measured due to its relatively low content. But this kind of gas should be concerned because it is harmful to people. Table 6.5 shows the results calculated by A. B. Ganqief, et al. in the Former Soviet Union, in accordance with thermodynamics model and hot metal conditions [17]. There are two computational conditions:

(1) Hot metal composition: C 4.4%, Si 0.7%, Mn 0.5%, S 0.015%.
(2) Hot metal temperature: 1300–1600 °C.

Table 6.5 Gas emitting in the process of tapping

Molecular formula	Emitting rate of harmful gas			
	1300 °C	1400 °C	1500 °C	1600 °C
g/tHM				
SO	0.166737	0.3536160	0.6870720	1.2450720
SO_2	0.1045824	0.1538944	0.2171008	0.2987968
S_2O	0.0226968	0.0289840	0.0356528	0.0424640
NO	0.0000442	0.0001849	0.0006666	0.0020881
S	0.1096192	0.2998496	0.7278721	1.5929920
NS	0.0014301	0.0036791	0.0084502	0.0175761
CS	0.0049293	0.0082302	0.0128088	0.0187280
CS_2	0.6224932	0.3815124	0.2428428	0.1582776
SiO	0.0030096	0.0401676	0.3958812	3.0080160
SiS	0.0086784	0.0818280	0.5860500	3.2756400
SiS_2	0.0153483	0.0709421	0.2685848	0.8415240
Mn	0.2199725	1.5224850	8.6845000	39.748500
MnO	0.0002353	0.0021659	0.0153800	0.0879760
Fe	0.1400336	0.9032800	4.7052320	20.517280
10^{-6} g/tHM				
NO_2	0.000001	0.000003	0.000021	0.000120
N_2O	0.004180	0.016700	0.059840	0.186600
CN	0.006300	0.028600	0.113620	3.788200
N_2C	0.132200	0.340800	0.788000	1.664000
C_2N_2	0.000234	0.000387	0.000598	0.000880
NCO	0.000533	0.002100	0.007050	0.020900
SiN	0.000001	0.000024	0.000752	0.155400
SO_3	0.001760	0.004720	0.011200	0.024800
Si	0.000004	0.000246	0.009240	0.233800

The calculation results are shown in Table 6.5.

Blast furnace smelting is a high-temperature process. Some elements or compounds gasify and then join the gas stream or circulate in the furnace. Some blast furnaces are used to blow through tap hole for clean up the hot metal and slag, which may lead to the spilling of slag and hot metal and bring difficulties in cast house operation. At the same time, a large amount of gas is emitted from taphole, which is harmful to the human body. Some of the burden contains high lead, then the white gas with higher lead emitted from tap hole, which is harmful to people's liver. Therefore, blowing through taphole should be abandoned in general tapping operation, to prevent the gas to harm the human body.

6.5 A Profound Lesson

Maintaining the normal state of tap hole is the core of the furnace cast house operation. The abnormal state of tap hole may lead to problems in tapping and serious accident. In 1960, the hot metal ladles were unprecedentedly scarce for Capital Steel, and the tapping was often delayed. Table 6.6 shows the actual situation at No. 3 blast furnace:

Due to the delay of hot metal ladles, hot metal and slag often remain in the furnace. There was a large amount of slag and hot metal stocked in the hearth, and tap holes were difficult to maintain normal state. The too shallow tap hole became a dangerous "normal" without seeing. To face the dangerous conditions, it was supposed to be shutting some blast furnaces down to relieve the problem of insufficient hot metal ladles. However, due to the requirement of high productivity pressure, it was afraid difficult to complete the production plan, then reluctantly insisted. At 11:00 on November 23rd, due to the inefficiency of hot metal ladles at No. 3 blast furnace, the amount of hot metal accumulated in the hearth was 198 t

Table 6.6 Tapping at No. 3 blast furnace of Capital Steel in 1960

Month	October			November	
Time (Ten days or day)	The first ten-day period	The second ten-day period	The third ten-day period	The first ten-day period	11th–22nd
Delay times of hot metal ladle	17	10	36	22	30
Delay times of slag ladle	7	7	20	12	22
Qualification rate of tap holes, %	90.4	78.5	43.4	62.5	13.3
High-pressure closure rate, %	93.6	71	58.6	77	53.1
Blowing and closure, %	1	17	24	–	26

without lower slag during three times of tapping. The tap hole had to be plugged because the ladles were full. The fifth tapping started at 15:49 with the depth of the tap hole of only 600 mm. According to the regulation, the depth of the tap hole of No. 3 blast furnace should be above 1.7 m. The manager was careful to release the blast. However, the three ladles were full, so tap holes were forced to plug at 16:03 with mud 350 kg. The manager worried that tap holes would be burned out, so he intentionally used more mud and kept the mud gun ready. When the blockage was added with 100 kg of mud at the tap hole plugging location, it was found that smoking emerged from the cylinder of mud gun and blowing off and clay gun at 16:17, trying to seal the tap hole. However, the electricity current was too high to add the mud. Then blowing off and discharging upper slag at 17:07, while the gun head had become red. The mud gun was removed away and watered for preparation for tapping.

There two hot metal ladles were prepared at 17:00, but the tap hole was very shallow. Two ladles were soon filled with soaring hot metal, but mud gun had been damaged. The continuously flowing of hot metal made these two ladles casting up with the rails. The No. 2 blast furnace was also forced to stop blowing, due to the closely casting for more than 40 meters of iron rails for hot metal transportation.

The company urgently mobilized workers to repair to dismantle two hot metal ladles. No. 3 blast furnace had suspended blowing off for 19 h 7 min. No. 2 blast furnace had suspended blowing off for 6 h 10 min. Two ladles were damaged. Therefore, there was more inefficiency in hot metal ladles in the plant, so all the blast furnaces had to be operated with reducing blast.

The lesson from the accident is that it is very important to keep tap holes in normal state, as well as the furnace cast house operation management!

References

1. Y. Yang, Iron-making Technology and Design, Book Five. (The Chinese Society for Metals, 1985), p. 127
2. Y. Liu, Effect of up-thrust on operation in blast furnace. Iron Steel **12**, 13–16 (1988)
3. You Wenquan, BF operation as slag and hot metal are incompletely discharged. Capital Steel Sci. Technol. **2**, 63–65 (1944)
4. V.W. Lobyz, H. Weber, Metall. Interpretation **2**, 14–18 (1956)
5. P. Campball, N. Standish, Diagnosing Blast Furnace Hearth Condition (preprint) (1984)
6. Y. Liu, The function of up-thrust on hearth of blast furnace. Iron Steel **12**, 1–4 (1995)
7. P. Huang, Y. Dong, Q. Hou et al., Study on hot metal and slag discharged from super large blast furnace in Jingtang, Capital Steel. Iron Steel **6**, 47–49 (2014)
8. Y. Morita et al., Viscosity of hot metal and molten steel. Iron Steel **2**, 54–61 (1982)
9. J. Chen, *Manual of Data and Diagram Commonly Used in Steel-making*, 2nd edn. (Metallurgical Industry Press, Beijing, 2010), p. 526
10. R. Qian, et al, Hot metal temperature measured in front of furnace, 1980 (internal data in Capital Steel)
11. W. Fu et al., *Work Corpus of BF Raceway and Hearth* (Metallurgical Industry Press, Beijing, 1986), pp. 1–19
12. N. Nakaruma, et al. Iron-Making Steel-Making **1**, 1–17 (1978)

13. B.I. Dakelish, BF operation without discharging upper slag. Steel **7**, 15–17 (1987)
14. J. Pan, Practice of BF operation without discharging upper slag. Iron Steel, 13–14 (1990)
15. J. Pan, Operation of small blast furnace without slag notch. Iron-Making **4**, 38–39 (1994)
16. X. Zhang, H. Li et al., The practice of cancelling discharging upper slag in blast furnace with a volume of 420 m^3 in Zhang Steel. Commun. Iron-Making Technol. **5**, 9–10 (2008)
17. A.B. Ganqief, et al. Ferrous Metall. Coll. Univ. **1**, 13–14 (1992)

Chapter 7
Basic Operation of Blast Furnace

Abstract Stable and smooth operation is the best status of blast furnace production process and the premise of obtaining the best index of blast furnace. Smooth operation is the best requirement for blast furnace operation. If the burden is the decisive factor (the first factor) for the smooth operation, the effect of operation on the smooth operation of BF is the second one. The wrong operation could destroy the smooth operation of the blast furnace, while the inferior charge would make it difficult to maintain the smooth operation for a long time. In fact, blast furnace smelting process changes constantly, which is accompanied by the production process instead of man-making. Mainly: (1) The burden changes, especially the change of coke quality; (2) Gas distribution changes; (3) The furnace profile changes; (4) Furnace temperature changes; (5) The blast furnace equipment faults in operation. Blast furnace workers should timely detect changes and take measures to keep the process running smoothly and stable. First of all, small changes should be detected in a timely manner, which requires blast furnace workers, especially foremen and furnace managers, to be able to skillfully identify changes in the blast furnace. The main method is knowing the data display provided by various instruments of blast furnace.

7.1 Burden Descending

Only when the burden descends steadily, the blast furnace could operate smoothly. The gravity of the charge is the only force of descent. Mahaneke has studied the power of burden descending [1]. Figure 7.1 shows a schematic diagram of measuring the effective weight of charging (pressure at the bottom of the furnace). There is a balance in Fig. 7.1. On the left-end bracket of the balance, there is a blast furnace model. The bracket is not connected to the model. To the right of the scale is placed the balancing weight. Figure 7.2 shows the measurement results. From Fig. 7.2, the weight of burden increases with the height of the blast furnace, but the pressure transferred to the bottom of the blast furnace increases to a certain extent and no longer increases. The falling force of the pillaring is affected by a variety of factors and the huge influence of "furnace profiles" and friction resistance of the burden from the bulk material model test only.

Y. Liu, *The Operation of Contemporary Blast Furnaces*,
https://doi.org/10.1007/978-981-15-7074-2_7

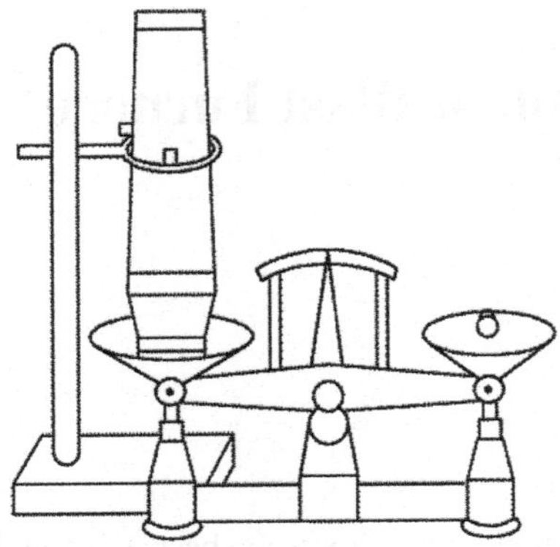

Fig. 7.1 Device for measuring hearth bottom pressure of bulk materials in model

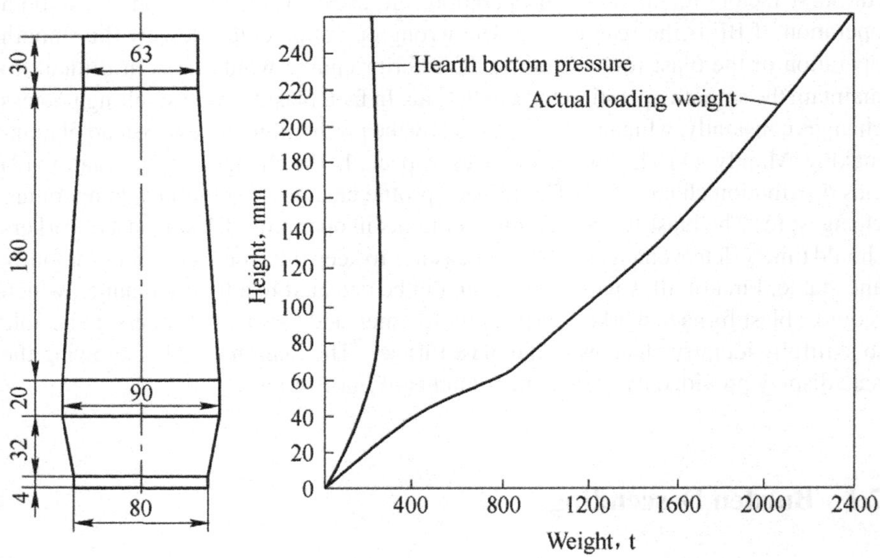

Fig. 7.2 Measurement results of hearth bottom pressure of bulk materials in model

Grugenov proposed that the vertical pressure of the burden in the furnace is equal to the weight of the burden minus the friction of the burden and the friction of the wall and the resistance of the gas [2]:

$$F = Q_{LL} - \left(F_{LQ} + F_{LL} + P_{MQ}\right) \tag{7.1}$$

where F is burden pressure (or residual weight of pillaring); Q_{LL} is burden weight; F_{LQ} is the friction between the wall and the burden; F_{LL} is the friction between burdens; P_{MQ} is resistance of rising gas to burden.

Barbarekin created the method of frictional resistance to measure the vertical force of burden descending. Figure 7.3 shows a device for measuring the falling force of pillaring [3].

At that time, the internal friction angle of burden (coke and sinter) in MMK Steel blast furnace of the former Soviet Union was about 43°. According to test of the device (Fig. 7.3), H. A. Janssen's formula was used to calculate the vertical force of stock descending. The results obtained by many scholars are generally consistent. Professor Yang Yongyi concludes that the weight of moving burden (pillaring) is about 40–45% of the weight of burden (pillaring) [4]. However, Mahaneke's static model test showed that the vertical descending force (the original author called the bottom pressure of the furnace) was only 10–15% of the weight of the pillaring. Obviously, the movement of the burden increased the descending force, which was an important guarantee for the smooth operation of the blast furnace.

To improve the movement ability of burden descent, the first step is to increase the residual weight *F* of the pillaring.

(1) Reducing wall resistance. Equation (7.1) shows that reducing the friction force of furnace wall against the burden *F*LQ could increase the residual weight *F* of the pillaring. Reasonable furnace profile design is an effective measure to improve *F*; maintaining normal furnace profiles and preventing furnace wall from scaffolding are also important measures to improve the ability of burden descent. Once scaffolding, the resistance of the wall would increase, which is the main reason why scabs destroy the running of the blast furnace.

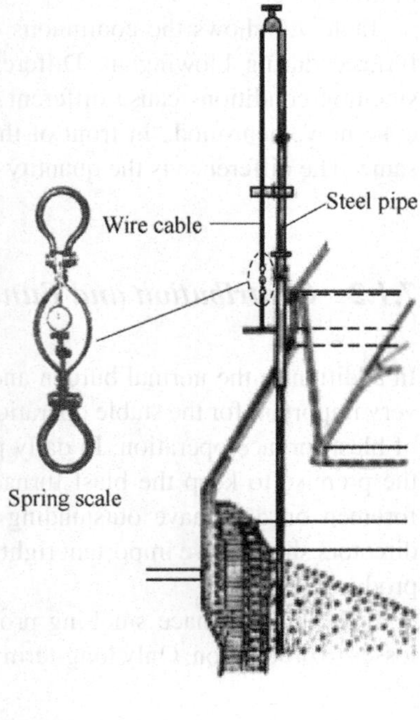

Fig. 7.3 Schematic diagram of measuring device

(2) Reducing gas resistance. The resistance P_{MQ} of rising gas to the pillaring depends on the gas velocity, distribution and permeability of the stock. The former depends on operation, and the latter depends on burden quality. The resistance of gas is directly proportional to the quadratic square of gas velocity. Therefore, the blast furnace should control reasonable smelting intensity and maintain reasonable gas velocity. Controlling gas flow distribution can make full use of gas energy and reduce resistance, which is an important contribution for BF top distribution.

7.1.1 Direct Observation and Manipulation (Operation) Curve Recognition

Basic operation of blast furnace refers to the analysis, judgment and treatment of blast furnace condition by the shift foreman in daily work. In Chaps. 2 and 3, direct manual observation has been explained. The following sections illustrate the identification method of operation curve through examples. Operation curve is the main "language" of the blast furnace process, and the foreman can understand the smelting process of the blast furnace through this language. This book will explain this language repeatedly. Here is a brief description of the working conditions of the tuyere. The tuyere is the only "window" for the blast furnace to directly observe the reaction in the furnace, which is very important for a deep understanding of the smelting process. Special instruments have been used to observe and measure the condition of the tuyere.

Table 7.1 shows the continuous observation of tuyere in a medium-sized blast furnace during blowing in. Different blast furnace equipment level, burden and smelting conditions cause different coke conditions in front of the tuyere, but the coke movement mode in front of the tuyere of each blast furnace is generally the same. The difference is the quantity and proportion.

7.1.2 Contribution and Function of Blast Furnace Foreman

In addition to the normal burden and equipment, the daily operation of foremen is very important for the stable operation of blast furnace. Foremen are the core figures of blast furnace operation. In daily production, the correct operation of foremen is the premise to keep the blast furnace smooth and stable operation. Blast furnace foremen on duty have outstanding effect and great responsibility. Smart factory directors should give important rights and treatment to their foremen in iron-making production.

If the blast furnace smelting process were damaged, it would inevitably bring losses to production. Only long-term smooth and stable operation blast furnace could

Table 7.1 Coke movement before the tuyere

Coke status before the tuyere	Blast volume	Tuyere		Schematic diagram	Explanation
		5	6		
		Blast volume			
	m^3/min	%	%		
Static		0	0	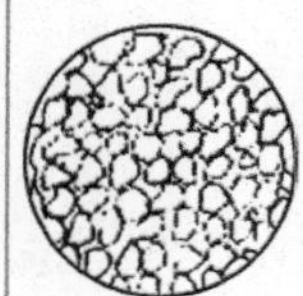	The tuyere is full of coke without moving
Accumulation	940	50	45	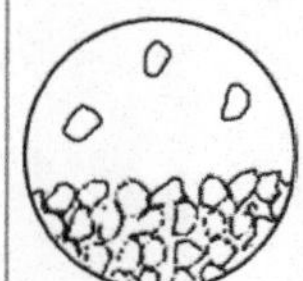	Coke accumulation under the tuyere but constantly displaced and updated
Trundle	1040	60	55	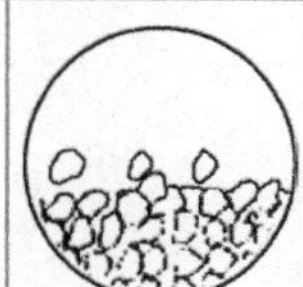	The coke falls into the tuyere from the upper part and rolls into the furnace after falling to the bottom
Pulsation	1134	65	60	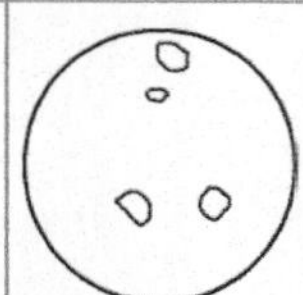	No accumulation of coke, only a few pieces of black coke jumping and flying in the tuyere
Flutter	>1134	>65	>60		Black coke was no longer visible. There is only bright coke fluttering

have good production technical indexes. The foreman's mistakes in operation sometimes lead to serious consequences. Similar cases are common occurrence. Figure 7.4 shows the blast pressure record of a 620 m^3 blast furnace for one day that the bridging was completely destroyed in 9 h by 5 consecutive times of checking from 7:03 to 16:00. The loss of the day due to bridging is conceivable. Figure 7.5 shows a foreman with superb skills, who not succeeded in once a checking, but also blowing in 90% of the blast after checking. The blast volume recovered to the level before checking in 30 min, and the full blast volume exceeded the total blast before checking was exceeded in 40 min. The results of both treatments by such two methods were quite different. Of course, the above differences were related to their respective furnace conditions at that time, and there was an incomparable side between them. Although

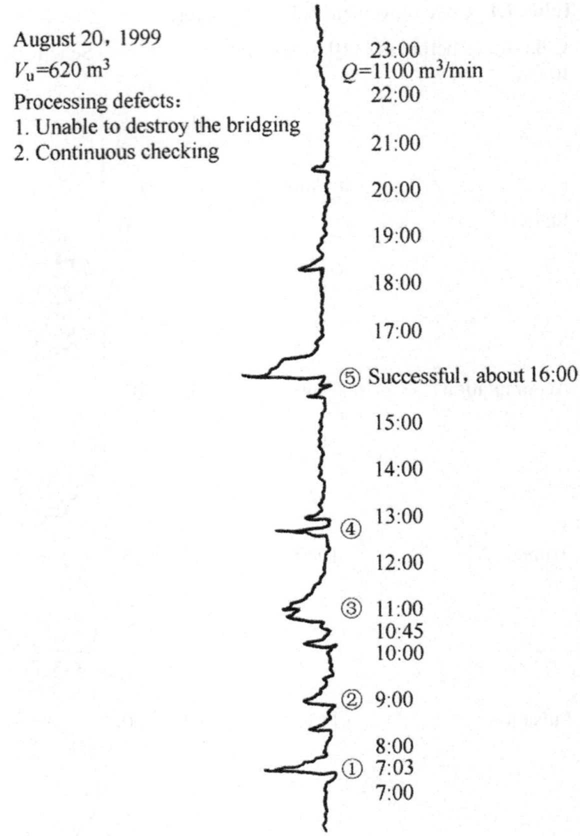

Fig. 7.4 Record of hot blast pressure for continuous checking

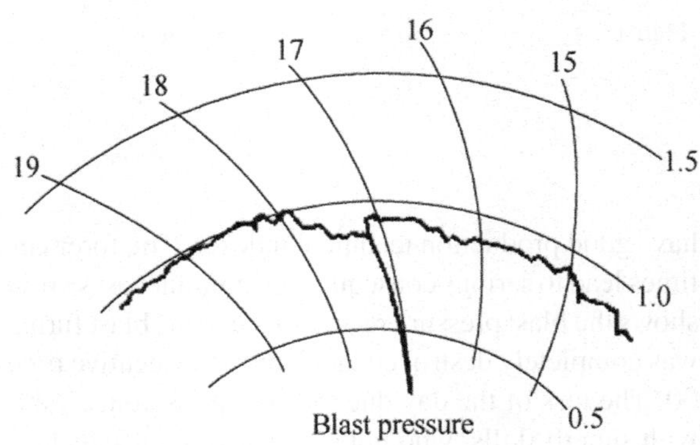

Fig. 7.5 Record of blast pressure on successful checking right the first time

both of them deal with bridging, the basic furnace conditions were different, and the burden conditions were different, there was a gap in operation technology anyway.

7.1.3 Reasonable Pursuit and Practical Operation

In a 1036 m^3 blast furnace in a factory, the night shift foreman did not deal with a 3.5 m slip and quickly charged burden to catch up the stock line and was anxious to maintain the charges to meet the requirement of administration. The foreman also did not charge coke blank for the deeply slipped stock line, but was anxious to catch the stock line with regular burden, which lost the greater for less. When tapping at 8:00 nest morning, tap hole opened, the hot metal could not tap out, for the furnace cooling with the risk of chilled hearth.

At the other plant, a 540 m^3 blast furnace, the night shift foreman wrote in the operating column of the daily journal. "The furnace operated smoothly. [Si] and basicity of slag were normal. The furnace temperature is assumed to be 0.45%." Starting from the morning of the day shift, the furnace condition was good. The slipping was at 13:10 and 13:30, with 3.5–4 m deep. After slipping, two charges of coke blank were added separately. Starting from 13:00, reduce loading, and from 14:00, 12 charges of coke blank were added, 108 tons in total. However, the furnace condition was irreparable and it was difficult to tap. From 13:30 to 13:55, the tap hole was opened twice in a row, and only 30 tons of iron was tapped. From 14:40 to 16:25, the tap hole was opened again, only tapping 12.5 tons of hot metal. At 16:55, the tuyeres are automatically filled with slag, and the blast furnace was forced to stop blowing. The operation log of blast furnace is shown in Table 7.2.

In fact, the night shift foreman wrote in the journal as "furnace operated smoothly, furnace temperature and slag base were in good condition. The presumptuous furnace temperature of 0.45%," which had contained misjudgment: The actual hot metal at 4:30 contained Si 0.57%, and at 6:00 it had been reduced to 0.32% (Table 7.2). According to the ruler record of the night shift on January 17 (Fig. 7.6), there were two slips from 4:00 to 6:00, one of which at 4:00 was deep, and about 10 batches were added to the blast furnace (Fig. 7.7). Although the foreman slipped and added coke blank during night shift, it did not help. Because of the wrong judge, he thought that the furnace temperature was not low (actually, furnace temperature was normal), and reduced the blast temperature from night shift continuously. According to the record, the blast temperature was 1080 °C at 6:00; 1070 °C at 8:00; 1060 °C at 10:00; and 1050 °C at 12:00 (Fig. 7.8). The wrong operation of two shifts resulted in the serious accident of chilled hearth.

The above example illustrates the importance of daily operation of the blast furnace. Every mistake in operation would bring losses, although each time the loss is different.

Table 7.2 Night and day shift of blast furnace operation log (excerpt)

Time	Batch	Blast volume, m^3/h	Oxygen enrichment, m^3/h	Blast pressure/kPa	Top gas pressure, kPa	Pressure drop, kPa	Coal injection amount, t/h	Real load	Blast temperature, °C	Tapping time(00:00)	Tapping amount, t	Si, %	S, %	Hot metal temperature, °C
1	6.5	1387		1.96	0.95	1.01	4.20	2.58	1080					
2	8.5	1401		1.97	0.89	1.08	4.20	2.67	1080	1:00–1:32	1052	0.54	0.02	1460
3	7.0	1397		1.96	0.90	1.06	4.20	2.61	1080	2:47–3:15	104.7	0.51	0.022	1455
4	7.5	1379		1.98	0.93	1.05	4.50	2.61	1080	4:10–4:30	57.3	0.57	0.019	1455
5	7.0	1405		1.95	0.96	0.99	4.20	2.61	1080	5:36–6:00	106		0.019	1450
6	7.0	1401		1.92	0.96	0.96	4.20	2.61	1080			0.32		
7	8.0	1399	1400	1.96	0.90	1.01	4.70	2.61	1075	7:03–8:00	104.1	0.35	0.027	1450
8	8.0	1402	1400	1.92	0.92	1.00	4.70	2.61	1070					
9	7.5	1369	1400	1.90	0.87	1.03	5.00	2.57	1065	8:55–9:25	66.5	0.32	0.023	1440
10	7.0	1352	1400	1.90	0.87	1.03	4.50	2.58	1060	10:30–10:50	55.2	0.35	0.025	1445
11	6.5	1353	1400	1.96	0.90	1.07	4.20	2.58	1055					
12	6.5	1185	1400	1.86	0.85	1.03	4.50	2.56	1050	11:55–12:33	107.7	0.33	0.026	1450
13	7.5	1478	1400	1.99	0.97	1.02	4.70	2.59	1030	13;30–13:55	29.9	0.32	0.023	1430
14	6.5	967	700	1.51	0.51	1.00	4.70		1015					
15	1.5	951	0	0.5	0.50	1.00	0		965	14:40–16:15	12.5	0.35	0.025	
16	0	0	0	0					950					

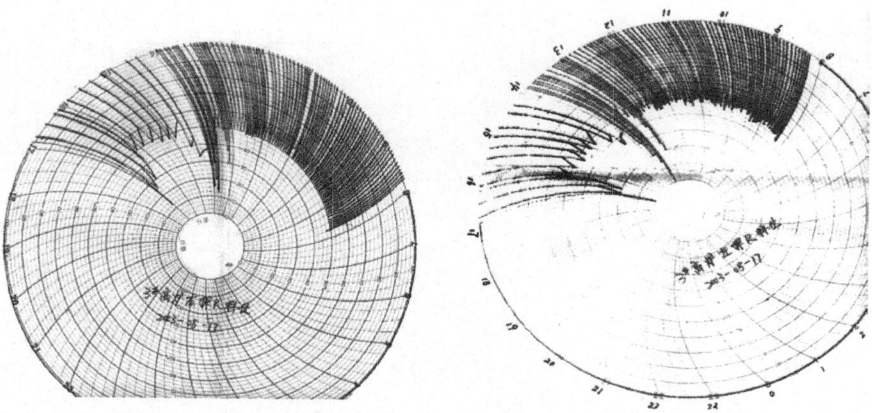

Stock level measured by right stock rod

Stock level measured by left stock rod

Fig. 7.6 Record of stock rod from day shift to middle shift on 17th

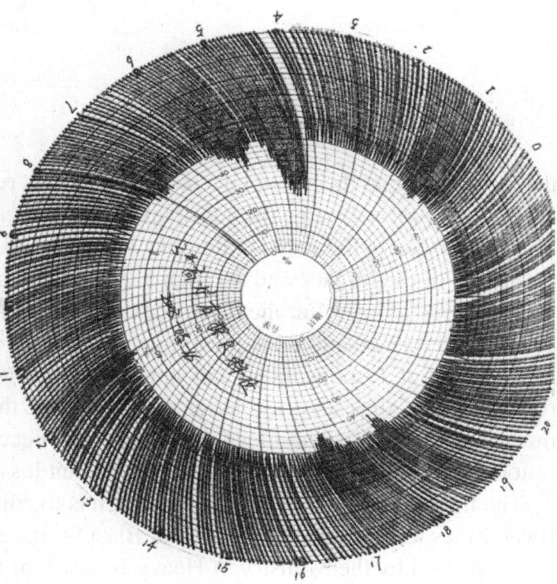

Fig. 7.7 Record of stock rod from day shift on 16th to night shift on 17th

7.1.4 Seize the Opportunity and Make Decision at the Right Moment

A channeling in the BF of a famous iron-making plant occurred on a weekend, while the production task was in tense situation. The foreman refused to reduce the blast and was afraid of taking responsibility. He asked the monitor on duty, who hesitated as well, then calling and looking for the leader but no response. While hesitating,

Fig. 7.8 Record of blast temperature from 16th to 17th

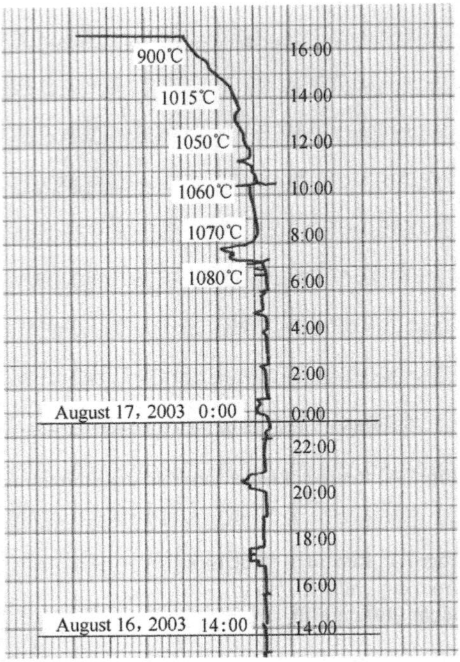

the foremen found the gas uptake had been turning red. When the top of the furnace caught fire, passersby saw the gas uptake like a red candle and called the duty room. Workers rushed to the scene to find that it was the oil leakage from hydraulic station on the top of the furnace and the high temperature on the top caused by the pipeline that triggered a fire. After the fire was extinguished, it was found that the gas uptake had been deformed and the top equipment had been seriously damaged.

The fear of responsibility and negligent management of equipment were the root causes of disasters, too strict assessment of the foreman, too heavy punishment, evasion and "request leadership" in case of accidents have restrained people's rationality and creativity, resulting in significant losses.

The important work of BF chief foremen is to adjust the furnace condition. "The Basic Rules for Technical Operation of Blast Furnace," formulated in 1955 in China, was approved by the Ministry of Heavy Industry of the People's Republic of China (the Ministry of Metallurgical Industry was not established at that time) and implemented throughout the country. "The task of adjusting the blast furnace process is to find out the harmful effects encountered in the smelting process and the germination of the abnormal process of the blast furnace, take effective measures correctly and restore the normal process of the blast furnace in time. Therefore, the adjusting method of the blast furnace process should be to prevent the normal operation to be destroyed and the normal operation should be restored in the shortest time from the destroying" [5].

In order to discover the abnormal process of blast furnace in time, the operators should be familiar with the characteristics of normal process of blast furnace first. This is the first BF operation regulation in China. Since then, all factories have formulated their own operation regulations based on this regulation and their own conditions. Now the analysis shows that the basic idea of the first national BF technical operation regulations is still correct. In the foreword of the regulation, it is proposed that "the single standard technical operation regulations of the blast furnace should be formulated to ensure that:

(1) To produce largest quantity of qualified pig iron.
(2) The unit consumption of raw materials is the lowest.
(3) Blast furnaces and their accessories have the longest service life.
(4) Safety of blast furnace workers.

The above four articles are the policy often mentioned now: "high yield, high quality, low consumption, longevity and safety."

Being familiar with the basic operation of blast furnace is not only limited to BF foremen, the chief workers, but also for the smelting engineers and the directors in charge of technology. Technical leaders of iron smelting plants should go to each blast furnace to understand the operation state before the dispatch meeting in the morning. Experts proficient in basic operations should stay at the monitor for a few minutes and observe the 24-h operation curve record, so as to have a clearer understanding of the actual operation and trend of the blast furnace.

Terms of Iron-making
Smelting process curve: The main parameters such as blast volume, blast pressure, blast temperature, top temperature, top gas pressure and stock line in the smelting process of blast furnace are continuously recorded and displayed on the operating platform of the duty room to provide reference for the operation of the foreman. The concentrated picture of these parameters is called smelting process curve, also called operation curve. It is the magnifier and microscope of the blast furnace smelting process, more important than the general engineering drawings, because the drawings are static and only provide physical characteristics and dimensions; while the process curve is dynamic, it tells the state and changes of the blast furnace process. The deeper you understand, the more comprehensive it "tells." Blast foreman, furnace chiefs and smelting engineers must have a thorough understanding and mastery of the "language" of the process curve. Only by mastering this "language" skillfully, the blast furnace could be dominated.

7.2 Channeling

Terms of Iron-making
Channeling: Channeling is a common abnormal phenomenon in blast furnace. The rising gas stream passes through the stock column, and the gas distribution is generally uniform. Once there is more gas in a certain area of the stock, it is so-called

channeling, and some people call it "gas flow." Since the gas is continuously flowing through the stock column of the blast furnace, the "gas flow" is defined as an abnormal state, which is easy to be misunderstood, so we refer to call the area with excessively gas passing through as "channeling."

7.2.1 The Cause for Channeling

Channeling is the result of inadaptability of rising gas and stock column permeability. When the gas passage is "crowded," the gas would automatically choose the way with lower resistance to pass through. Once a "channel" formed, the powder of the stock would be blown away and the channel would be developing automatically. This is also the reason why there are more chances for channels to occur at the furnace wall, because the furnace wall is smoother and gas is easy to pass through. It is customarily called the channel passing near the furnace wall, which is called the edge channel.

The "crowding" of the gas passage in the stock is the result of burden strength decreasing or powders increasing in the burden. The powder reduces the permeability of the stock and direct cause of the channeling.

Sometimes, because the furnace temperature is too high, the volume of the rising gas expands, and the FMQ increases (Eq. (7.1)), channels could also occur. All factors affecting gas volume increase might contribute to the channel.

Main reasons for the formation of channeling are:

(1) Poor burden strength or more powder;
(2) Defects of equipment, resulting in directional channels, such as uneven blast intake at tuyeres, scaffolding or erosion of furnace wall, etc.;
(3) Channeling caused by improper or incorrect operation. For example, the unreasonable distribution of materials, resulting in edge or center channels; furnace temperature control disorders resulting in channels and so on.

7.2.2 Treatment on Channeling

Normally, channeling is not destructive, but cannot be ignored, for it is often the root of metallurgical accidents.

Recurrent channels must be dealt immediately; otherwise sometimes they would develop automatically and more seriously. It is a precursor to the sticky of the blast furnace process. In view of the causes of channels, there are some corresponding active treatments:

(1) Peripheral or central channels have been discussed in Chap. 3 "The Structure of Stock Column and the Control of Gas Distribution."
(2) The blast tuyeres should be adjusted for channels caused by uneven blast intake.

(3) If the strength of coke or sinter is too low, besides strengthening screening, the leaders should be advised to improve the strength; at the same time, the blast volume should be reduced to suit the permeability of the stock; or the distribution of gas at the edge or center should be opened appropriately, and the load should be retreated, if necessary, to eliminate the channeling.
(4) Furnace wall scaffolding should be eliminated by flushing furnace or other methods.
(5) The edge channels formed by local erosion of the furnace body can be controlled by changing the distribution of materials or tuyeres, in addition to suggest for the leaders to arrange treatment.

Nonrecurrent channels should be dealt with resolutely. By checking, the original burden distribution should be destroyed, the existing gas passage should be changed, and the furnace cooling and even the chilled hearth caused by the channeling should be avoided.

7.2.3 Rare Real Channeling

The stock column near the channeling is relatively loose with large voids where more gas is flowing through, generally no space forming without burden that is the real "tubular voids." Real channel used to occur on a "short and fat" blast furnace. The volume of the blast furnace was 576 m^3 with hearth diameter 6.1 m, the belly diameter 7 m, the effective height 18.1 m and working height 16.9 m, which was 4–5 m shorter than the same diameter blast furnace. Because of the short and fat furnace, the channeling happened frequently in production process. In 1987, there was a biased stock line before the channeling, but it did not be dealt with timely and thoroughly, then leading to the emergence of malignant channels very soon. After that, the left rod was not visible and the right rod was zero. The top temperature was over 800 °C, so the dial cannot record automatically beyond the limit. Furnace temperature drops sharply, 15 tuyeres were full blooding, and slag and hot metal could not tap from tap holes, only gas spraying. Forced to blow out and open the furnace head manhole and found that the coke held on one side of the large bell, the other side is inclined to the direction of the channel (the large bell rod had been bent). The inner wall of the channel melted to form a glass-like slag shell about 100–200 mm thick without bottom. The large bell looked like a red lantern. The channel was surrounded by bulk materials, and the coke was red. This is the real "channeling."

Channeling is the foreboding of abnormal blast furnace. Many changes in blast furnace can cause channeling. Because channels occur often, so it is easy to be neglected. The above examples show that channeling might be the foreboding of blast furnace operation accident and the root of major accidents, and it should be vigilant!

7.2.4 The Types and Features of Channeling

Features of Channeling:

(1) The blast volume increase. Figure 7.9 shows a recurrent little channeling. The blast volume rises with the blast pressure dropping instantaneously. Figure 7.10 shows a run hot channeling;
(2) The blast pressure decreases;
(3) The permeability index increase;
(4) The top furnace temperature record is usually installed at the lower part of gas uptake. If the top temperatures rise in four directions at the same time, the channeling would be at center; if the temperatures in four directions are bifurcated, it would be a local edge channeling, and the direction of the highest temperature should be the direction of channeling occurrence.

The curves from left to right refer to top furnace pressure, channeling and blast volume, respectively.

(5) Because of the occurrence of channeling, the utilization rate of gas decreases, the speed of burden decreases for a long-time channeling, which is reflected in the stock rod record, and the speed of burden decreases; the channeling within short time has less influence.
(6) In the case of recurrent channeling and the amount of dust increasing, there is no dust curve picture in the general duty room, so it can only be known from the records of dust discharging.

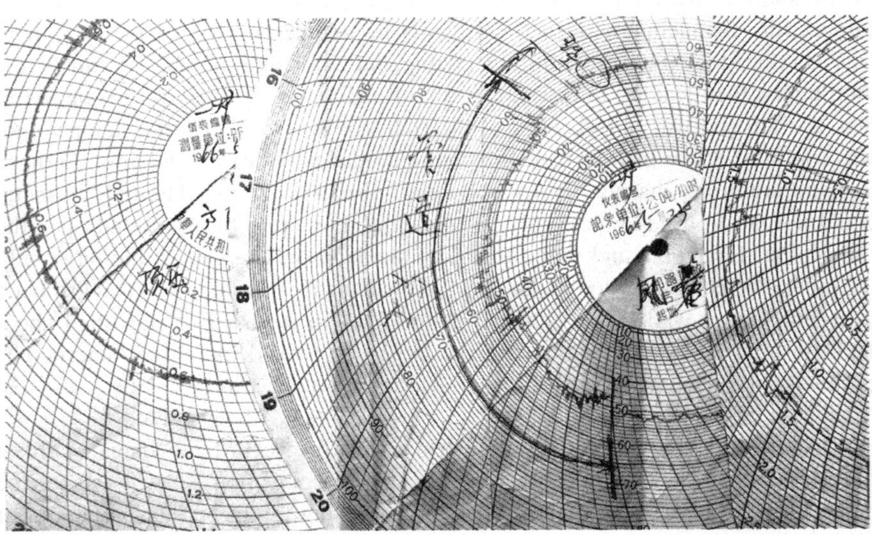

Fig. 7.9 Recurrent channeling

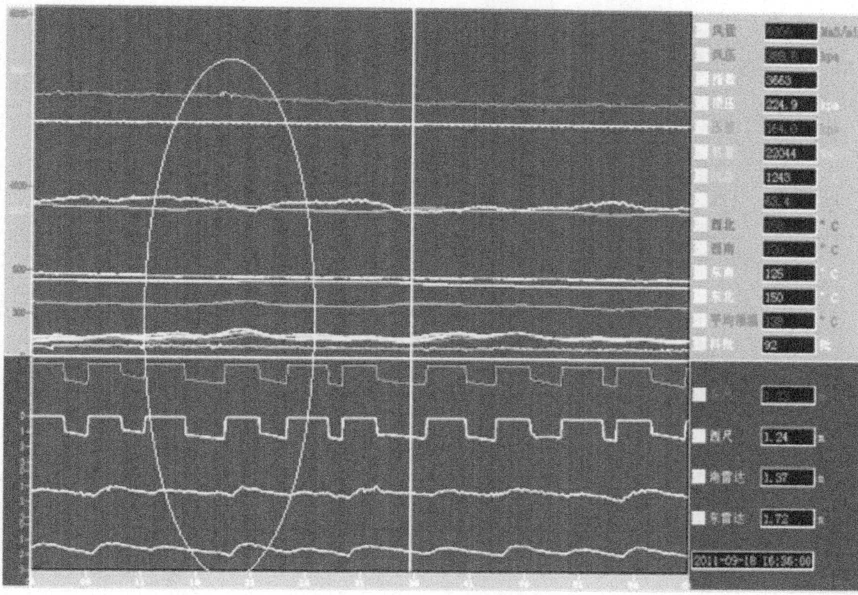

Fig. 7.10 Little channel raised by run hot

Figure 7.11 shows the continuous channeling (in the circle) after the slip. During the period of 19:00–20:00, the channeling continued for nearly an hour.

Channeling can be divided into two categories according to their causes:

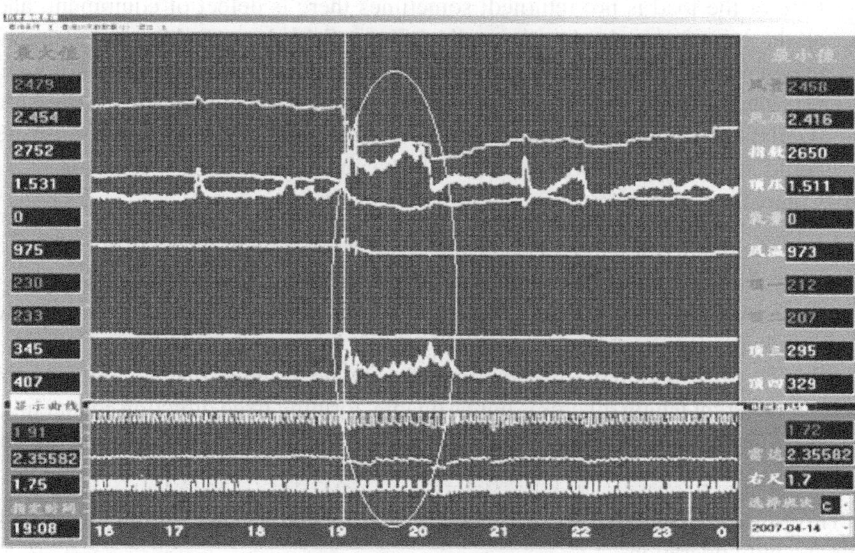

Fig. 7.11 Picture of continuous channeling

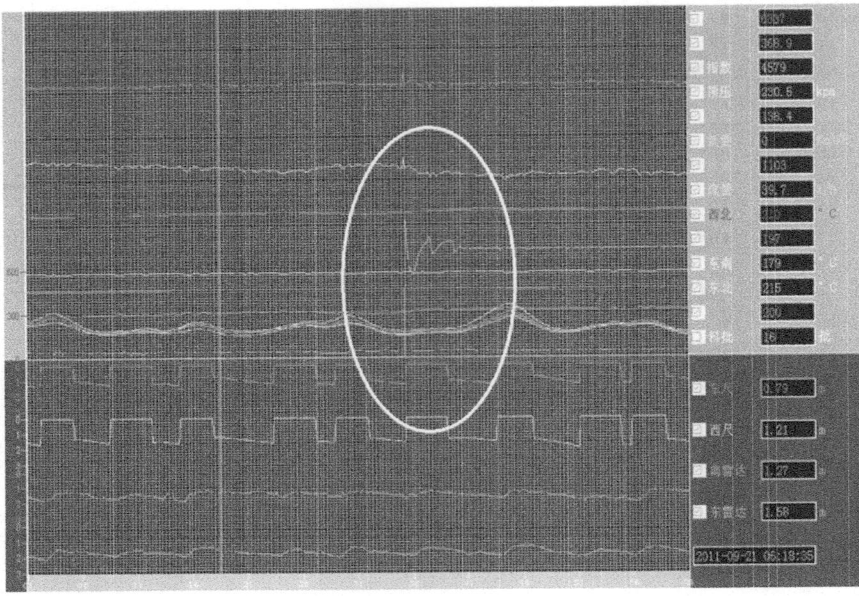

Fig. 7.12 Oxygen flushing channeling

(1) Because of wrong operation or improper regulation, the temporary channeling occurs. The most common one is caused by the rise of furnace temperature, the rise of blast pressure and the tight furnace. Sometimes, coke from bin clearing or dated sinter is used, so the permeability of stock is poor and not treated in time or the load is not returned; sometimes there is defect of equipment, and all these conditions lead to channeling. Figure 7.12 shows sudden oxygenation. Oxygen was added from 0 to 1200 m^3/h. The control valve did not work well. Oxygen oscillates and rushes out of the "channel." Sometimes, the top pressure regulating valve group did not work well, the blast is added, the top pressure could not be adjusted well, and then the channeling occurred (Fig. 7.13).

(2) Channeling formed by non-operational reasons:

1. The quality of burden is very poor, and it is not appropriate to pursue blast volume; that is, blast volume could not accommodate to the burden.
2. Some of them often have edge or center channels because of unreasonable charging system.
3. The distribution of different diameter tuyeres is unreasonable, or the use of different length tuyeres is unreasonable, forming fixed direction channeling.
4. The furnace profile is abnormal, such as scaffolding, erosion and so on, forming a more directional fixed channel.

This kind of channeling should be dealt with the reasons and is difficult to work only by daily operation.

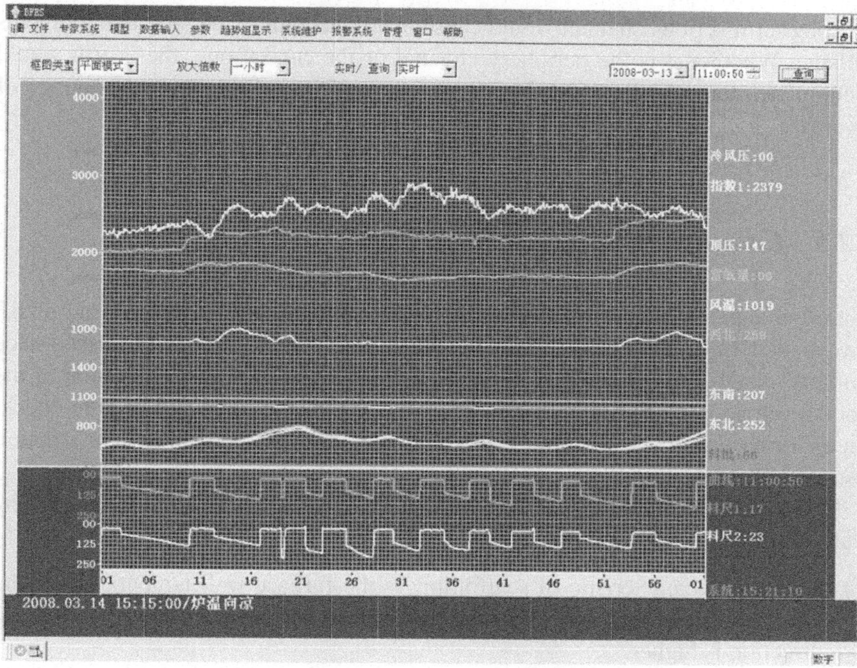

Fig. 7.13 Channeling with poor top pressure adjusting when blast adding

Observing channels, short-time records are easy to be observed clearly, longtime records for regular channels might be observed, and contingency channels could not be observed easily. Figure 7.14 shows a record of one hour and twelve hours, and an hour record was a part of twelve hours. During one hour's record, the state of the channel could be observed clearly. In a longtime record, it was difficult to see the details of the channeling.

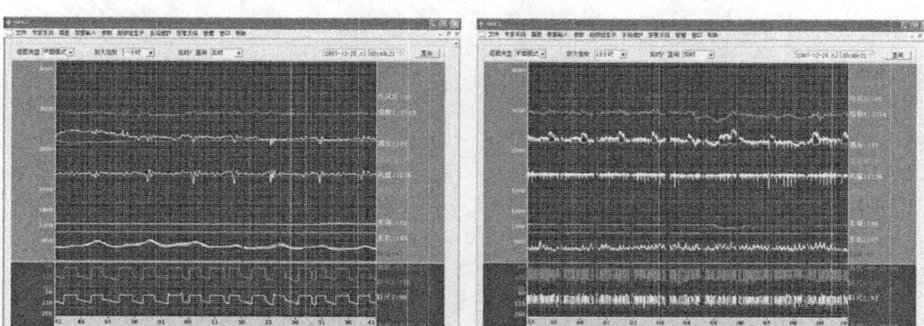

1 hour 12 hours

Fig. 7.14 Record of 1 and 12 h

Long-term serious channel would cause hearth accumulation, and many tuyeres damage. Whether it is the periphery channel or the central one, the hearth would accumulate over a long period. Serious stacking of hearth might cause damage of tuyeres.

7.2.5 Channeling and Bias Stock Rod

Not all channeling would lead to stock surface inclination, but phenomena of BF bias stock surface are mostly caused by channeling (Fig. 7.15). Some blast furnace workers mistakenly believe that the wall scaffolding leads to the inclination of the stock level, so they often stop the furnace, lower the stock line and tried to check, but the furnace wall was mostly smooth without scaffolding, resulting in production loss from the misjudgment.

Key points for judging whether the stock surface inclination is caused by scaffolding are:

(1) The basic characteristic of scaffolding is the furnace wall temperature of the scaffolding location lower than the adjacent area, and the stock line at scaffolding direction is higher.
(2) The stock surface is inclined for the channeling, generally on one side of the channeling. The stock surface is lower, the stock line depth is deeper than the opposite one, and the furnace wall and top temperature are higher. Figures 7.16

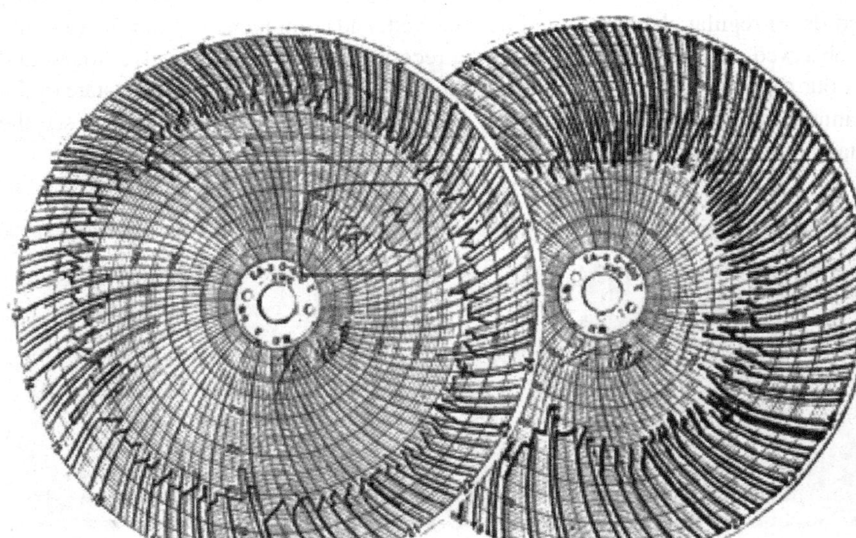

Fig. 7.15 Blast furnace stock surface depth difference caused by channeling (bias stock line)

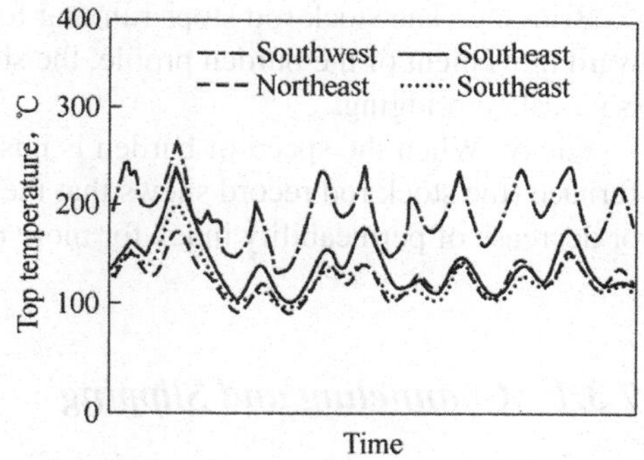

Fig. 7.16 Temperature distribution on furnace top

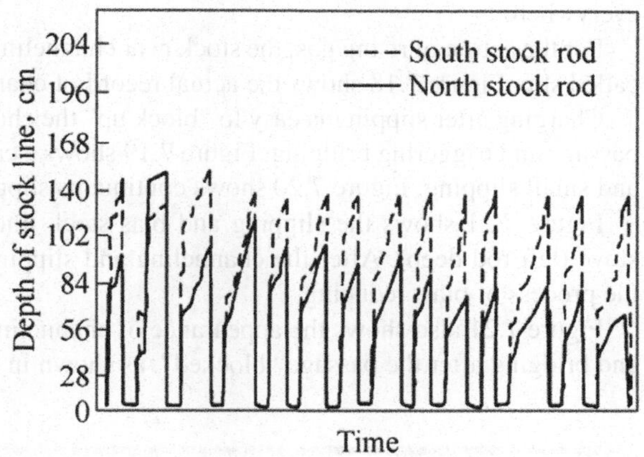

Fig. 7.17 Depth of stock rod

and 7.17 show the deviations in gas flow distribution, resulting in bias burden [6]. The blast furnace had a capacity of 450 m^3. Because of the strong gas flow in the south side, the charge scale in the south side was 0.8–1 m deeper than that in the north side, and the top temperature in the south side was 100–150 °C higher than that in other directions.

7.3 Slipping and Bridging

Terms of Iron-making

Slip: The stock rod falls down with the charge level. If the burden profile suddenly falls down (the stock rod suddenly and rapidly falls down) more than 0.5 m above the normal depth, it is called slip. Channeling is produced by uneven density of charging column. Under the action of rising gas, the burden can easily slip into the loose area (channeling). Channeling often leads to slip. In this book, it is generally called slip; sometimes, it is also called slide or collapse to keep the original citation.

Bridging: The stock rod stops running for more than 1 min following the downward movement of the burden profile, the stock rod records "stop" for 1 min, and it is so-called bridging.

Sticky: When the speed of burden is unstable in the production process of blast furnace (the stock rod record shows that the interval is not uniform), the fluctuation or decrease of permeability index for more than ten minutes is called "stick."

7.3.1 Channeling and Slipping

Slip often occurs in channeling. The density of the stock column is different everywhere.

For the action of rising gas, the stock near channeling is easy to collapse, which is called slip. Figure 7.18 shows the actual record of channeling and slip in the circle.

Charging after slipping is easy to "block up" the channel, obstructing the smooth passage and triggering bridging. Figure 7.19 shows a case study of small channeling and small slipping. Figure 7.20 shows continuous slipping.

Figure 7.21 shows the slipping and bias stock line after the switchover of hot stove (left rod deep). After the channeling and slipping, there was still slipping in the process of blast reducing.

Figure 7.21 also shows the appearance of channeling, followed by deep slipping and bridging after the passage "blocked", as shown in the circle above.

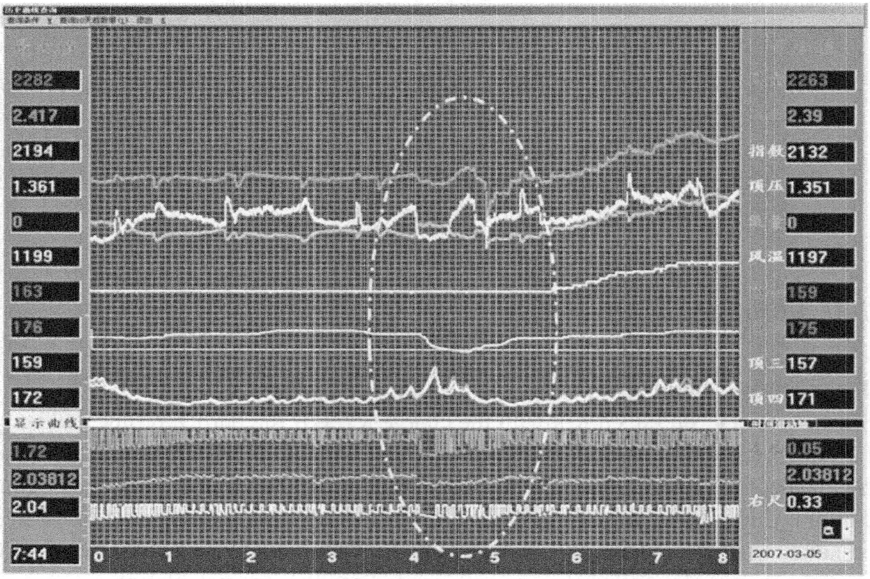

Fig. 7.18 Slipping after channeling

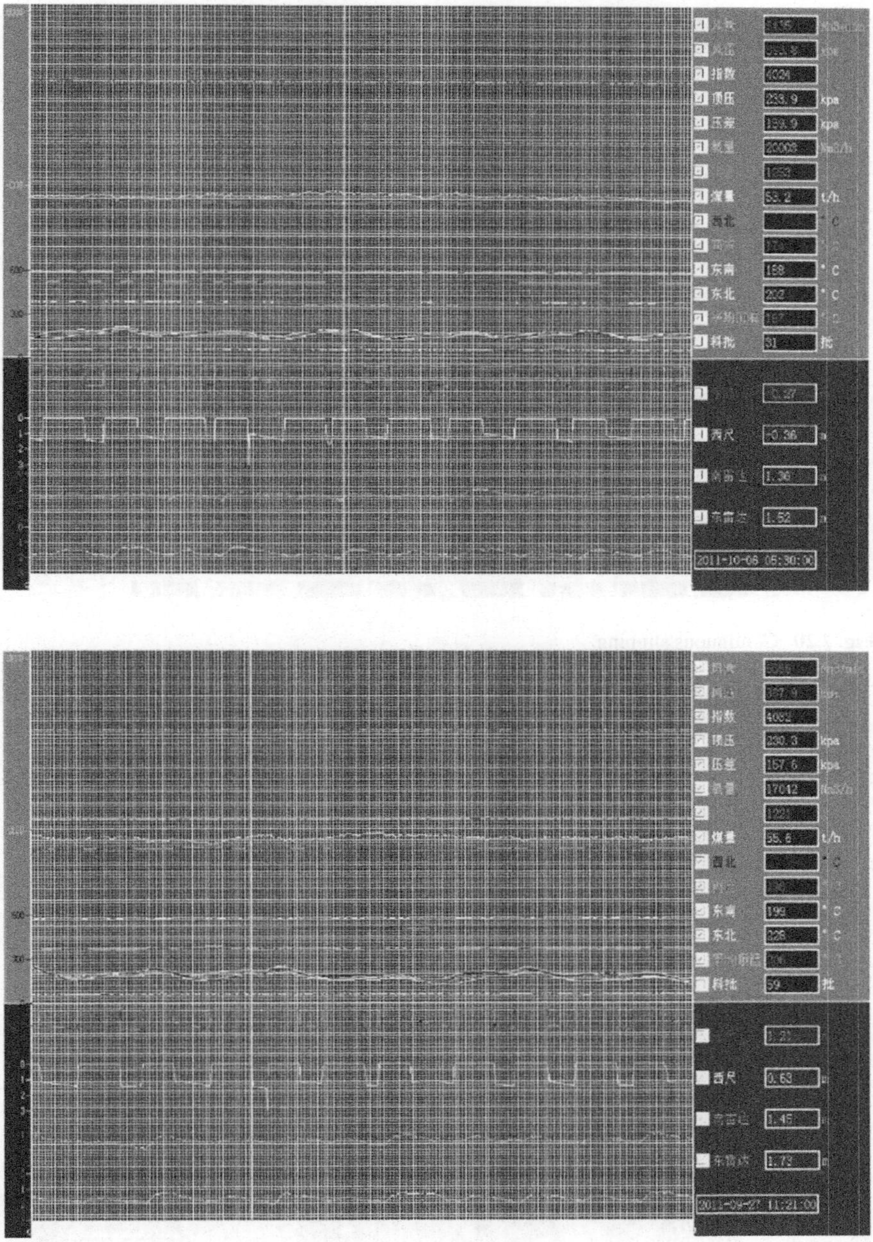

Fig. 7.19 Channeling and slipping

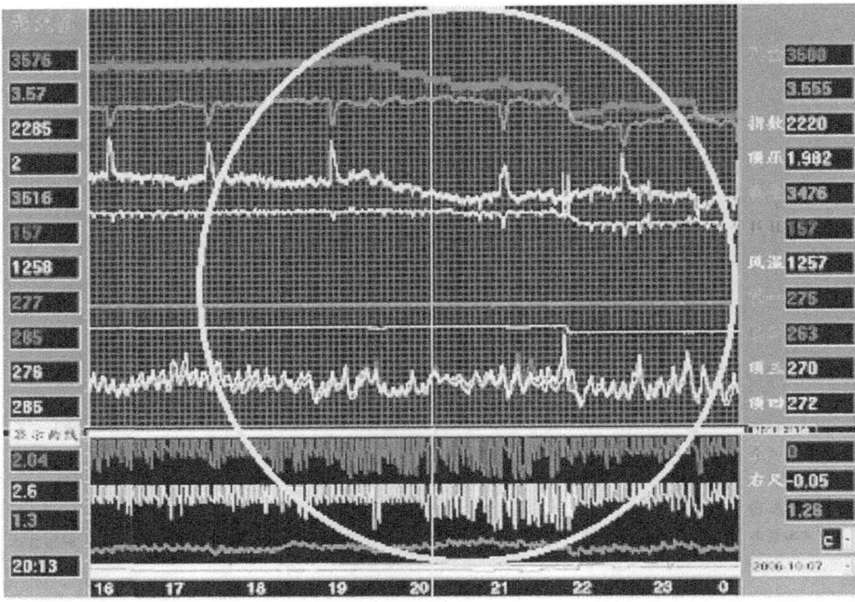

Fig. 7.20 Continuous slipping

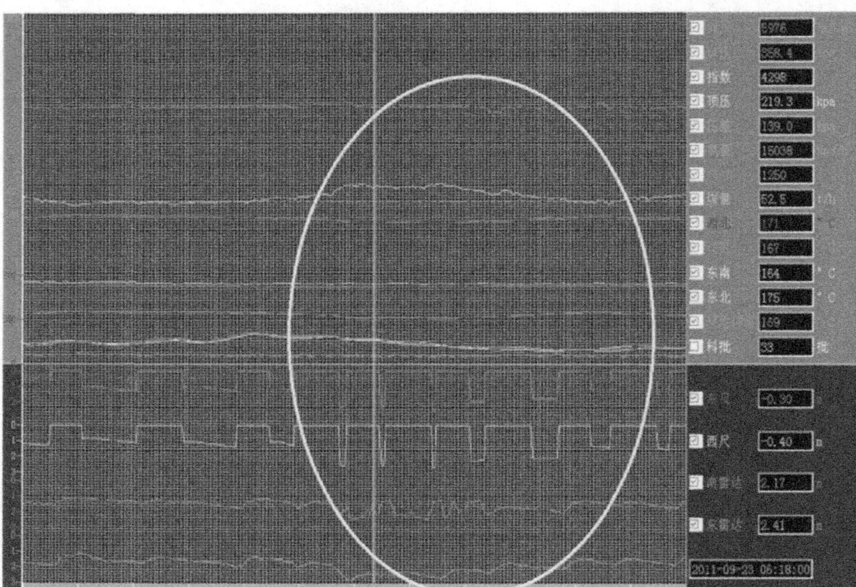

Fig. 7.21 Channeling–slipping–bridging

7.3.2 Bridging Mechanism

Since blast furnaces came into world, there was bridging in the process, although it was not clearly indicated the name, bridging.

The first blast furnace bridging in the world was recorded in China. "The Book of Han" recorded: "In the second year of Emperor Pinghe of Han Dynasty (27 B.C.), the blast furnace of the official iron-making plant in Peijun (now in the eastern part of Peixian County, Jiangsu Province) occurred a bridge and sounded like thunder and drum. Thirteen iron makers escaped away in panic. After the sound, a big pit by the blast furnace was formed with several feet deep. The blast furnace was broken into ten pieces, and a hot metal splashing like meteors, which was as similar as an accident happened 64 years ago." This record shows not only the blast furnace bridge, but also the slip, the hearth broke, explosion and hot metal flying to the sky. The Book of Han is a famous historical work in China. The author Ban Gu died in 92 A.D. This book contains important events in 229 years from 206 B.C. to 23 A.D. Two blast furnace bridges, the rupture and explosion of the blast furnace in the book, whatever the motivation, show that the incident is sensational, and even historians wrote as national events.

Just like the study of blast furnace production, the study of bridging has been carrying out. Since the 1940s, scholars in the former Soviet Union had systematically studied and experimented on the relationship between the stock descending force and resistant force. The results indicated that the descending force of moving stock equaled 40–45% of the weight of stock column. There are three main parts against the force of stock descending: the frictional force of the wall, the resistance of rising gas and the frictional force inside the stock column (see Eq. (7.1); (Fig. 7.22)).

With the model test and theoretical analysis of blast furnace bridging, Yang Yongyi found that the sum of the above three forces is far less than the force of stock column descending. He suggested that the bridging was the result of "sub-weightlessness." When there is a layer of burden with poor permeability at a certain position of the stock column, bridging would occur, although the effective weight of the whole column is much greater than the resistance. Figure 7.23 depicts the sub-weightlessness in detail.

Sub-weightlessness indicates that bridging would occur in the case of larger local resistance and more local burden powder in the stock column.

The pressure changes in different parts of the blast furnace reflect the location of bridging. For upper bridge, the upper pressure increases while the lower pressure decreases. For lower bridge, the lower pressure increases, as shown in Table 7.3 and Fig. 7.24.

Xiong Yao put forward the theory of "supporting force," which regards the stock column as "piston flow." In addition to the three forces proposed by former Soviet scholar, the "dead zone" of hearth pushes the stock column upward with the cohesive zone. Therefore, active hearth plays an important role in reducing bridging [7].

The above viewpoints can explain the causes of bridging and the ways to prevent it: (1) Improve the quality of burden; (2) the blast volume should be compatible with

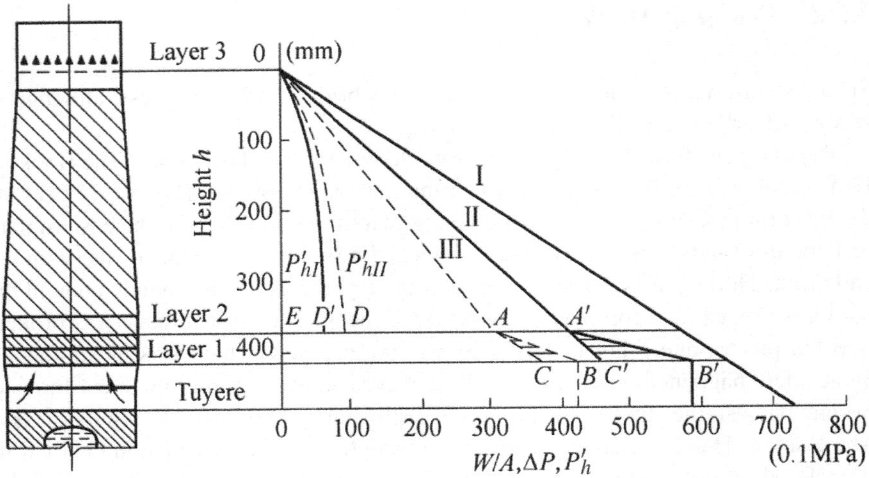

Fig. 7.22 Graphic of sub-weightlessness. Tuyere—layer 1: non-burden zone Layers 1–2: super-weightlessness zone (pressure drop is more than burden weight)
Layers 2–3: sub-weightlessness zone (pressure drop is less than burden weight)
I—net weight of pillaring, $W/A = h_{\gamma M}$;
II—distribution of pressure drop of bridge above critical point (disjointed, blast volume 1.9 m^3/min)
III—distribution of pressure drop of the moment before checking (blast volume 1.48 m^3/min)
P'_{hI} and P'_{hII} are pressure variations along height of stock column (calculated value);
BC (parallel to bottom edge) in $\triangle ABC$ and $B'C'$ (parallel to bottom edge) in $\triangle A'B'C'$ are equal to upward super-weightlessness

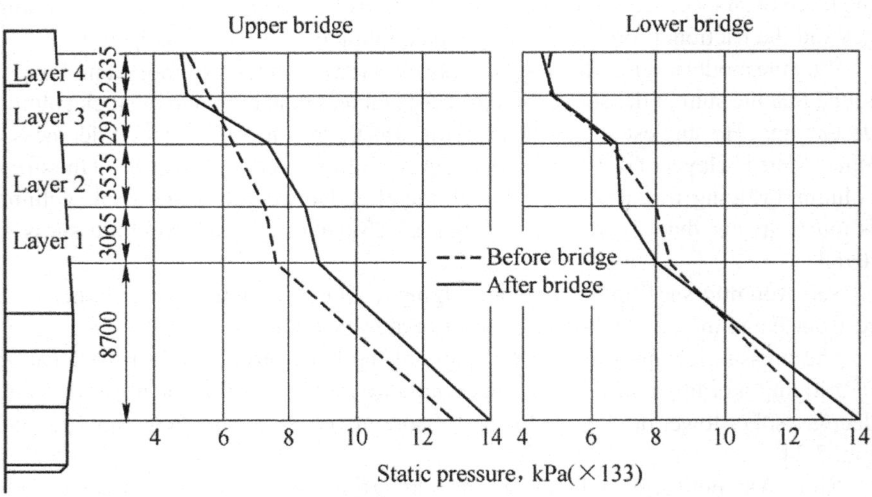

Fig. 7.23 Variation of static pressure before and after bridge in a certain factory

Table 7.3 Variation of static pressure in each layer before and after bridge

Bridge position	Time	Hot blast, °C	Static pressure, kPa (× 133)			
			Layer 1	Layer 2	Layer 3	Layer 4
Upper	November 16, 1959					
	Before	1280	770	716	610	548
	After	1400	880	828	722	485
Lower	October 12, 1959					
	Before	1290	845	783	675	470
	After	1410	820	703	680	480

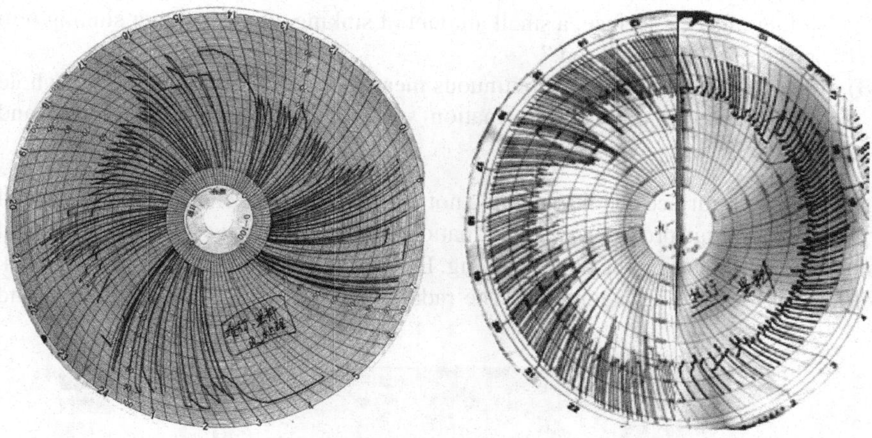

(a) (b)

Fig. 7.24 Record by mechanical stock rod

the quality of burden and does not exceed the permissible limit of permeability of stock column; (3) rational distribution of gas; (4) the distribution of burden and blast volume should be suitable; and (5) active hearth.

7.3.3 Classification of Stock Rods

In practice, stock rods can provide important blast furnace information. There are three types of stock rods for blast furnaces:

(1) Mechanical stock rod: a cylindrical iron hammer hung on the sprocket by a 4–6-m-long chain. Put mechanical stock rod on the surface after charging, and follow the "stock column" down. It has the advantage of tracking the material

surface most of the time, but attention should be paid to the difference between mechanical failure or stock surface changing causing curve deformation, sometimes the chain hammer partially melted, the weight reduction will also cause stock rod oscillation, sometimes sprocket failure lead to a "small step" of the stock line, giving the false impression that the burden is unstable decline.

Figure 7.24a shows the record of tight furnace, bridging and treatment of blast furnace, the left side of Fig. 7.24b is the record of frequent slip, and the right side is the record of temperature rising and hot bridging.

(2) Scanning stock rod: The mechanical stock rod is scanned regularly by computer, and the depth of the stock rod is used as the test data to draw the record (Fig. 7.25). The detection time interval has a great influence on the detection accuracy. Generally, it is scanned once in 3–5 s, without continuity. It might ignore slight sliding of stock surface, a small amount of sinking, chain hammer sinking into "groove" and "buried rod."
(3) Radar stock rod: the only continuous measuring stock rod, tracking the whole process. It can provide information such as stock surface fluctuation and "writhe."

Some stock surface conditions could not be reflected by mechanical stock rod, and radar stock rod can provide more information. Figure 7.26 shows that the mechanical stock rod raised during the coke charging. Figure 7.27 shows the fluctuation of stock surface in the case of channeling. The radar stock rod reflects the fluctuation and

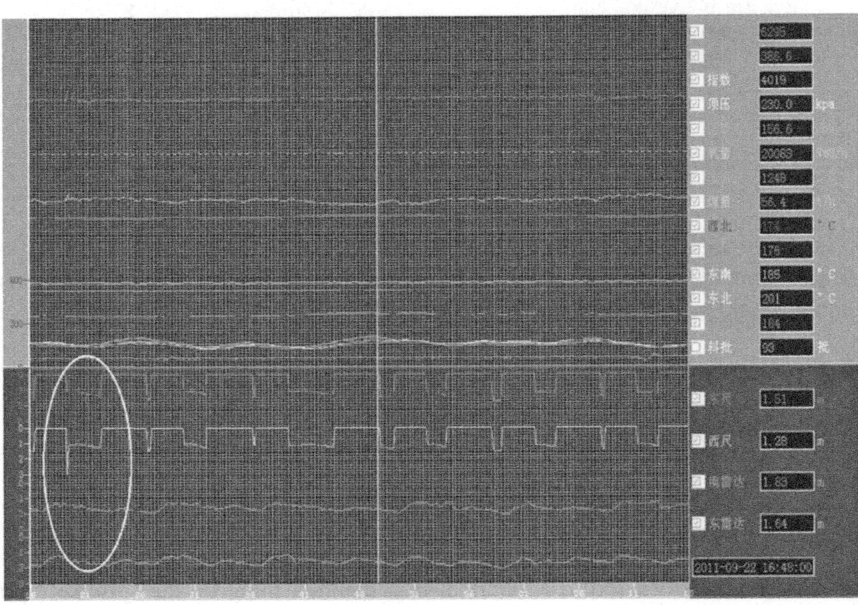

Fig. 7.25 Local slip recorded by scanning stock rod

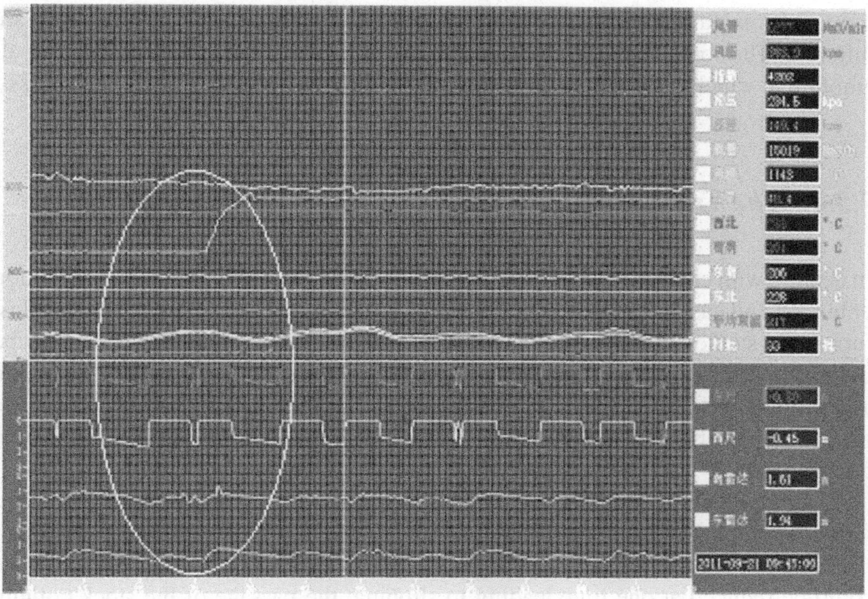

Fig. 7.26 Variation of south stock rod curve before and after coke charging after channeling

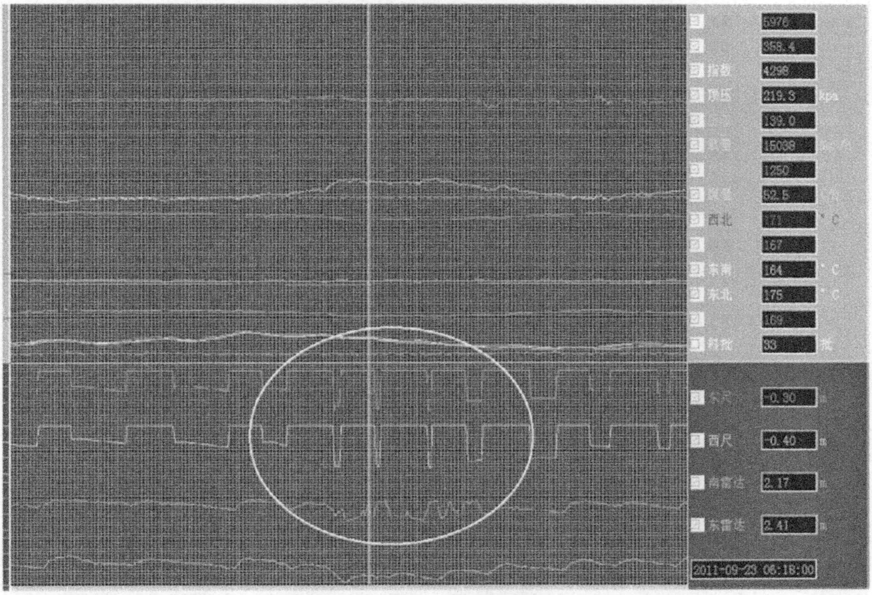

Fig. 7.27 Stock surface fluctuation reflected by south radar stock rod after deep slipping

writhe of coke falling on the stock surface. The results of the two radar stock rods are different. The south (radar) rod fluctuated violently, while the north one had a small amplitude, which truly reflects the stock surface condition in the charging process.

7.3.4 The Treatment and Lesson of Bridging

In dealing with channeling and slipping, it is essential to keep reasonable distribution of burden and gas, so as to keep burden and gas moving steadily. Only when the blast furnace is stable and smooth operation, the energy of the blast furnace could be utilized to the maximum extent. Bridging generally is the continuation of channeling and slipping. Without channeling or slipping, the examples of sudden occurrence of bridging might happen, but very few. There are always special reasons for sudden bridging.

Figure 7.28 shows an example of poor handling of bridging: Too slow snorting failed to completely destroy the passage in the stock column after bridging. Although the blast was reduced a lot, it delayed for more than an hour. If such a volume of blast could be reduced at one time leading to slipping, the passage of the stock column would be completely changed and made the process of adding blast acceptable.

Figure 7.29 shows a very rare example. At a 1580 m^3 blast furnace, the top

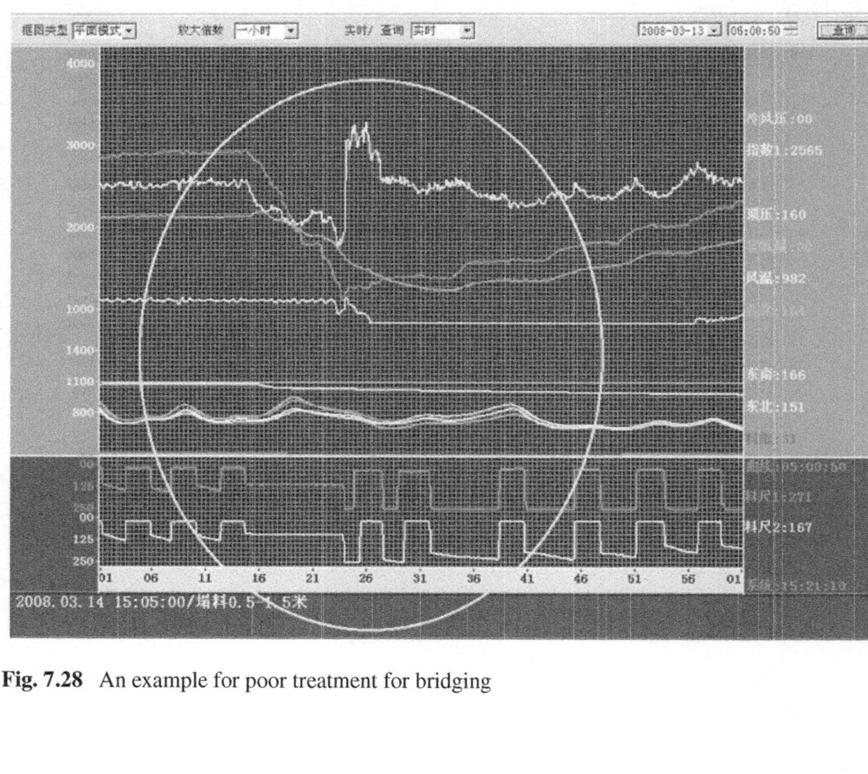

Fig. 7.28 An example for poor treatment for bridging

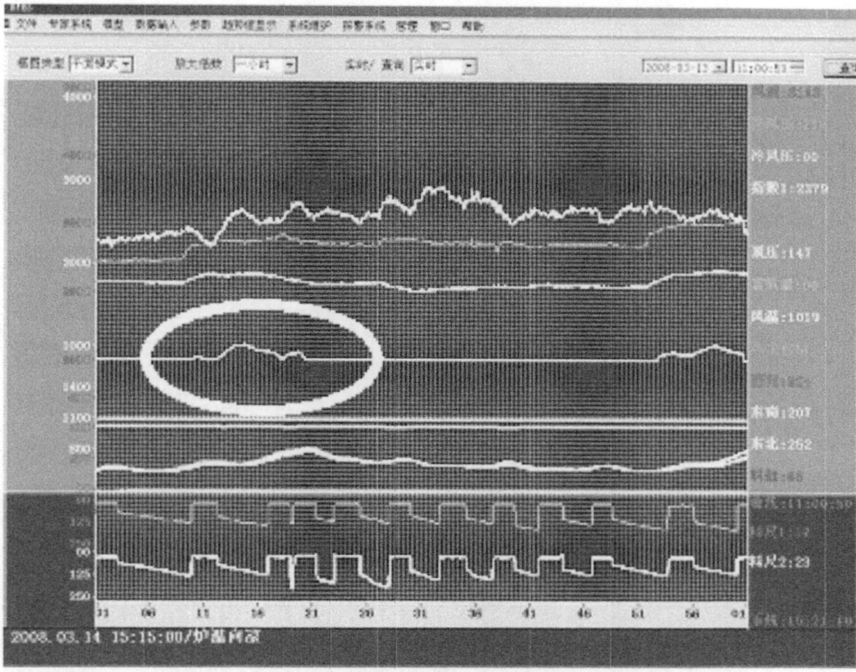

Fig. 7.29 Example for top pressure regulation failure

pressure automatic regulator facility did not work when blast adding, resulting in unstable top pressure (see the record of the top pressure in the circle of Fig. 7.29) and in channeling. The blast volume and permeability index fluctuate and the top temperature bifurcated. The operator neglected the failure of top pressure regulation and the channeling, which could be avoided actually.

The primary principle of treatment for bridging is to maintain adequate furnace temperature. Figure 7.30 shows a record of one shift of a 1200 m^3 blast furnace with sufficient furnace temperature, but it was into bridging state with zero blast volume. After shift, the blast was added six times in two hours, only the blast pressure increased rather than blast volume. After the seventh time, the blast volume suddenly increased, and the blast and permeability index increased simultaneously. Afterward, there were 12 times of blast adding in more than two hours, with each average interval of 10 min. During this period, the blast volume and pressure increased synchronously. At the 12th blast adding, the blast pressure dropped rapidly with the permeability index increasing significantly, the stock slipped 5 m, and the stock line began to move. Then, blast is added three times again. The blast pressure rose, and the blast volume decreased at the third time. The "permeability index" dropped sharply, and the blast pressure curve intersects with the blast volume curve. The blast furnace continued to add blast 2 times, although the blast pressure increased, the blast volume dropped dramatically, and it was forced to reduce the blast at 23:00.

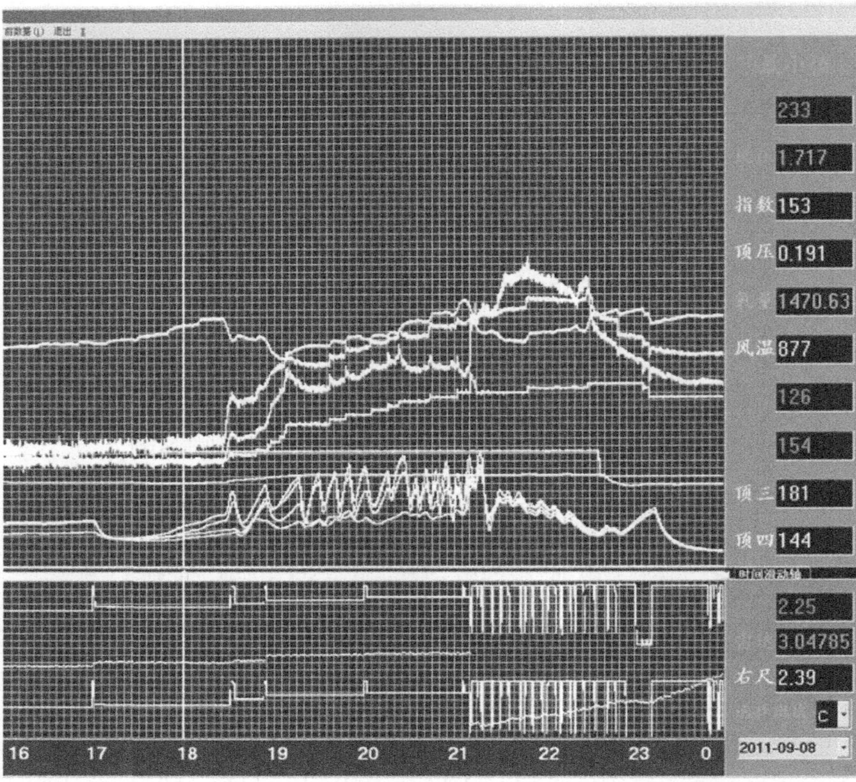

Fig. 7.30 Treatment on bridging

When there is bridging with no blast or small blast volume, adding blast is indispensable, but to pay attention to:

(1) Adequate furnace temperature;
(2) During the process of adding blast, the increase of blast pressure must be higher than or equal to the blast volume, and the "permeability index" should be properly guaranteed. During the third blast adding, the blast pressure increased faster than the blast volume, and the "permeability index" decreased, indicating that it was not the right time for blast surge.

Figure 7.31 shows the bridging in the process of blast. The bridging was treated by checking before 1:00, and the relationship between pressure and volume was improved after checking. Because of the low blast volume, the effect of checking was not good. After 3:00, the "permeability index" of each blast was first high and then low. In this case, the opportunity should be waited for. But after 4:00 in fact, the blast adding once again caused another bridging. After long-term bridging, it should be avoided bridging as far as possible. At this time, tapping was abnormal. It was difficult to accurately determine how much iron is in the hearth. Once bridging, it

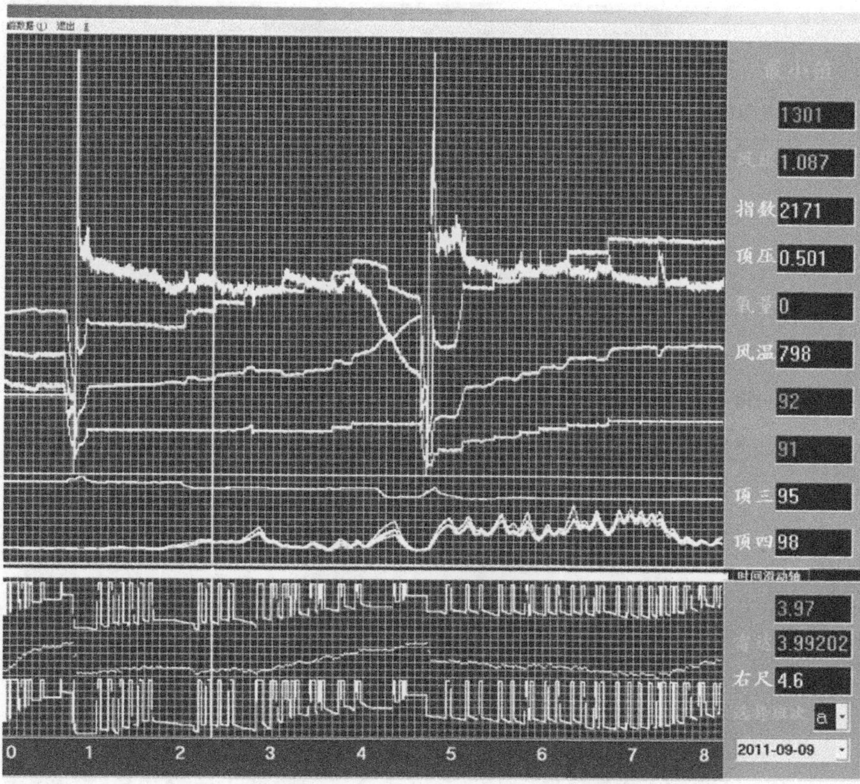

Fig. 7.31 Treatment on bridging and checking

would be difficult to predict whether slag blooding will occur or not. It was necessary to keep blast supply when the iron in hearth is unknown. Every effort should be made to make the bridge slipping down automatically under the condition of blast supply to prevent slag blooding. Checking at 4:50 was last resort, though a little risky. After checking, to add blast following the principle of "permeability index" made progress smoothly.

The bridge (Fig. 7.32) was caused by furnace converting to running hot.

Figure 7.33 shows that the furnace was not in good condition. The foreman was afraid of the furnace cooling and then made the blast temperature continuously raise after 2:00 and raise by 50 °C again after 7:30. During this period, the blast volume shrunk continuously, the blast pressure rose continuously, the bridging at 8:50 was forced to drop the blast temperature by 100 °C, and the pressure was reduced at 8:55 in an attempt to destroy the bridge. After 2 min, the checking was carried out as stock rod moved reluctantly, and the improvement of pressure relationship between blast pressure and volume was limited. Although the blast increased slowly in 3 h, the blast volume still could not restore the level before checking. In the next 4 h, the

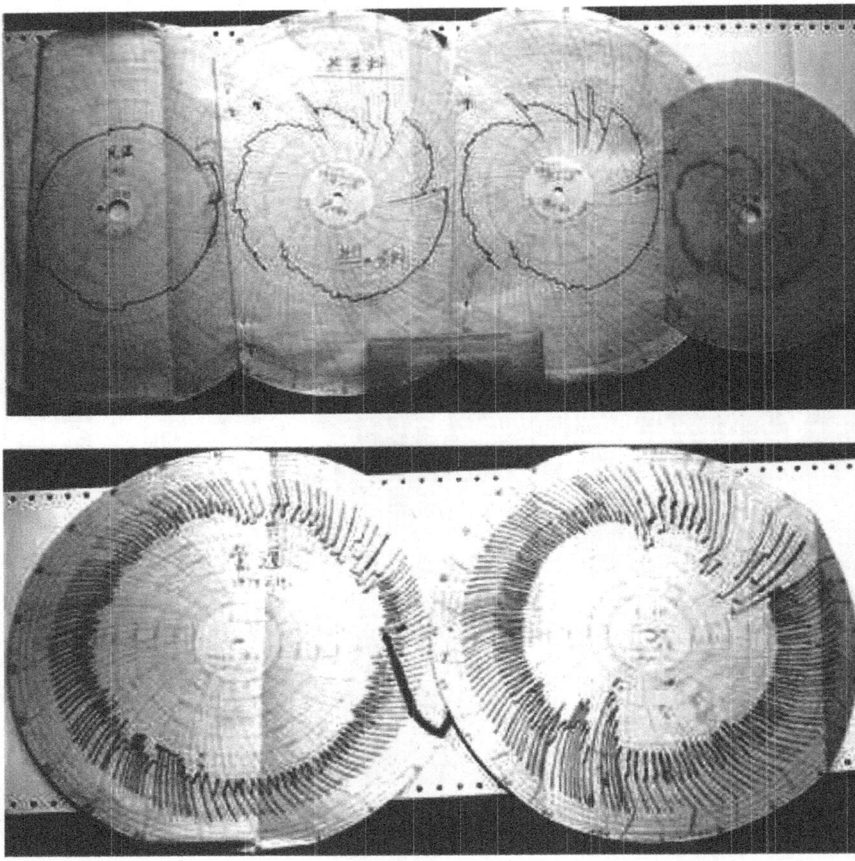

Fig. 7.32 An example for hot bridging

blast was forced to increase again, and then 4-time checking, from 16:00 to 23:00, after seven hours, the blast volume was restored to the previous normal level.

This example was not a desirable way to handle bridging. Comparing with Figs. 7.32 and 7.33, it can be seen that:

(1) The foreman was timid, afraid of furnace cooling and blind to raise blast temperature. The wrong elevation temperature and reverse operation bring about serious consequences.
(2) The bridge was caused by the reverse increase of blast temperature. The first bridge could be avoided actually.
(3) Undetermined handling of bridge, insufficient blowing off and failed to thoroughly solve the relationship between blast pressure and blast volume, so that the burden descending was not smooth, and the stock rod slid and stagnated. Because of the unswerving checking, four times of continuous checking made

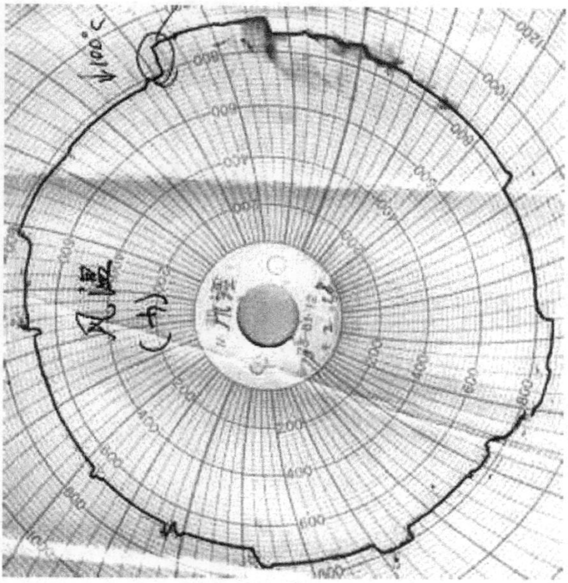

Fig. 7.33 Amplified blast temperature record

the stock column more compact, which was not conducive to the recovery of furnace condition.

(4) The blast volume was smaller, so it was forced to increase, and then started checking. The last three times were operated by this way. Although the idea was desirable, the time of blasting was too short, which was not as effective as imagination.

In Fig. 7.34, bias stock line was caused for the channeling. The material sometimes slips and the channeling is destroyed timely due to the wind reduction twice. At that time, the condition of burden was poor, and it was difficult to completely avoid the channeling and slip. At 2:00 before and after the blasting process, there was a slip, but it did not affect the addition of full blast. After 2:00, it reached full blast, and the furnace condition was normal. At 8:00, the speed of burden slowed down, and after a short time (about 20 min), it was added back. Such mistake caused the burden to bridge after 25 min. At 9:00, the blast was sharply reduced, the south stock rod fell to 4 m, and the north stock rod fell about 2 m. Although the blast was slowly added afterward, the stock rod went very hard and it was difficult to add blast. At 11:00, the blast was again reduced, and it took another 3 h to get close to the full blast.

This treatment was evaluated as follows:

(1) The bridging, which might be prevented actually, occurred only for the missed opportunities in undeserved mistake to get a small amount of blast.
(2) For the first time, there was too little blast reduction to achieve the purpose of checking. After blasting again, the stock rod did not work well. After blasting again, the stock rod did not work well and blast volume increased little with the increasing blast pressure. It showed that the treatment was insufficient.

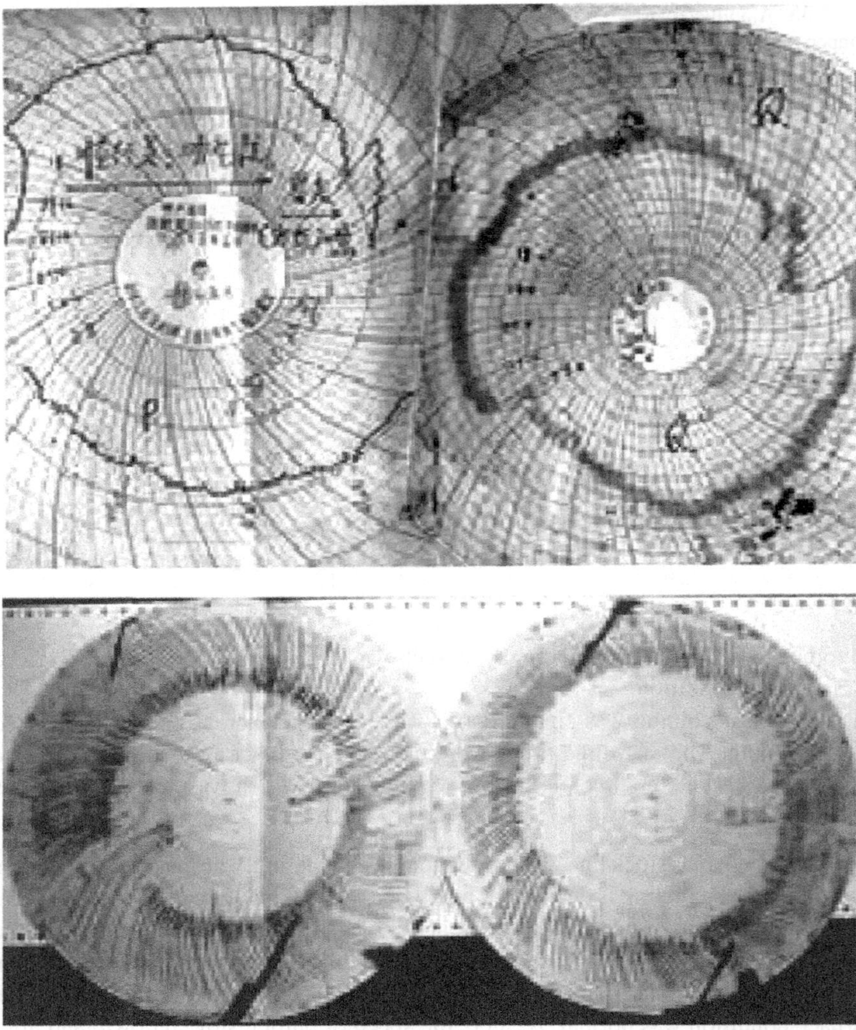

Fig. 7.34 Bridge and treatment

(3) The above shortcomings lead to long-term slow blast operation and huge loss of blast furnace production.

Figure 7.35 shows the slag blooding caused by improper bridge treatment. The threat of blooding comes from two aspects:

(1) The furnace temperature is low or cold, so it is easy to blood.
(2) The hot metal and slag tapped not well, and the liquid level of the metal in hearth is close to the tuyere, so easy to slag.

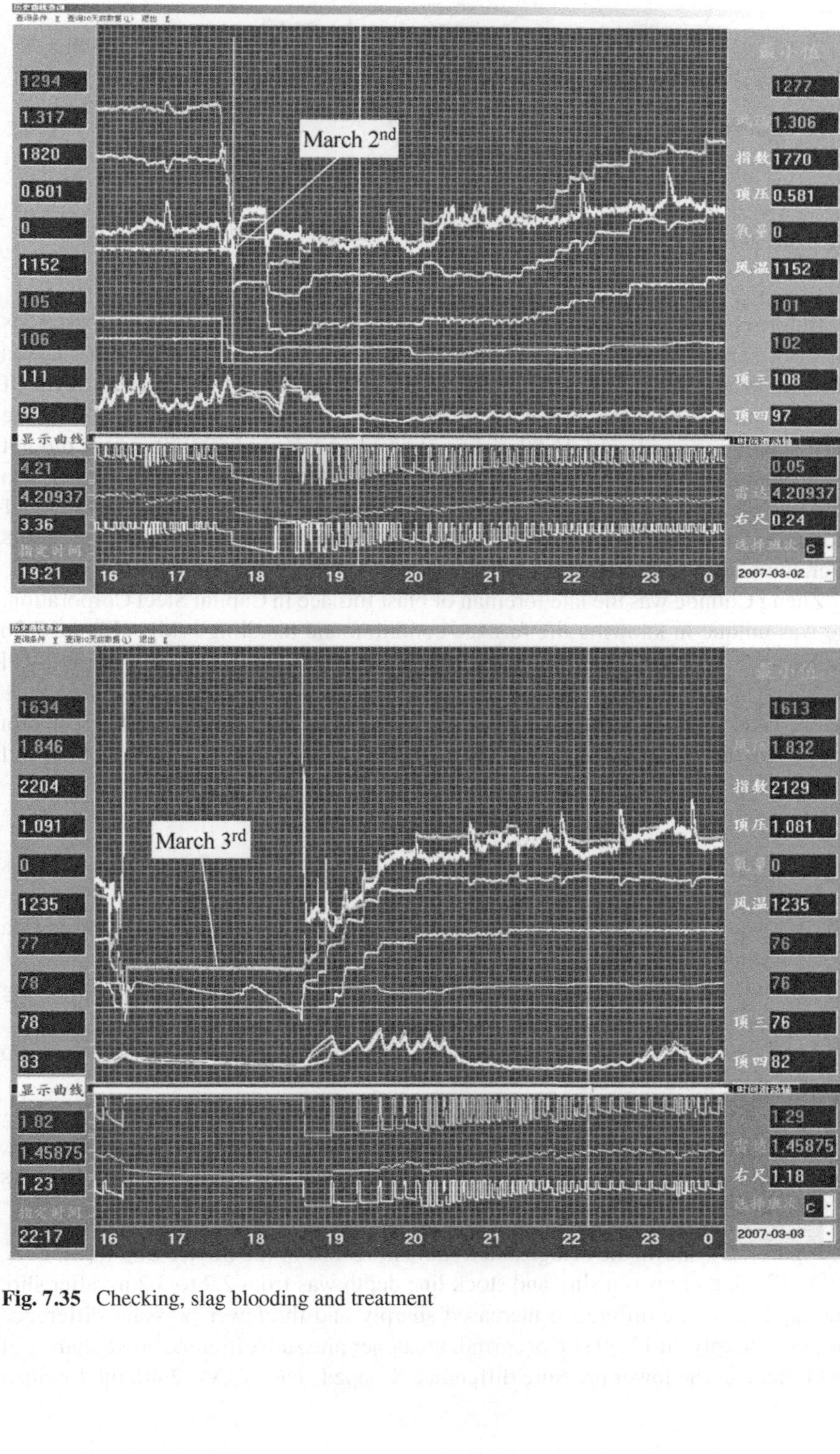

Fig. 7.35 Checking, slag blooding and treatment

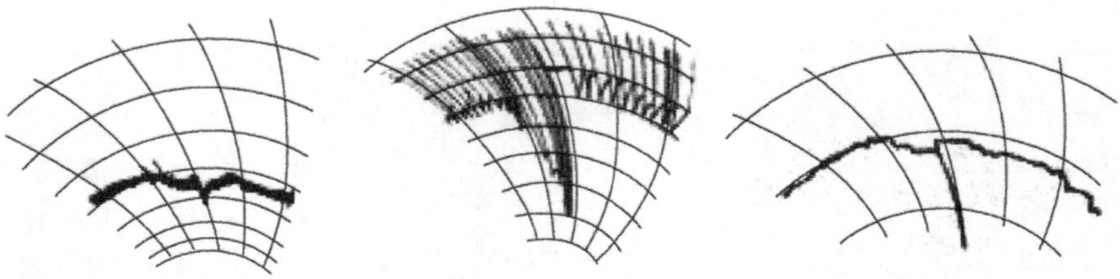

Fig. 7.36 Zheng Chunde's checking

Before checking, the foreman should considerate carefully, no check with a serious threat of slag blooding, unless it is forced to do so. Figure 7.35 shows that it was difficult to recover after blooding of tuyere. The original blast volume level of 1290 m^3/min on March 2 was already under blowing. After checking at 17:40, the blast supply resumed until 21:00 and reached 1277 m^3/min at 24:00, which did not reach the blast volume level before checking. A long-term reducing wind leads to heavy losses. Besides, after preparation for treating the burned coal lance and 11 blowpipes, the blast stopped at 16:22 for 138 min the next day, and nearly two hours later, the blast volume became normal.

Zheng Chunde was the late foreman of blast furnace in Capital Steel Corporation. He was unique in knowing the furnace condition and handling bridge. Figure 7.36 shows his classic work on bridge treating at No. 2 BF (1327 m^3) of Capital Steel on June 8, 1976. He blew out dramatically and then quickly blew in, reaching 90% before checking; 30 min later, blast volume reached the level before checking; 40 min after checking, blast furnace blast volume exceeded before checking, reaching full blast.

The key points of his operation are:

(1) Careful study of furnace conditions and full estimation of iron tapping status and furnace temperature level.
(2) Furnace burden quality situation, careful comparative analysis, so as to have a good idea.
(3) Decide the operation scale according to the actual situation, the level of blowing out and blowing in, to ensure the success of checking at one fling.
(4) Only success at one fling is the most effective checking and the least loss. So he always did his best.
(5) Pay attention to observing the tuyere.
(6) Pay attention to safety and protection, and avoid slag blooding. Before checking, always arrange the foremen, and the hydraulics and coal-injecting workers making good preparations for preventing burnout and slag blooding.

Figure 7.37 shows the bridge and treatment record of the Soviet Union [8].

At 10:20, there was a slip, and stock line depth was from 2.2 to 3.2 m; after slip, the upper pressure difference increased sharply and the lower pressure difference dropped sharply; at 12:30 slip occurred, the upper pressure difference rose sharply at this time, and the lower pressure difference dropped sharply. At 12:40, the foremen

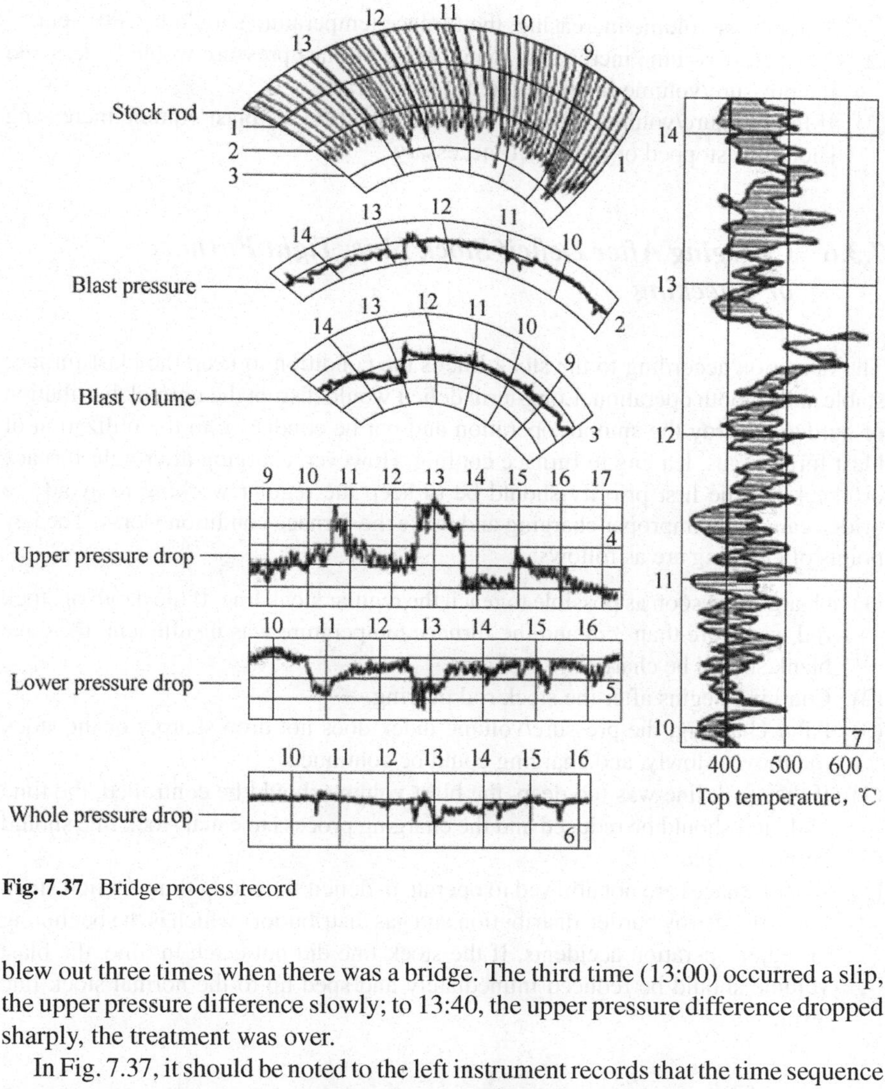

Fig. 7.37 Bridge process record

blew out three times when there was a bridge. The third time (13:00) occurred a slip, the upper pressure difference slowly; to 13:40, the upper pressure difference dropped sharply, the treatment was over.

In Fig. 7.37, it should be noted to the left instrument records that the time sequence for upper three instruments is from right to left, but for the lower three instruments is from left to right.

7.3.5 Blasting Under Tight Furnace

Blast volume, which is the basis of blast furnace production, should be restored first at the tight furnace. It is difficult to increase blast volume with tight furnace. The timing and quantity of blast volume increasing are very important. Based on the past lessons, some operating points are put forward:

(1) When blast volume increasing, the furnace temperature should be sufficient.
(2) After blast volume increasing, the increase of blast pressure would be less and the pressure/volume index would not deteriorate.
(3) If the pressure/volume index deteriorated sharply, the blast volume increasing should be stopped or reduced if necessary.

7.3.6 Charging After Deficit Stock Line, Tight Furnace or Checking

The operation according to the stock line is the condition to keep the blast furnace stable and smooth operation. Long-term deficit would disrupt the normal distribution of burden, destroy the smooth operation and not be conducive to the utilization of blast furnace gas, leading to furnace cooling. However, charging after tight furnace or checking, the first priority should be to keep the regular working to avoid the bridge caused by improper charging and make the furnace condition worse. The key points of charging are as follows:

(1) Charging as soon as possible to reach the regular stock line. If the depth of stock rod was more than 3 m and the furnace temperature was insufficient, the coke blank should be charged.
(2) Charging begins after the stock rod moving.
(3) After charging, the pressure/volume index does not drop sharply or the stock rod moves slowly, and charging could be continued.
(4) If the stock line was too deep, the blast volume should be controlled, the time of deficit should be reduced and the charging process to catch stock line should be accelerated.
(5) Blast furnaces are not allowed to operate in deficit for a long time. Deficit stock line will destroy burden distribution and gas distribution, which is the beginning of major operation accidents. If the stock line did not catch in time, the blast volume should be reduced immediately and sped up to the normal stock line level.

7.4 Lessons from Operation

7.4.1 The Checking (Snorting) Controlling

Foremen, especially the green hands, are inexperienced in operation, but the most important lesson is to keep calm in case of trouble. In 1995, the coke quality of Capital Steel was very poor, while there were many accidents in blast furnaces. At 5:58 on February 25, the No. 1 blowing pipe was burnt out suddenly, and the foreman checked in a hurry. Because of the excessive speed of checking, all 28 tuyeres were slag blooded, although it took 25 min to re-blow slowly (Fig. 7.37), the slag which

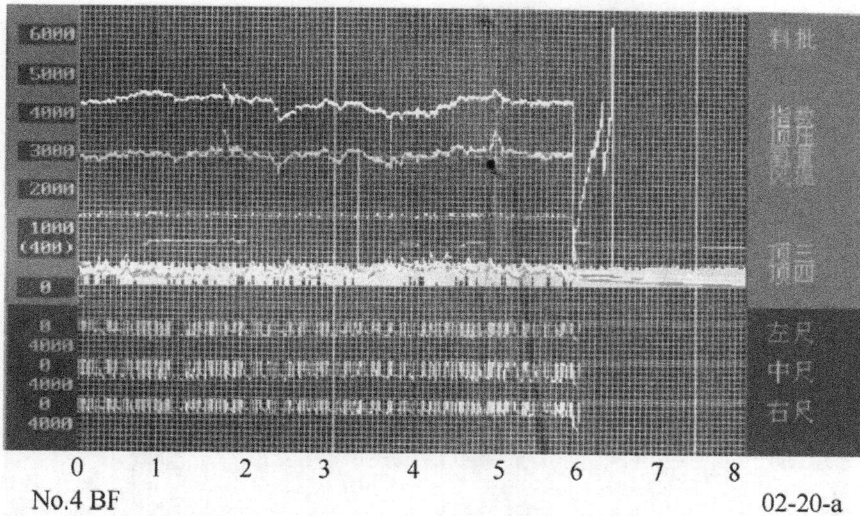

Fig. 7.38 An example for excessive speed checking

had entered the blowing pipe was returned immediately and the slag blooding of all tuyeres had become a fact and had to stop blast in 10 min (Fig. 7.38).

If the checking was not too fast, while observing the burning out of the blowpipe and coke injection, and while slowly checking to control the ejection, it might not slag blooding every part. Fortunately, the checking was in time, and most of the blowpipe blooding was not serious.

7.4.2 Re-blast After Checking

It has been mentioned before that checking is a common technology used for blast furnace operation and also often makes mistakes. Figure 7.39 shows an example of excessive re-blast after checking, while 90% of the original blast was reached after checking. If the checking conditions were good, it would be completely possible, and sometimes it can be reached to 95%. The coke quality of this blast furnace was very poor, and the continuous channeling appealed. Under this condition, blowing to the original level of 90% must be failed. From the figure, the secondary re-blast started at 4000 m^3/min, which was equivalent to 75% of the original blast volume before checking, and reached the original level in an hour.

This example indicated that the burden condition and the furnace condition at that time must be considered.

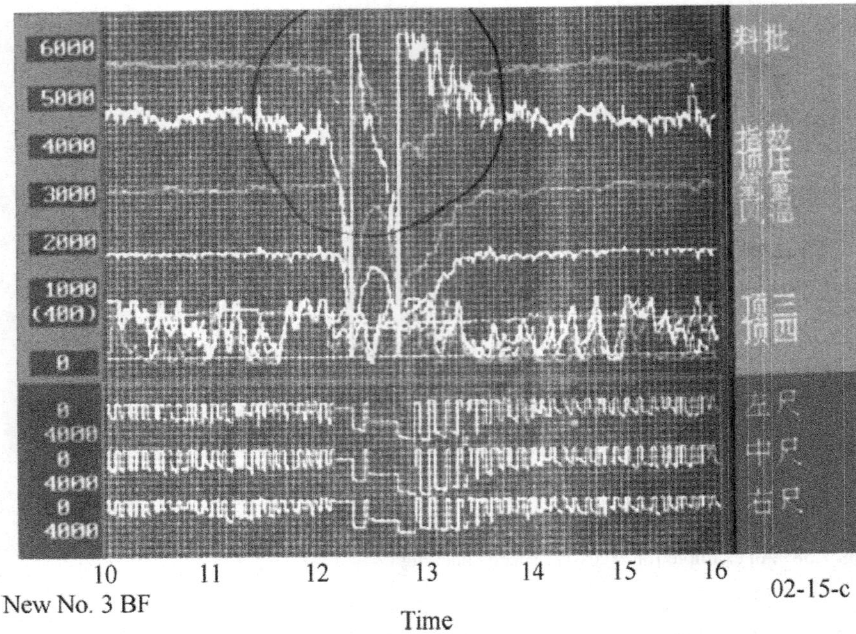

Fig. 7.39 Re-blast too much after checking

7.4.3 Checking at Low Furnace Temperature

It is easy to slag blooding when the furnace temperature is low, especially when the quality of coke is poor. Figure 7.40 shows an example of checking and re-blasting. At 14:55, hot metal contained 0.08% silicon; at 16:10, try to stop the blast and renew the tuyere, because the furnace temperature was too low, and the quality of coke was very poor, worrying about blowing pipe slag blooding, so operation was very difficult. Figure 7.40 shows that from 16:20 to 17:00 (40 min) there were 5 times of checking to completely stop blast operation. It was also very difficult to re-blast after renewing the tuyere.

7.5 The Smooth Operation Index of Blast Furnace

7.5.1 The Furnace Condition in Smooth Operation

The degree of the furnace condition in smooth operation is the main indicator to measure the progress of blast furnace smelting. When the furnace condition is in smooth operation, the stock descends uniformly and stably, which is reflected on the stock rod. The interval between adjacent charges is approximately equal, and the

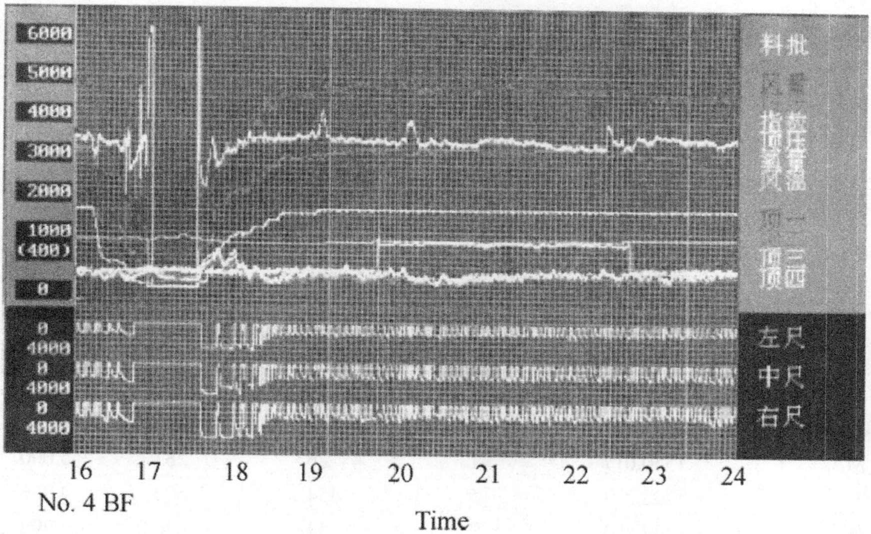

Fig. 7.40 Checking at low temperature

depth of the stock line is substantially uniform. The variation of smooth operation must be acted out on the stock rod. When furnace condition is not smooth, the motion of the stock rod might divide into two cases: One is that the stock rod suddenly falls sliding to the depth over the normal depth of 0.5 m, which is customarily called "slip"; in another case, the stock rod suddenly stops to move and its record also does not drop any more showing a horizontal line, which is called "bridging." When the bridging happens, the normal movement of the stock would be destroyed, the stock cannot descend normally, and the smelting process of blast process would greatly slow down. Based on the above analysis, the times of the slip and bridging could be used as indicators to measure the smooth operation condition of the blast furnace. And according to their influence on the smooth operation of the blast furnace, a different "weight value" could be used to calibrate. According to practical experience, the weight value of the bridging is determined to be 3, and the weight value of the slip is determined to be 1. Therefore, the proposed smooth operation index of the blast furnace can be expressed by Eq. (7.2):

$$S_x = \frac{10}{1 + \frac{T_L + 3X_L}{\mathrm{nd}}} \tag{7.2}$$

where *SX* is smooth operation index; *TL* is slip times within calculating days; *XL* is bridge times within calculating days; *n* is number of BF; *d* is calculating days.

Table 7.4 shows the production data of 27 ironworks in China in 1997. According to the results calculated by Eq. (7.2), we can see the smooth operation conditions of blast furnaces in different ironworks in China from Table 7.4. If the smooth operation index is greater than or equal to 8, the blast furnace smooth operation is in

Table 7.4 Comparison of smooth operation among 27 iron-making plants in China in 1997

Company name	Annual output, kt	BF number	Bridge times	Slip times	S_X
P	3366.465	4	35	4	9.036
S	7463.451	5	0	142	9.278
X	166.669	2	22	12	9.102
M	1642.025	3	7	111	8.924
B	7857.401	3	0	147	8.817
T2	1711.5	2	14	118	8.203
BT	4071.158	4	46	210	8.075
A	8302.108	11	230	444	7.798
T	2001. 985	4	111	200	7.326
J	1960	6	284	116	6.935
MA	1550.017	1	26	287	5.000
T1	1728.273	5	454	502	4.966
JU	1344.803	2	54	629	4.799
H	2552.355	6	432	1300	4.576
XN	764.803	4	945	448	4.447
K	1083.726	4	96	1547	4.431
N	1077.466	4	524	230	4.403
SM	772.225	3	350	447	4.225
AN	1869.487	5	449	1171	4.202
TH	1328.224	5	598	546	4.098
B2	2418.518	3	493	101	4.093
LENG	265.485	4	773	384	3.507
W	246.557	3	309	1227	3.370
L	1008.895	4	832	382	3.365
CN	752.122	4	1004	628	2.863

good condition. If it is less than 5, the blast furnace smooth operation condition is not good. If it is between both, the blast furnace smooth operation condition is medium. The data in Table 7.4 may have some errors, probably because some ironworks have inaccurate statistics. In general, the possibility of missing notes might be more than multiple notes. Therefore, the actual smooth operation condition may be worse. For these ironworks, not smooth operation might be also a special "wealth" or a "gold mine" to be developed. Some of them are committed to the development of such "gold mines" while improving and strengthening the smelting process. As a result, the production conditions rapidly improved, both the production and consumption harvested, and the smelting costs had fallen sharply [9].

7.5.2 To Compare the Smooth Operation Level of the Ironworks with Index S_X

There were 11 blast furnaces in AN Steel in 1997. Although workers have the same technical level and management experience, the smooth operation conditions of these blast furnaces varied differently, due to different raw materials and equipments. Figure 7.41 shows the smooth operation index of each furnace in 1977. The most of the blast furnace smooth operation was in good condition, and only a small part of the furnaces were in bad condition.

The development of Anyang Steel was very fast in recent years, but its smooth operation condition had a downward trend. Figure 7.42 shows its smooth operation index in recent years [10].

Table 7.5 shows the smooth operation index of blast furnace in the ironworks of

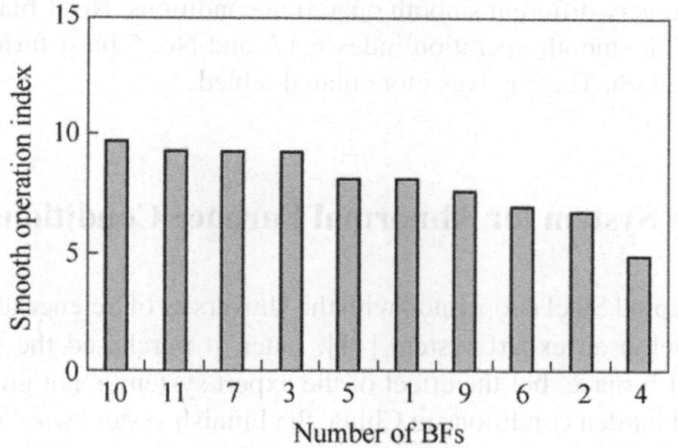

Fig. 7.41 BFs' S_X of AN Steel

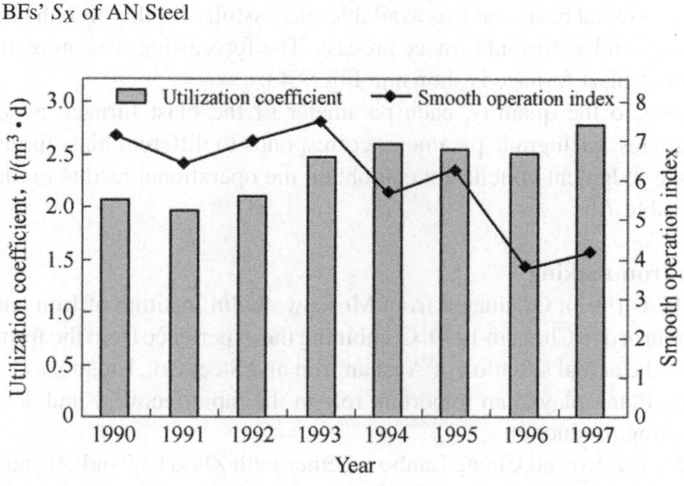

Fig. 7.42 Trend of S_X of Anyang Steel's BFs

Table 7.5 BF S_X of iron-making plant of Nansteel

Company	Number	Time	Days	Index	Slip	Bridge
Nansteel	Total	1997	365	4.403	230	524
	1	January–September, 1998	243	6.47	23	36
	2	January–September, 1998	243	5.97	14	50
	3	January–September, 1998	243	5.32	31	61
	4	January–September, 1998	243	5.94	37	43
	5	January–September, 1998	243	3.06	76	113
	Total	January–September, 1998	243	5.27	181	303

Nansteel. The smooth operation index of blast furnaces was 4.403 in 1997; from January to September in 1998, the average index was 5.27, which was a big step of progress, although there was a big gap with 8.817 in Baosteel. The five blast furnaces of Nansteel had very different smooth operation conditions. No. 1 blast furnace was the best one with smooth operation index 6.17, and No. 5 blast furnace, the worst one with index 3.06. The gap was more than doubled.

7.6 Expert System for Abnormal Furnace Conditions

In 1990, the Capital Steel cooperated with the University of Science and Technology Beijing to develop an expert system [11]. Later, it purchased the Finnish expert system of blast furnace, but the effect of the expert system is not good, because of the unqualified burden conditions in China, the Finnish system was "unacceptable." Afterward, in Capital Steel a new system was independently researched and developed on the original basis and was available successfully for the smooth operation of blast furnaces and abnormal furnace process. The forecasting reasoning of bridging sub-system of blast furnace is shown in Fig. 7.43.

According to the quantity, each parameter of the blast furnace is sorted into different grades, each grade parameter corresponds to different blast furnace states, and then the judgment conclusion can obtain the operational results in that year as shown in Table 7.6.

Expert in Iron-making

Cai Bo (1924–1991): Graduated from Moscow, Stalin Institute of Iron and Steel in 1948, he returned to China in 1949. Combining the experience from the former Soviet Union with the actual situation of Anshan Iron and Steel Co., Ltd, he, as Director of Ironmaking Plant, played an important role in the rapid recovery and development of iron-making production.

"In 1956, Cai Bo and Cheng Lanbo together with Zhou Chuandian and Liu Zhen made experiments in No. 5 and 6 blast furnace, respectively, to raise the temperature of dry air", by which time vapor blast had been wildly used in China. At that time,

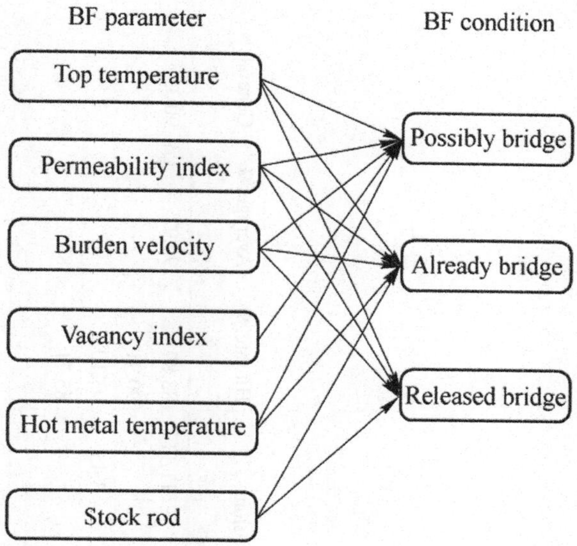

Fig. 7.43 Subsidiary system structure of bridge

Ye Zhupei, a famous metallurgist, put forward the theory of "Three High," which was recognized by blast furnace experts. Ye's papers on high humidity are published not only in China but also in the "Collection of Commemorations of Academician Balkin's 70th Birthday" published by the Soviet Academy of Sciences. The result of the experiments had profound implications for the operation of the blast furnace.

Marshal Nie Rongzhen (in charge of China's scientific development program) commented on Cai Bo's paper: "It is very important that you summarize the scientific research on iron-making in the past 30 years. This is the first systematic scientific research summary of industrial departments that I have seen" [12].

After 1984, together with academician Wang Zhixi and other predecessors in the metallurgical field, he studied the development strategy of China's iron and steel industry, proposed the construction principle of "Big Ship, Big Port and Big Factory," and found the correct development direction of China's iron and steel industry.

Cai Bo deserves to be called "the founder of new China's modern iron-making technology" [13].

Table 7.6 Hit rate statics of BF condition prediction

Abnormal	Month	Prediction times	Hit number	Mistakes		Omissions		Hit rate, %	Average, %	General, %
				Times	%	Times	%			
Blockage/bridge	March	9	8			1	11.11	88.89	90.97	90.35
	April	11	10	1	9.09			90.91		
	May	29	27	1	3.45	1	3.45	93.10		
Abnormal-flow channel	March	7	6	1	14.29			85.71	89.73	
	April									
	May	16	15	1	6.25			93.75		

References

1. H.I. Mahaneke, Regularity of mixture ingredients. Steel **10**, 874–880 (1948)
2. B.K. Grugenov, Steel **1**, pp. 20–23 (1952); **9**, pp. 771–773 (1956)
3. H.H. Barbarekin, Steel **4**, 289–291 (1959)
4. Editorial Board of Yang Yongyi Papers Collection, in *Yang Yongyi Papers Collection.* (Metallurgical Industry Press, Beijing, 1997), pp. 66–78
5. The National Conference on Blast Furnace Production Technology was formulated in 1955, in *Basic Rules for Technical Operation of Blast Furnace in China.* (Heavy Industry Press, Beijing, 1955), p. 16
6. Z. Zhang, D. Liu, Treatment and operation practice of blast furnace inclined burden for Tongcai No. 2 BF. Iron-Making Exchange **3**, 36–40 (2011)
7. X. Yao, Views on the "supporting force" in the process of blast furnace charging and bridge. Iron-Making **5**, 47–51 (1990)
8. M.I. Ostroukhov, L.I. Snapbil, in *Exploitation of Blast Furnace Utilization.* (Metallurgy, Moscow, 1975), 69–129
9. Y. Liu, Suggestions on analyzing the progressive state of blast furnace by smooth operation index. Iron-Making **4**, 50–51 (1998)
10. Q. Dou, et al., Iron Steel **32**(Supplement), 277 (1997); Iron-Making **3**, 55 (1998)
11. T. Yang, Y. Liu, et al., Development and application of smelting expert system for No. 2 blast furnace at Capital Steel Corporation. Iron-Making **4**, 50–51 (1998)
12. B. Cai, Thirty Years of Iron-making Research in Central Iron and Steel Research Institute (1982)
13. C. Zhou, B. Cai, *Founder of Modern Iron-making Technology in New China, Zhou Chuandian Collection*, vol. 3 (Metallurgical Industry Press, Beijing, 2001), pp. 545–558

Chapter 8
Precaution and Disposition of Furnace Breakage

Abstract The damage of cooling devices is the barometer for people to observe the damage of furnace body. Once the cooling devices are burned, the surface is deformed and cracked, gas and flame will be ejected if it occurs in the stack; if it is in the belly of the furnace, coke and slag will be ejected. At that time, if water leaks into the furnace in an instant, a large amount of water gas would be produced and thus leading to explode, which is enough to destroy the furnace itself. If it occurs in the hearth, the hot metal will outflow from the hearth, leading to destructive, unpredictable explosion when meets with water.

8.1 The Breakage Is a Kind of Process

For the blast furnace, the process from corrosion to burning through is different in time. Here are two typical cases: A 4550 m^3 blast furnace in Japan was in major capital after 28-year safe operation. The average tapping volume of the blast furnace first generation is 21,000 t/m^3. The longest time for blowing out is seven days [1]; another blast furnace met with an accident of breakage only after two-month operation. Blast furnaces in China now are able to maintain a good operation for 10–20 years. The realization of furnace longevity involves systematic engineering with multi-process and disciplinary and is also a shared goal for iron-making workers. It is not only a prerequisite for a stable operation to maintain a good furnace type or good operation type, but also a necessity for obtaining excellent production technical indicators.

8.1.1 The Location of Breakage

It is a process for furnace to be burnt. Times are needed for lining to be corroded. As the lining being eroded, the hot metal gradually approaches stave cooler. The temperature difference of inflow and outflow water becomes larger and larger. Therefore, the heat flow intensity that goes through the stave cooler also becomes stronger. Figure 8.1 shows the actual measured result of Capital Steel No. 4 blast furnace

Y. Liu, *The Operation of Contemporary Blast Furnaces*,
https://doi.org/10.1007/978-981-15-7074-2_8

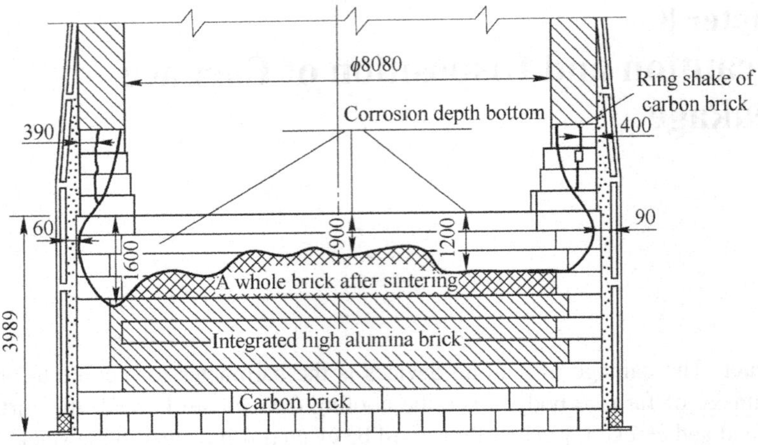

Fig. 8.1 Corrosion of Capital Steel No. 4 blast furnace (first generation)

which has been overhauled. After ten years and three months' operation, the bottom brick is eroded for four layers with the erosion speed of 13 mm/months. The worst is that it is only 60 mm distance away from the furnace second-layer stave cooler.

The No. 4 blast furnace (the second generation) was put into operation on June 4, 1983. After 30 months, the hearth was seriously eroded. Local heat flux intensity is higher over 15,000 kcal/(m^2 h) (62,800 kJ/(m^2 h)). On March 5, 1986, the second layer of stave cooler was burnt.

The tapping speed of an iron notch of furnace No. 4 is very quick (the frequent utilization coefficient is between 2.3 and 2.5 t/(m^3 d), resulting to a server corrosion in hearth. Figure 8.2 is the second generation of the No. 4 measuring furnace, which is actively prepared after burning and repairing, after a year of major repairs.

Wu Steel No. 5 blast furnace (3200 m^3) was in operation since October 19, 1991, and was stopped on May 17, 2007. The operation lasted for 15 years and 8 months. The erosion rate of hearth bottom is about 9.6 mm/month. The thinnest part of the ring carbon is 186 mm from the stave cooler of 2 segments, as shown in Fig. 8.3.

Fukuyama No. 5 blast furnace was in operation in 1973 and was ended in 1983. During that time, the world went through two times of oil crisis. In the downward direction of the No. 3 vent, hearth erosion was serious (Fig. 8.4). Solid iron penetrated into the furnace shell steel plate about 400 mm. Brick lining became metamorphic brittle with an elephant-foot-type erosion shape.

Figure 8.5 is the structure of Hamborn No. 9 blast furnace ceramic cup. It produces higher than 15.5 million tons of furnace cylinder. The corrosion shape is also like elephant foot. Figure 8.5 shows the calculation of the mathematical model of the furnace erosion.

Elephant foot erosion in hearth is also called as "garlic shape" erosion. This area is the thinnest area which is very easy to be burnt through. The weakest point here is where the cooling intensity is the lowest. The cooling intensity is the weakest between the two stave coolers, and therefore, this place is prone to be the breakpoint.

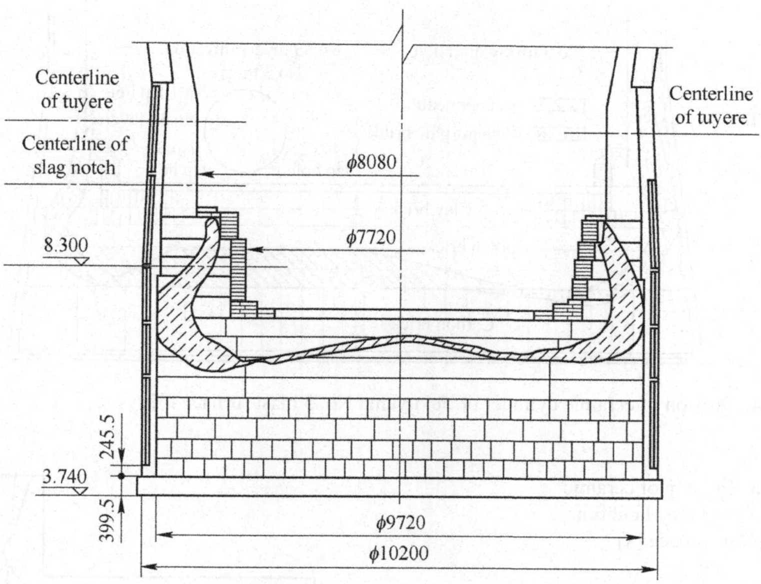

Fig. 8.2 Measured hearth condition of Capital Steel No. 4 blast furnace (1983–1987) in 1987

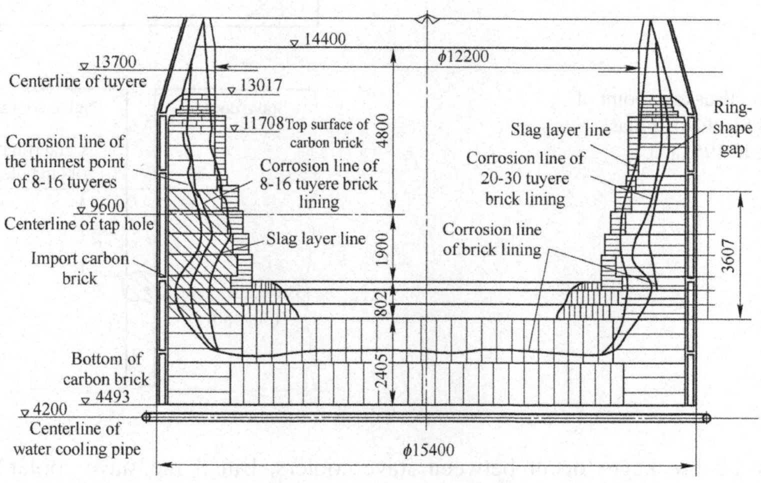

Fig. 8.3 Actual measurement of Wu Steel No. 5 blast furnace major capital [2]

Figure 8.6 is an example of blast furnace production in Capital Steel. In the figure, this point is the "farthest" one from cooling water pipe of stave cooler. It is obvious that with an enough cooling intensity, solidification of permeable hot metal can be realized, and burning out can be avoided; if cooling intensity is insufficient, deep hot metal will go deep and burn through the stave cooler, expanding the range of burning.

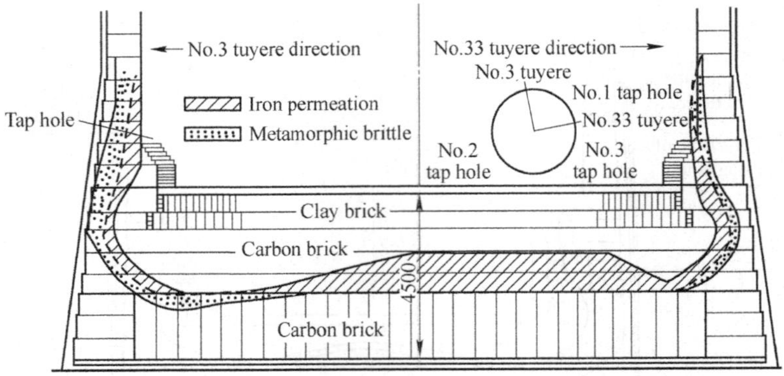

Fig. 8.4 Erosion of ceramic cylinder of Fukuyama No. 5 blast furnace [3]

Fig. 8.5 Erosion of ceramic cylinder of Hutte Hamborn No. 9 blast furnace [4]

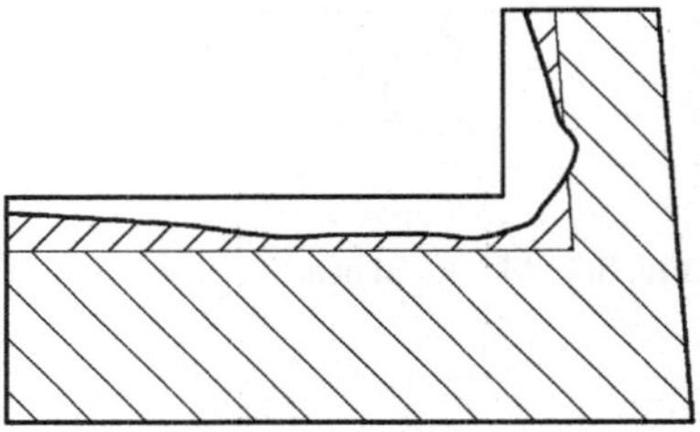

Fig. 8.6 Breakage point of Capital Steel No. 2 blast furnace in 1955 [5]

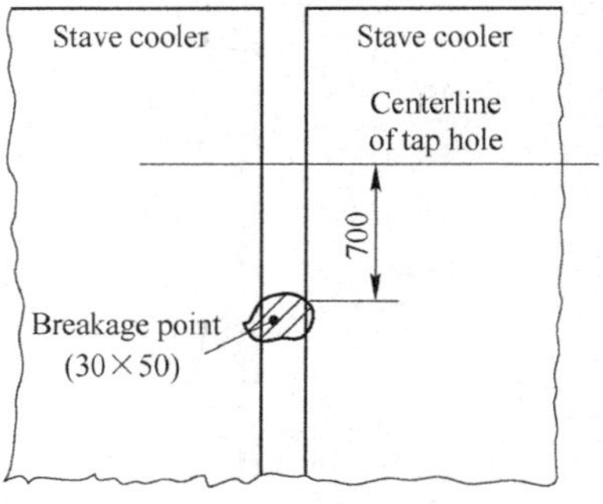

Not all breakages occur between stave coolers. But if the stave cooler has a relatively low cooling intensity, it is most likely to burn through.

According to researches of many blast furnaces, the breakage sites are generally in the elephant foot area of the second section of the stave cooler, correspondingly lies down to the 1–2 m elevation range of the middle line of tap hole. In the past, the breakage point was usually in the tap hole area of comprehensive hearth and bottom blast furnace and the blast furnace with high aluminum resistance and tap hole composite brick.

8.1.2 Reasons for Sudden Breakage

After serious blast furnace breakage, we always feel on edge during operation. Any minute changes in water temperature of stave cooler or temperature measurer will remind us of a comprehensive observation. Whereas, there could also be no sign for the accident breakage in Chinese plants just as the breakage of An Steel No. 3 blast furnace on August 25, 2008.

An Steel No. 3 blast furnace was put into operation on December 28, 2005, with a volume of is 3200 m^3.

The thickness of breakage in hearth is 1914 mm. That means even for a local breakage, a relatively long time is needed. In March 2010, An Steel No. 3 blast furnace blew off for a major capital, which was necessary as the downside area of hearth was in serious corrosion. According to the analyses of An Steel, the reason why there is no indication for this breakage is that there are less temperature measurement points in hearth. Only 24 points were set in the second and third sector of stave cooler, which are the key sectors. In the circumference direction, four points are set in three layers, with a horizontal distance of 12 meters (vertical distance is 1 m between two layers). It is less like to be measured by the points no matter local corrosion or breakage. That is the main reason for obscene of signs.

Outer layer of blast furnace was made by UCAR small piece of carbon brick, famous American product. Insides, “Ceramic Cup” is used for a low thermal conductivity with 15 years designed life. Again, it is designed by Design Institute. It is hard to think that the hearth is burnt through in less than three years. The corrosion process of the lining is obvious, as summarized by An Steel that UCAR high graphite carbon brick is easy to be melted by hot metal, and the cooling ratio area of stave cooler of the furnace is small (0.603), leading to the low cooling intensity and the fast corrosion speed. The flaw in the design structure is another reason for the short life.

Against the accident, timely observation is need, and more emphasis is required during the whole operation. Otherwise, disaster is inevitable.

8.1.3 Reduce Elephant Foot Corrosion

Elephant foot corrosion is proved by a great number of experiments. The reasons lie on the circulation, stress and others matters of hot metal in the hearth. Deepen the depth of salamander is an effective method to reduce elephant foot corrosion. Figure 8.7 is the recommended salamander depth and actual salamander depth based on the thrust view [6] of the blast furnace [7]. From the figure, huge difference can be found in the actual salamander depth if blast furnace in different countries. Even the blast furnace with same volume, the depth varies greatly. Blast furnace with a volume above 1000 m^3 has different salamander depth from 1 m to more than 3 m. Table 8.1; Fig. 8.8 show the ratio of the recommended salamander depth to the diameter of the hearth. As you see from Table 8.1: The ratio of the recommended salamander

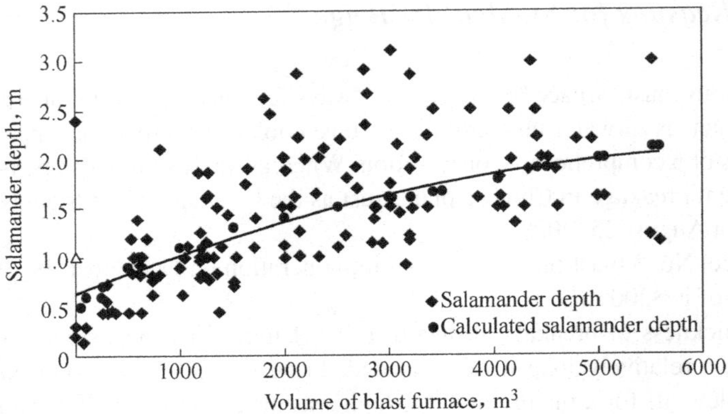

Fig. 8.7 Recommended salamander depth and actual salamander depth

depth of small blast furnace to the diameter of the hearth is greater than or close to 0.2; for blast furnace with a volume larger than 500 m^3, the ratio of the salamander depth to the diameter of the furnace cylinder is only 16–14%. Nowadays, the blast furnace is often designed in accordance with a salamander depth of 20% diameter of the furnace, which is suitable for the small blast furnace. However, for blast furnace with more than 500 m^3 volume, the depth is bigger, and a more reasonable depth is about 15%, and many experts advocate 20%. The load in the figure is K/J (the road of burden put into the furnace). It will affect the weight of the pillar but slightly affect the salamander depth.

The hot metal in the position of the salamander depth is relatively calm except for the disturbance brought by slagging, which has an influence on the height of compact deadman. It is confirmed by the test that using isotopes to measure the hot metal of the hearth [8]. Barbarekin and others studied the hot metal drained out of salamander depth of three blast furnaces of MMK Steel located in the former Soviet Union. The result shows that the composition and temperature of hot metal is different in different depths: the lower in position, the lower in temperature. The temperature of hot metal near the tap hole is about 1460 °C, and the temperature of the hot metal below the tap hole (the bottom of the tap hole) for about 2.5–4 m (one blast furnace is about 2.5 m and the other two furnaces about 4 m) is only 1200 °C [9].

The lower the hot metal dropped, the less carbon content is in the iron. A larger amount of carbon is precipitated and formed as a layer of deposited carbon, which acts as a protective layer (Fig. 8.9). Because of a low carbon content, the melting point of hot metal increases, together with a decreased fluidity of hot metal. The protective layer also offers protection for hearth bottom.

The content of hot metal of three blast furnaces aforesaid (before blowing off) is shown in Table 8.2. In hot metal blast furnace, A, when hot metal flow closed to the bottom of the furnace, so the content of Si in blast furnace B and C is improved for a better fluidity. Three blast furnaces then were blow off in order to collect sample, which were used to test composition and temperature, as shown in Fig. 8.10.

Table 8.1 Ratio of the recommended salamander depth to the diameter of the hearth

Volume of blast furnace, m^3	50	100	255	500	1000	1500	2000	2500	3000	4000	5000	5500
Recommended salamander depth L_d, m	0.5	0.6	0.6	0.9	1.1	1.3	1.4	1.5	1.6	1.79	2.0	2.1
Hearth diameter d, m	2.2	2.8	4.1	5.8	7.4	8.6	10	11	11.6	13.2	14.6	15.1
L_d/d	0.23	0.21	0.17	0.16	0.15	0.15	0.14	0.14	0.14	0.14	0.14	0.14

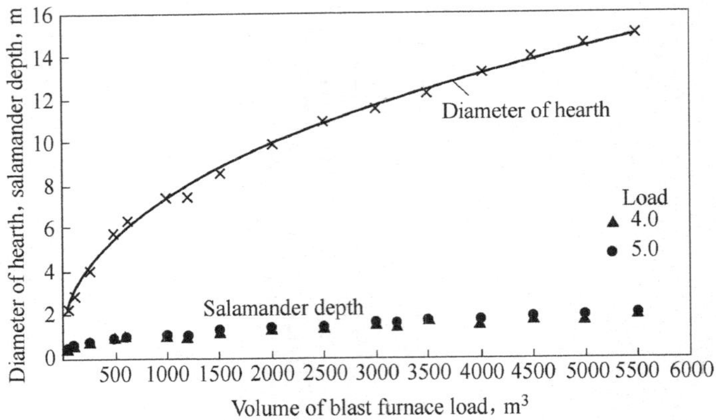

Fig. 8.8 Hearth diameter and recommended salamander depth

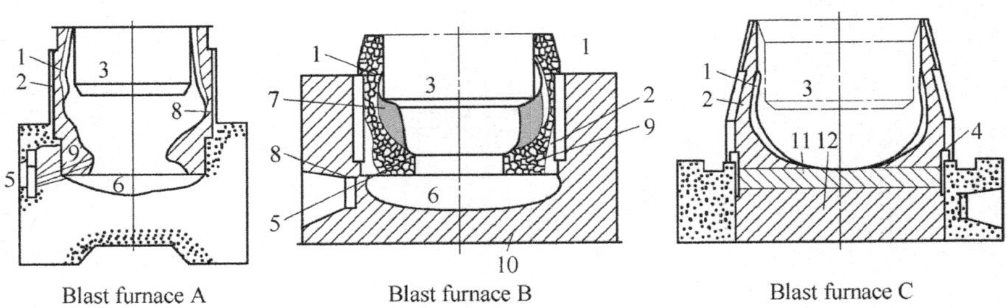

Fig. 8.9 Inner shape of hearth and the position of iron in blast furnace. 1—Slag crust; 2—Brickworks; 3—Middle line of hasp; 4—Upside residual tap hole; 5—Downside residual tap hole; 6—Solid residual iron; 7—Graphite sediment; 8—Carbon brick; 9—Deteriorated and agglomerate brick; 10—Cement; 11—Clay brick laid on end; 12—Clay brick laid flat

Table 8.2 Temperature and content of hot metal

Furnace		A	B	C
The content of hot metal, %	Si	0.85	2	1.51
	Mn	0.4	0.18	0.79
	S	0.037	0.037	0.026
	C			4.35
The temperature of hot metal, °C			1395	1398

When residual iron in the Chelyabinsk No. 2 blast furnace was ejected out from the furnace in 1950, the temperature of the hot metal at the lower tap hole was only 1160–1180 °C [10].

The hearth bottom of tested furnace is not chilled, which means the hot metal in the salamander depth has another function: hot metal at the depth of 2–3 m can low temperature of itself. Therefore, a deeper salamander depth is able to reduce the

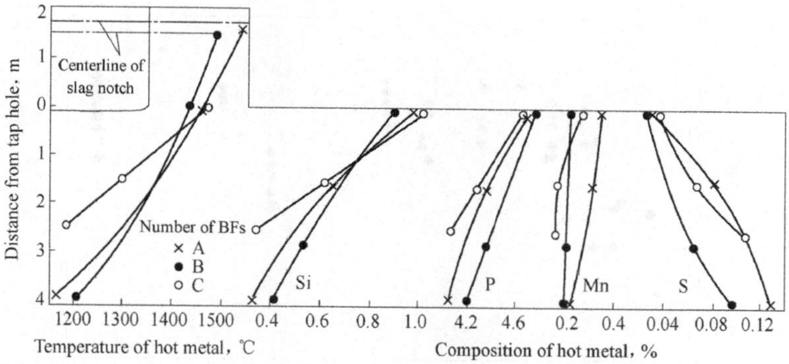

Fig. 8.10 Actual measured temperature of hot metal and the contents

corrosion during circulation of the metal and low the temperature of hot metal as well, and the degree of damage in the bottom of hearth is also reduced.

Nevertheless, the author still recommends a salamander depth with a depth less than 2.5 meters. The reason lies on the corrosion: several-year operation will bring a stronger hot metal corrosion on salamander depth in furnace. Blast furnace today is equipped with cooling facility. Therefore, a serious corrosion which is typical in 1950s is not possible nowadays. It is a waste for the volume to be built with an excessively deep salamander depth. Especially for the large blast furnace, another 1-m thickness in salamander depth means the expansion of blast furnace volume for dozens or hundreds cubic meter. Blast furnace with a salamander depth around 15% completely meets the request of smelting and is also able to protect the hearth for a long-time life. The cooling intensity in elephant foot area and the quality of refractor need to be increased.

8.1.4 Tapping Speed

In the daily operation, operator needs to control the tapping speed. For some large blast furnace, a quick tapping speed leads to "circulation," which will damage the hearth and the wall. Although a thicker salamander depth is able to relieve ring circulation, it still cannot be avoided with a fast tapping speed. Figure 8.11 shows the tapping speed of some blast furnaces.

8.1.5 Which Area Should Be Focused on and Be Tested?

Generally, the breakage of hearth concentrates on the elephant foot area, especially in the hot metal container under the second sector of salamander depth of the hearth. Therefore, the designer should attach importance to that, and more monitoring points

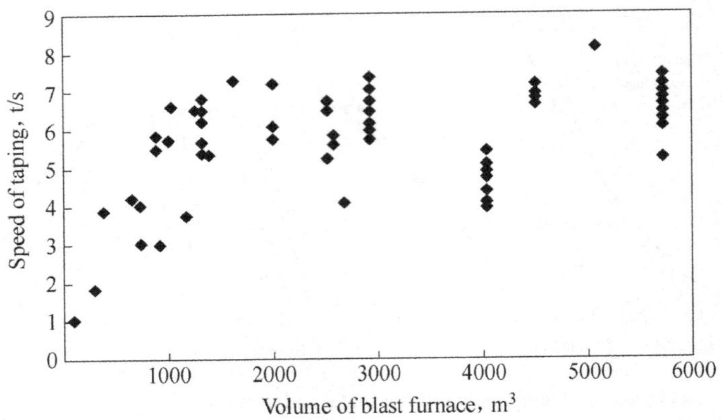

Fig. 8.11 Tapping speed of some blast furnaces

are needed there. It also needs a thermocouple with high-quality, and the process of erection should also be high-level. Operator of the blast furnace should keep records of this area frequently in order to avoid any anomalous changes

We have a variety of methods to monitor hearth bottoms, but the temperature detection is still the major way. The change of water temperature difference of stave cooler in this area is able to reflect the corrosion of hearth bottom or the active degree of hearth in time. The heating process generally is the process of corrosion. The thermocouple of the lining here is more sensitive with special design. Alongside the radical direction from the hearth to the center or alongside the radical direction of the furnace wall, 2–3 points are settled with different distances. Therefore, the depth of corrosion can be calculated with the known distance and different temperatures. The galvanic couple of ring type carbon brick of the hearth is buried with the depth of 200–400 mm in the stave cooler heat wall, and the 2–3 point is settled, resulting in a easy calculation of the erosion change. The number of galvanic couples in elephant foot area is different according to different diameter. In Bao Steel 4000 m^3 level BF, there are $16 \times 2 = 32$ points with a height difference of 500 mm or so. It is available for reference.

8.2 Breakage of Hearth and Disposition

8.2.1 Foreshadow of Breakage

Figure 8.12 shows the record of water temperature difference and intensity of heat current of the Capital Steel No. 4 blast furnace on March 6, 1985.

From Fig. 8.12, it is obvious that the water temperature difference of two adjacent stave coolers ($32^{\#}$-1 and $32^{\#}$-2) reaches 0.9 °C and 1.1 °C. Intensity of heat current is

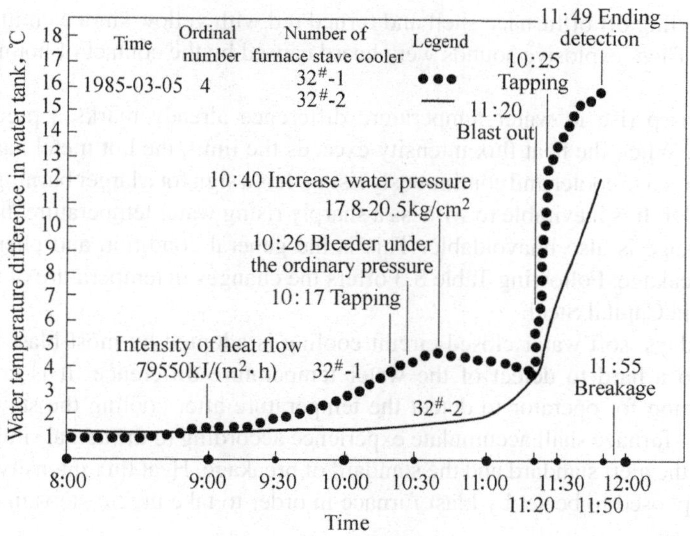

Fig. 8.12 Record of No. 4 blast furnace before breakage

37 kW/m^2 and 45.4 kW/m^2, respectively, (15,500 kcal/(m^2 h) and 19,000 kcal/(m^2 h)). According to the experience of Capital Steel (integrated hearth bottom with hearth bottom cooler), heat flow strength reaches to 35.8 kW/m^2 (15,000 kcal/(m^2 h)). Stave cooler is in a dangerous state and reaches the limit of safe production: the warning temperature; heat flow strength reaches to 43 kW/m^2 (18,000 kcal/(m^2 h)), which marks the limit temperature, and the hot metal reached the surface of the stave cooler. Breakage of stave cooler would occur at any time if no resolute measures are taken.

At 9 o'clock, the water temperature difference of 32$^{\#}$-1 continues to rise and exceeded the limit temperature to 1.2 °C. The intensity of heat fluxes up to 49.5 kW/m^2 (20,720 kcal/(m^2 h)). A high heat flow intensity means that the hot metal has already invaded into the stave cooler. Cooling water is no longer possible to lower the temperature, leading to a continuously rising temperature in stave cooler. Breakage can occur at any time. Urgent measures must be taken at this time. The water pressure of 32$^{\#}$-1 stave cooler should be increased from 17.8 to 20.5 kg/cm^2, which is completed at 9: 45. Whereas, the temperature difference continues to rise, resulting in the early organization of tapping, and the blast is ready to stop. At 10:17, the process of tapping begins, and 32$^{\#}$-1 water temperature difference continues to rise. At 10:20, the temperature is up to 3.3 °C, and operator immediately changes the pressure to the normal pressure and starts the process of bleeder and is also ready to stop the blast. The blast pressure gradually decreased from 2.63 to 1.1 kg/cm^2. Due to a slow flow speed of hot metal, the blast pressure cannot be continuously decreased. As a result, the process of tapping has maintained until 11:25.

The water temperature difference is increased continually, together with a larger breakage expansion of stave cooler. At 11:20, water tender found that a large section of rust peeling off of furnace shell and turned red with yellow smoke emitting from the base. Then, explosive sounds were heard caused by the contact of hot metal and water.

The sharp rise in water temperature difference already marks a precursor of breakage. When the heat flux intensity exceeds the limit, the hot metal has already burned the stave cooler and continuously rising. It is a sign for a larger burning areas of stave cooler. It is inevitable to witness a sharply rising water temperature difference, and breakage is also unavoidable. This is the general condition and phenomenon before breakage. Following Table 8.3 offers the changes in temperature of two blast furnaces in Capital Steel.

Nowadays, soft water closed-circuit cooling is adopted by most blast furnaces, leading to a hard to detect of the water temperature difference. It is convenient and common for operator to detect the temperature after cooling the stave cooler. Each blast furnace shall accumulate experience according to the actual situation and establish the alert standard and the standard of breakage. Heat flux intensity warning line is supposed to be set by blast furnace in order to take urgent steps in the event of breakage.

To avoid breakage, two points are found according to the temperature change of lining or stave cooler:

(1) The continuous increase of temperature indicates that the lining is constantly eroded;
(2) If there is a sudden extraordinary speed of heating, reason should be found and take resolute measures to prevent the erosion of the lining.

8.2.2 Corrosion Speed

The erosion speed of different parts of blast furnace varies greatly with many influencing factors. For the erosion of the furnace bottom, it has a few influencing factors, such as the bottom structure of furnace, lining material, cooling strength, masonry and burden quality, daily maintenance, smelting intensity and other major factors. The following part will mainly discuss the hearth corrosion in respect of corrosion speed.

Capital Steel No. 1 blast furnace was went into production on May 8, 1955, it has a clay brick hearth bottom with no cooling device. Breakage happened on April 14, 1958. It only worked for 1071 days. Hearth bottom corrosion depth was 1640 mm after demolishing with speed of 1.689 mm/d, and the first generation of iron was 1548 t/m^3.

No. 1 blast furnace of the eighth generation was put into operation on the August 8, 1962, and the breakage happened on April 5, 1965. Corrosion speed of hearth bottom of No. 1 and No. 3 blast furnace is shown in Table 8.4. The bottom structure and corrosion condition of the furnace are shown in Figs. 8.13, 8.14 and 8.15.

Table 8.3 Water temperature difference changes of breakage in stave cooler

Date	Order	Volume, m^3	Number of stave cooler		Water temperature difference changes, °C						
Nov. 16, 1989	3	963	$18^{\#}$–1	Time	10:10–11:15	11:19	11:35	11:37	11:38	11:40	
				Temperature, °C	1 (tapping)	Go out of blast	6.6	7.6	8.4	Breakage	
March 5, 1985	4	1200	$18^{\#}$–1	Time	8:00	9:00	10:00	10:30	11:00	11:49	11:55
				Temperature, °C	0.9	1.2	2.8	3.8	3.7	16.2	Breakage

Table 8.4 Corrosion speed of hearth bottom

Blast furnace	Date of blowing on	Date of blowing off	Operation times	Corrosion depth, mm	Corrosion speed, mm/day	Production of first generation, t/m^3	Utilization coefficient, t/(m^3 d)	Brick of hearth bottom
No. 1	May 8, 1955	April 14, 1958	1071	800	0.747	1548	1.445	Chamotte
No. 1	August 8, 1962	April 5, 1965	971	1640	1.689	1544	1.59	Ramming mix
No. 3	May 23, 1959	February 23, 1970	3733	1400	0.375	5479	1.468	High alumina brick

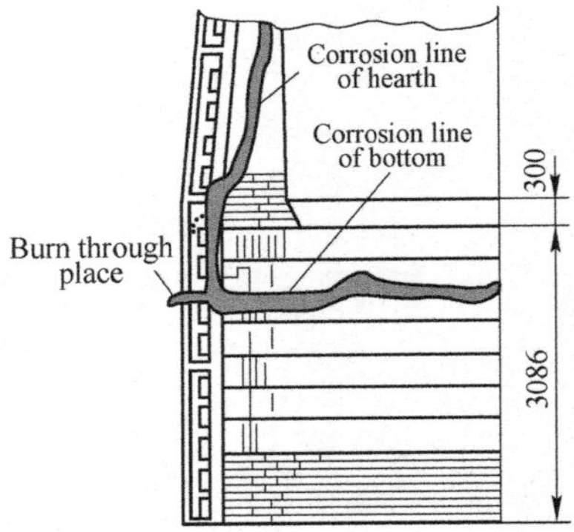

Fig. 8.13 Corrosion condition of Capital Steel No. 1 blast furnace (seventh generation)

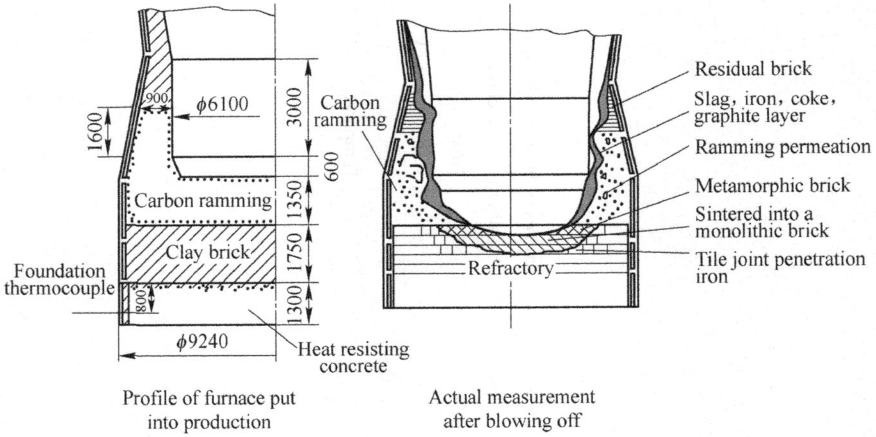

Fig. 8.14 Corrosion condition of Capital Steel No. 1 blast furnace (eighth generation)

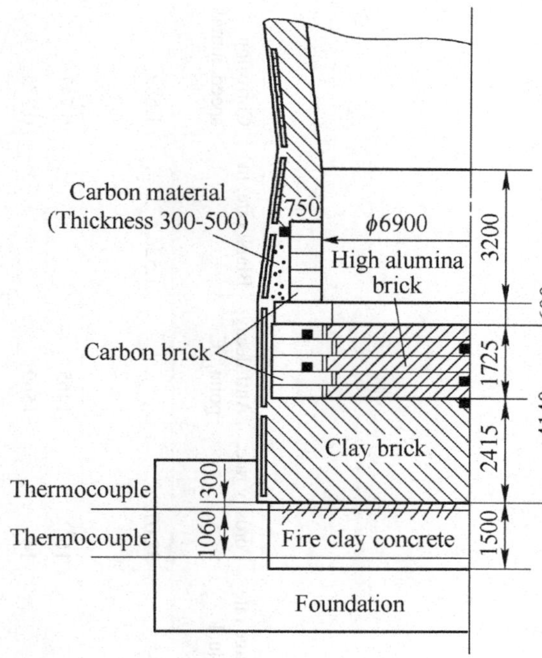

Fig. 8.15 Blowing on condition of Capital Steel No. 3 blast furnace (first generation)

Both two blast furnaces do not have hearth bottom cooling device. The seventh generation No. 1 blast furnace uses chamotte, and after breakage, carbon mash material is used to repair (Fig. 8.14). No. 3 blast furnace was put into operation in 1959. It has a comprehensive hearth bottom of high aluminum brick and chamotte. The specific structure is shown in Fig. 8.15. Table 8.5 shows the properties of three types of hearth bottom materials.

From the table above, in the three types of materials, ramming lining is the worst with the quickest corrosion, which is faster than that of chamotte; high aluminum brick has the best quality, and its life is four times longer than that of ramming lining. The above facts show that the quality of brick lining plays an important role in preventing corrosion.

The corrosion of the lining of the blast furnace above the hot metal is much more complicated than that of the lower part. In the 1960s, Capital Steel has widely adopted isotope detection in order to detect lining. Figure 8.16 is the testing results of furnace. After operating on August 8, 1962, each iron measures the radioactive radiation intensity of hot metal. The first peak occurred on August 17, that is, ten days after the operation as the protecting brick of hearth has been corroded. The second peak occurred on September 7—one month after the operation. The isotope of the bosh was eroded, and the temperature difference of the stave cooler was remarkably increased. The third peak was a severe corrosion of the lining near slag notch. The highest peak occurred on October 19 with iron ray intensity reaching 719 times/min, which conveying the information of the corrosion at the first layer of hearth bottom

Table 8.5 Properties of hearth bottom materials

Category	Al_2O_3, %	Fe_2O_3, %	Refractoriness, °C	Compressive strength, kg/cm^2	Shrinkage rate of residual bricks, %	Porosity rate, %	Anti-decay point, °C	Blank rate, m^3	Corrosion speed, mm/d
Carbon ramming hearth bottom				207 (Room temperature)	0.0149	32.07		1.52	1.689
Chamotte	44.35	1.35	1755	780	0.01	16	1465		0.747
High alumina brick	80–85	1.39	> 1790	> 600	0.12	14	1560		0.375

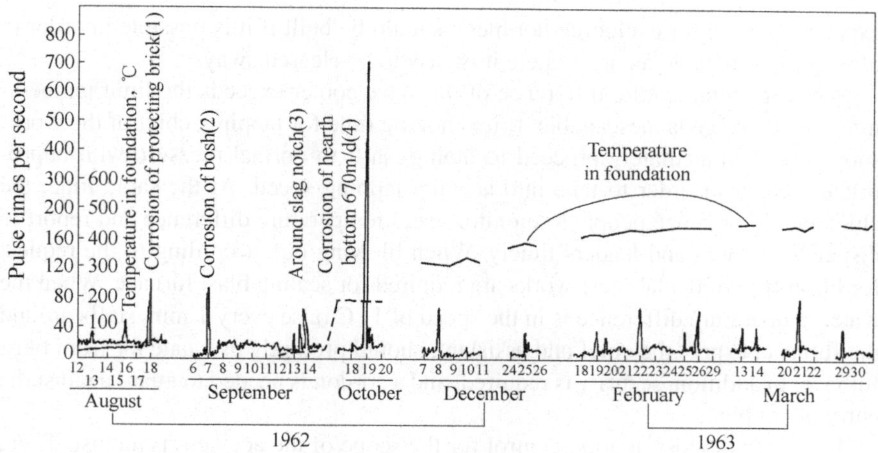

Fig. 8.16 Testing result of Capital Steel No. 1 blast furnace

brick. The isotope buried under the brick might also be corroded. Before September 15, the bottom temperature of the furnace was only 40 °C. But in 17, it jumped to 200 °C. Since then, small spikes of the intensity of hot metal rays often appeared, and the temperature of the bottom also saw a gradually rising.

Isotope detection is timely and reliable, but safety protection and measurement are cumbersome. Therefore, it is no longer in use.

8.2.3 Cut the Loss of Breakage

When the hearth is under the threat of breakage, measures need to be taken in order to prevent breakage. At the same time, in order to reduce losses, preparation should be made in case the breakage happened.

First of all, all obstacles need to be removed. Furnace foundation and the open space below the iron house needed to be cleaned up. And build the retaining wall around the inlet and cable ditch of the groundwater channel around the entrance. Equipment should better be removed, and protective measures need to be taken for another equipment that cannot be moved. Cables hanging under the iron house shall be protected with fireproof and adiabatic materials. The furnace foundation and underneath part of the iron house shall be kept dry. If there is water, it shall be excluded. For blast furnace at the last phase, it is difficult to avoid water nearby as plenty of spray water flowing along the furnace body, bringing water everywhere around the foundation. Under the circumstances, the water should be removed in the direction away from breakage with pipes or through dark ditch. Try to avoid the encounter of hot metal and water, which may result in explosions. A ditch which is exclusively used for excluding hot metal should be built if it is possible in order to blow off hot metal to an area where it is easy to be cleared away.

When the temperature difference of the stave cooler exceeds the limit temperature, the breakage is inescapable. If it is not the time for tapping, chief of the workshop should immediately proceed to taping, shift to normal pressure with appropriate bleeder in order to take into account tapping speed. At the same time, the chief should organize people to monitor water temperature difference and report to dispatching room and leaders timely. When blowing out, according to the request for blastoff period, elaborate works are required for sealing blast furnace. When the water temperature difference is in the speed of 1 °C raise every 1 min, staffs around the furnace, expect the chief and assistant, should promptly evacuate far from blast furnace. In addition, security is required, and a cordon is needed to guard against the entry of people.

To reduce the loss, a good control for the scope of the accident is a must. Then, the range of stave cooler needs to be reduced, and an important method is going out of blast as soon as the breakage is detected, especially for the modern large blast furnace. It has a high blast pressure which is in strong destructiveness. If people do not blow down the blast immediately, the area for breakage will easily expand. Due to an expanding area of burden, a large amount of hot metal, slag and burden of high temperature would be expelled like volcanic eruptions, and they may spread to everywhere and destroy everything. So that the disaster is much heavier from a simple breakage to a huge fire and produces incalculable losses. Excellent foreman, in the face of breakage, should be calm and prudent. He or she should organize the evacuation of personnel timely and try to shrink the breakage area of stave cooler and minimize the outflow of hot metal and slag. If the area of breakage is very small, it is impossible that the burden would flow out from the breakage hole of the hearth containing hot metal, only hot metal and a small amount of slag would flow out.

The condition that the burden sprayed underneath the iron notch is a result of bucking and blastoff, which should be used as a lesson and be avoided resolutely.

8.2.4 Recovery After Breakage [11]

Furnace corrosion is generally in the shape of "elephant foot" and occurs mostly between two staves coolers. Here is the weak point for cooling, from which breakage usually starts. Therefore, the breakage generally happens on two staves coolers simultaneously. Figure 8.16 is the schematic diagram of breakage of Capital Steel No. 4 blast furnace on March 5, 1985. Capital Steel the second-generation No. 4 blast furnace, with volume of 1200 m^3, has 72 block-staves coolers in the two layers of hearth. Each stave cooler has two inlets and two outlets (two in and two out) in order to have a better performance of cooling. It has two slag notches and one tapping hole. It was put into operation on June 4, 1983.

Ten months after operation, the furnace wall near the tap hole is corroded because of the tap hole with low position. The water temperature difference of stave cooler underneath the tap hole increased significantly. Up to April 1984, the water temperature difference exceeded 1 °C, which was close to Capital Steel's warning heat flow intensity of 15,000 kcal/(m^2 h) (62,800 kJ/(m^2 h)). Cooling water pressure should be enhanced immediately from 3.4 to 10 kg/cm^2, and the water temperature difference should be reduced to 0.8 °C. But the hearth wall has already been corroded. After another three months, water temperature difference of 47 outlets exceeds successively more than 1 °C, accounting for 33% of the second layer of stave cooler water outlet. It meant that the lining in the second-layer stave cooler has been corroded in a wide range.

After that, water temperature increase phenomenon becomes more common in the second stave cooler. Although a lot of measures have been adopted, stability cannot be maintained in a long-term round. The lining continues to be corroded. By the end of 1985 (about 30 months after operation), the corrosion of the lining near the second-layer stave cooler was in a quite serious situation. The temperature difference of the stave cooler fluctuated with tapping. The water temperature difference dropped in part of stave cooler during tapping. About 30 min after tapping, the temperature began to rise again. That indicates the hot metal has been close to relative the stave cooler. In the January 1986, the water temperature difference did not decrease in some stave coolers even during the process of tapping, and it took 1 h for the temperature too low to an acceptable level. Finally, breakage occurred on March 5. It started between the stave cooler of the $32^{\#}$ and $33^{\#}$. In the middle of these two stave coolers, it was an area in the shape of elephant foot. According to overhaul judgment in 1987, a 500 mm wide charcoal brick has been very thin or completely corroded. Breakage started from the No. 32nd and expanded to the 31st in a horizontal direction. Part of $31^{\#}$ near the side of the $32^{\#}$ was burnt.

During this breakage, 100 t of iron and a small amount of slag flowed out from the hearth. Although there were some kinds of sounds of explosion near the furnace foundation and the fact of breakage, other equipment was in good condition, and no casualty was caused due to a seasonable evacuation of personnel (Fig. 8.17).

After the breakage of No. 4 blast furnace, people were organized to clear up the site, while the burning stave cooler was judged on the basis of the inlet and outlet conditions. The burnt area of shell contained three stave coolers, including No. 31, 32, 33. Therefore, it is concluded that all these three stave coolers should be replaced. Spare parts of these three stave coolers should be prepared, while fire clay should be organized to be made in order to produce shell, fire clay brick and filling stuffs.

Cutting the shell according to position of the three stave coolers, and then pulling down the burnt stave coolers. In general, the burnt stave cooler is hard to pull down as a whole. And most of that need to be separated into few pieces and then be taken off separately.

After removing the stave cooler, the residual brick and coke are exposed, which should then be watered slightly on the coke surface in order to clean up the surface and outside of the coke. Clean up furnace along the residual brick and to maintain a depth of a brick. Put new bricks in absence area, then install on the new stave cooler, and sooth it with plastic fire clay. Finally, the shell is welded well.

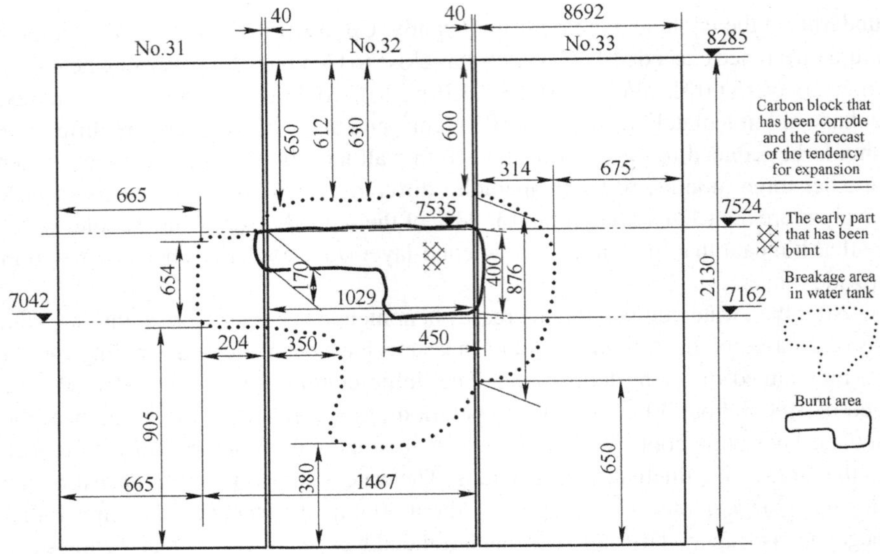

Fig. 8.17 Schematic diagram of breakage of Capital Steel No. 4 blast furnace

In Japan, a 3850 m^3 blast furnace, put into operation on January 29, 1973, was burnt through on June 3, 1977. Breakage position and recovery situation are shown in Fig. 8.18 [12].

It took 13 days to repair, and on June 14, it started blasting. A more detailed report is shown Table 8.6.

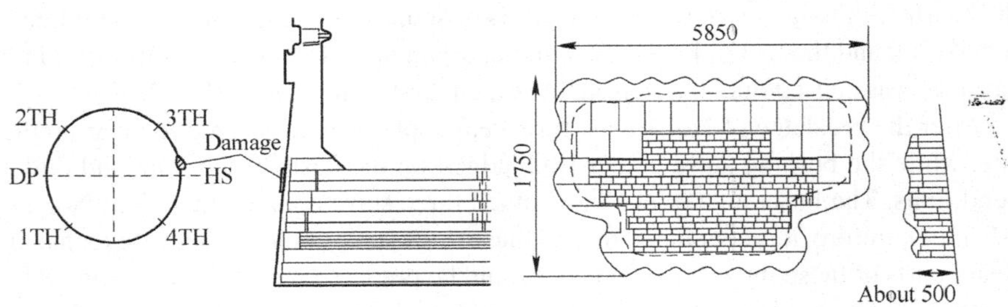

Fig. 8.18 Breakage point and reparation work of wielding

Table 8.6 Schedule of reparation and work description

Date	3	4	5	6	7	8	9	10	11	12	13	14
Content	1	2 3		4	5	6	7	8 9	10			
	16:00 breakage 1. Repaired the water pipe 2. cleared up ejection		3-4 cleared up ejection	4-5 opened the shell 5-6 started the brick work in the lower end		6-7 lower end of the shell was wielded		7-8 started the brick work in the middle part	8-9 started the brick work in the upper part		9-10 the shell was wielded and the water pipe of hearth bottom was repaired on June 14th, it started blasting	

8.3 Breakage of Hearth Bottom

8.3.1 *Description of the Accident*

The accident happened in September 1962. It was the No. 2 blast furnace of Shijingshan Steel Plant (predecessor of Capital Steel). The volume of it is 516 m^3. In 1953, it accepted major capital, and in 1960, it was stopped for reparation due to accident during the operation: The bottom was not cooled with the lower part of clay brick and the upper ramming (Fig. 8.19).

After the night shift, it was found that compared with calculated tapping volume, the amount of tapping iron was less and fewer. Then, the foreman asked the oncoming chief and furnace chief to pay attention to observation. By 11:00, there had been a cumulative 254 t of iron loss (Table 8.7). Gas was found from the cracks of foundation. The gas contained more than 95% of CO_2 after sampling analysis (Table 8.8).

Judging from the gas composition, it is clear that the gas is not from the inside of furnace. It is the CO_2 of the pebble after decomposition in the concrete of furnace foundation. At this time, the concrete of furnace foundation and white smoke blew out from cracks (Fig. 8.20), indicating that the hot metal has entered the foundation.

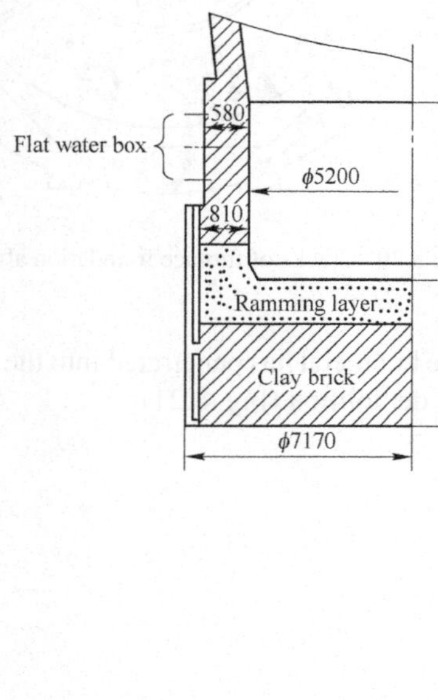

Fig. 8.19 Ramming hearth and bottom of No. 2 blast furnace during middle repair

Table 8.7 Actual tapping volume [13]

Tapping time	Batch of interval materials	Calculated tapping volume, t	Actual tapping volume, t	Difference, t
0:44	23	72.45	59.75	−12.7
3:30	23	72.45	43.95	−28.5
6:00	23	72.45	41.05	−31.4
8:41	26	92.4	29.6	−52.8
11:20	25	79.25	40.2	−39.05
14:02	24	76.1	46.5	−29.6
11:46	40	94.5	33.9	−60.6
Sum	174	549.4	295.25	−254.25

Table 8.8 Gas composition (%)

Position	CO_2	O_2	CO	H_2	CH_4
No. 6 plinth	95.4	0.1	2.0	4.0	0.3
No. 9 plinth	95.5	0.3	1.5	3.8	0.5

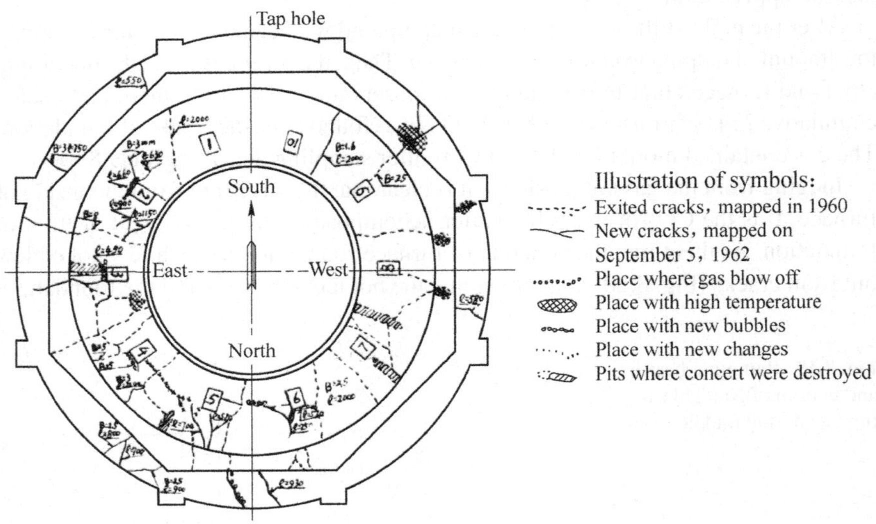

Fig. 8.20 Cracks of furnace foundation after the penetration of hot metal

The hot metal has penetrated into the foundation of 2.5 m, weighing about 250 t after the demolition (Fig. 8.21).

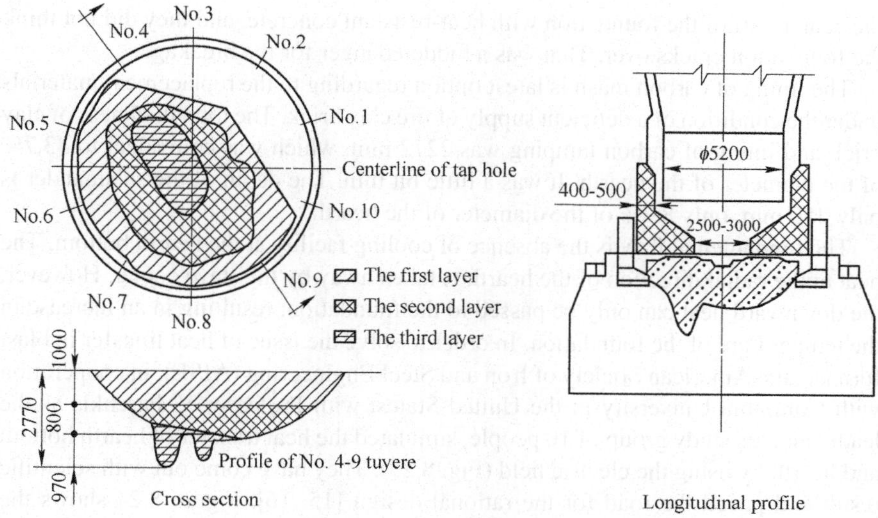

Fig. 8.21 Cross section and longitudinal profile of the breakage area in the bottom of the furnace

8.3.2 The Reasonable Structure of Furnace Foundation

During 1950s–1960s, breakage at the furnace foundation occurred frequently [14] because the foundation did not be cooled. Ten years before the first breakage in China, it also usually happened in the former Soviet Union [15]. Some blast furnaces were heavily destroyed (Fig. 8.22).

There are three cracks in the above No. 2 blast furnace before the reparation. During the middle-time reparation, operator only replaced the ordinary concrete in

Fig. 8.22 Corrosion condition of the foundation in former Soviet Union

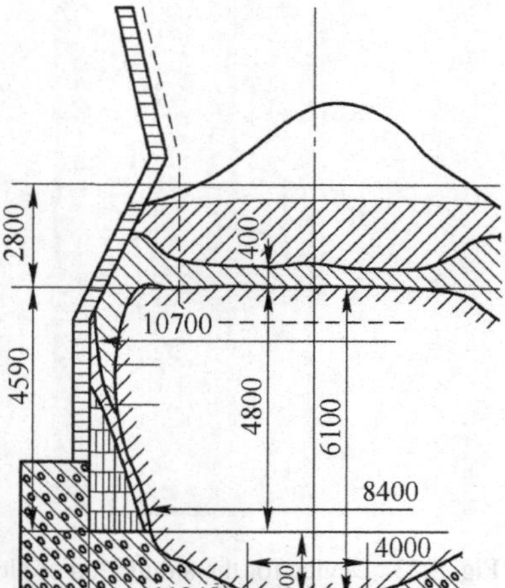

the center part of the foundation with heat-resistant concrete, and they did not think the foundation cracks over. That was a hidden danger for the breakage.

The lining of carbon mash is latest option regarding to the replacement materials under the condition of a deficient supply of fire clay brick. The total thickness of clay brick and lining of carbon tamping was 2272 mm, which was equivalent to 43.7% of the diameter of the hearth. It was a little bit thin. The depth of the salamander is only 450 mm, only 8.7% of the diameter of the hearth.

The fundamental flaw is the absence of cooling facility at the hearth bottom. The heat at the radial direction of the hearth is taken away by the stave cooler. However, the downward heat can only be passed to the foundation, resulting in an increase in the temperature of the foundation. In order to solve the issue of heat transfer in blast furnace, the American Society of Iron and Steel Engineering (AISE), in cooperation with Columbia University in the United States, with Professor V. Paschkis as the leader and his study group of 10 people, simulated the heat transfer of hearth bottom and hearth by using the electric field (Fig. 8.23). They have come out with scientific result and paved the road for the rational design [15, 16]. Figure 8.24 shows the temperature distribution at hearth bottom when equipped with and without cooling facility. The difference of temperature isotherm at the bottom is obviously seen from the figure. Without cooling device, the isotherm still extends downward even it was 600°F and 10 feet below the bottom; the same isotherm has become horizontal (Fig. 8.24) for a bottom with cooling facility when it is 6 feet below the bottom. The simulation results show that the temperature distribution of the bottom with cooling device is flat, while it is in the pot shape for bottom without cooling facility. Since then, the process to cool down the bottom began to prevail, which is an outstanding contribution for the researches.

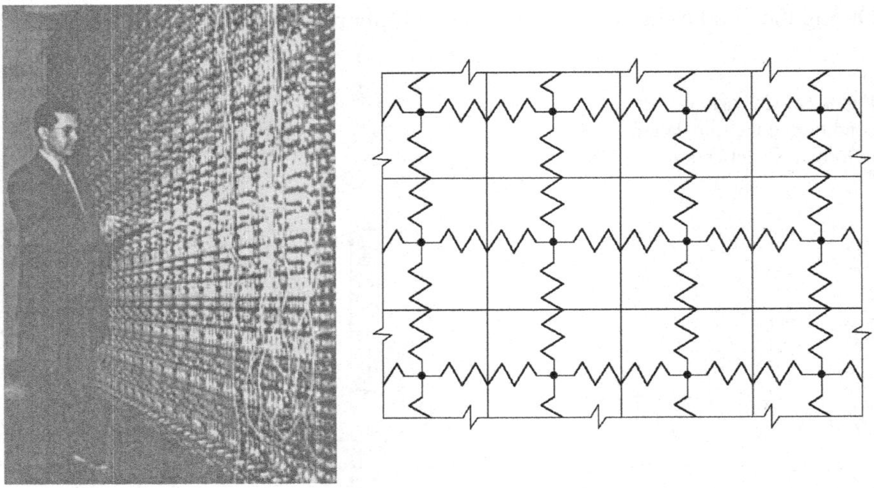

Fig. 8.23 Device for the simulation of electric field at the hearth bottom and its theory

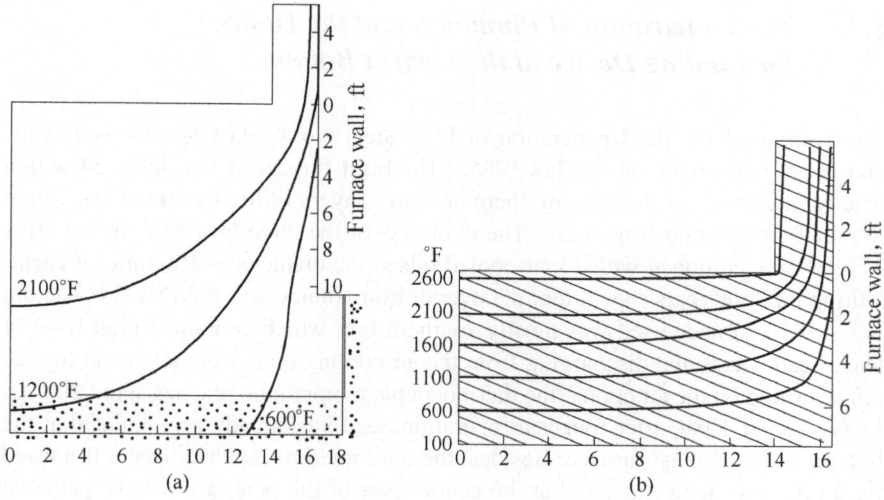

Fig. 8.24 Temperature distribution at hearth bottom when equipped with and without cooling facility

From the point of view of heat transfer, they gave three kinds of bottom design drawings (Fig. 8.25), but none of these have been popularized. Professor V. Paschkis and his group thought too much about heat transfer efficiency but ignoring the inconvenience in construction. This is the difference between physics professors and engineers.

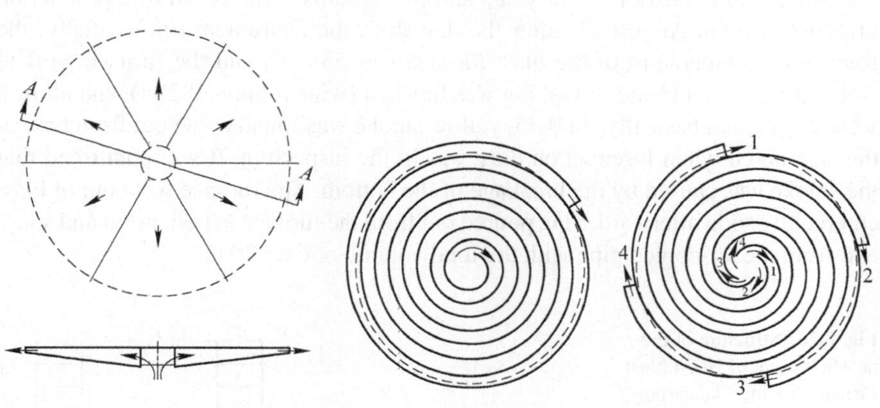

Fig. 8.25 Cooling system (left), single-screw and four-screw cooling system of the No. 6 tunnel of hearth bottom

8.3.3 The Penetration of Plumbum and the Destroy for Cooling Device at the Hearth Bottom

The volume of the third generation of Lian Steel No. 4 blast furnace is 323 m^3 and is put into operation in May 1985. "The blast furnace is not designed with a lead-remove hole; at the bottom, there are three layers of high alumina brick with the thick of 845 mm (Fig. 8.26). The thickness of the three layers of carbon brick is 1208 mm, equipped with 13 air-cooled tubes (the diameter is 150 mm for each). During the four years' operation, the overall iron volume was 6,260,000 t. The ore No. 4 blast furnace used was mostly southern ore, which contained high level of lead. Lead was found discharging from the air cooling pipe, iron notch and tuyere. After turning on the air cooler, the thermocouple temperature reaches about 200 °C. In September 1989, after four-year operation, lead has already accumulated at the bottom of No. 4 blast furnace. Besides, the air-cooled pipes have been left unused for a relatively long time, and at the central part of the pipe, a few have gathered rust. A small amount of lead was expelled from the air-cooled pipe. At that time, the temperature reflected by thermocouple at the foundation was the same as usual (about 200 °C). On September 17, hot metal and slag flowed into the air-cooled pipe along the lead ("channel") because the workers used oxygen to melt the lead in the air-cooled pipe. After breakage, slag flowed to the foundation (a total outflow of 40 t). But it did not cause a large explosion" [17]. In the September 1997, this kind of breakage also happened to the hearth bottom of No. 1 blast furnace.

"On August 11, 1993, breakage occurred on No. 3 blast furnace (100 m^3) at Tang Steel First iron-making plant. Since June 9, 1989, the furnace has been put into operation continuously for four years and two months, with a total iron production of 424,400 t. On August 11, after the day shift, the instrument run normally; the foundation temperature of the blast furnace was 550 °C, and the furnace worked well. At 8:25, first round of tapping was finished (with volume of 29 t), and all iron were discharged basically. At 9:35, yellow smoke was found come out from base at the slag notch when foreman on duty was in the inspection. It was confirmed that the smoke was caused by the breakage of the bottom. The furnace went out of blast emergently soon afterward. Slag poured out from the furnace in both north and south and from the air-cooled pipe with the total volume of over 20 t.

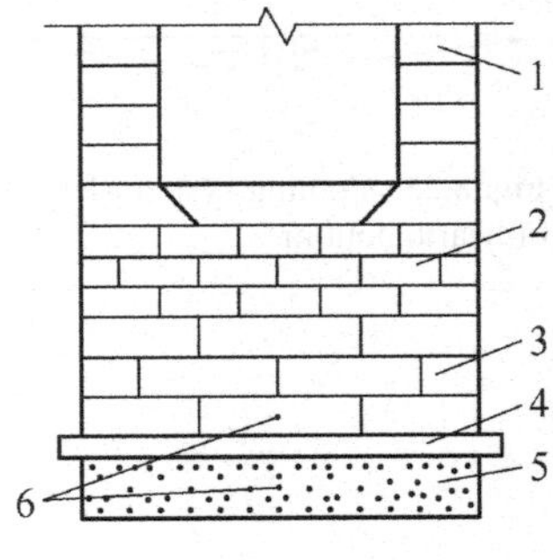

Fig. 8.26 Structure of the hearth bottom of No. 4 blast furnace 1. Ring like carbon block; 2. High alumina brick No. 1; 3. Carbon block; 4. Air-cooled pipe; 5. Foundation; 6. Thermocouple

Fig. 8.27 Diagram of the breakage at the bottom

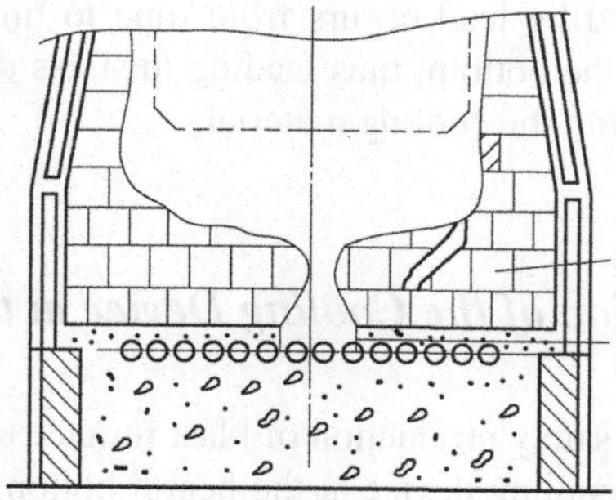

After the middle-time repair, the furnace went in a good condition. A few months before the accident, the blast furnace went normally with no collapse and suspension. At the beginning of August, several production indicators were even in better performance than that of other factories of similar scale. The utilization coefficient was 3.0, coke ratio was 579 kg, and the blast temperature was 1023 °C. Measurement thermocouple at the hearth bottom of No. 3 blast furnace was sensitive and in a normal operation. Before breakage, the bottom temperature was only 550 °C, which was far from the limit value of the plant's technical regulations" [18].

The No. 3 blast furnace is equipped with a self-baking carbon brick integrated air-cooled bottom. The top layer was high aluminum brick, and the bottom is four layers of self-baking carbon brick. The total thickness of the hearth bottom was 1734 mm. Fourteen air-cooled pipes (with diameter of 95 mm × 10 mm) were built between the bottom and heat-resistant base. It also was equipped with draft fan-forced air-cooled device. A thermocouple for temperature measurement was installed above the air-cooled pipe.

"The breakage channel basically locates in the center of the hearth bottom along the vertical direction. Its shape is shown in Fig. 8.27. The size of upper notch is about 130 mm × 80 mm, while the lower mouth about 560 mm × 440 mm. The channel is filled with gray-white residue. When the bottom is burned, the slag in the furnace is gushing out of the chaps from the eighth, the ninth and the tenth pipe, and all air-cooled pipe is burnt out except the above-mentioned three air-cooled pipes (the eighth is left with a 1 m residual).

The density of lead is 11.34 t/m^3, which is much higher than that of iron 7.8 t/m^3. Therefore, once the leaded mineral enters into the furnace, the lead, which is heavy, would easily pass through the hot metal and sink to the hearth bottom. For example, if the brick work in hearth bottom is in poor quality or the bricks have micro gaps, the lead will pass through the gap and destroy the bottom. Clay brick with low-quality and non-roasted carbon block or ramming furnace bottom, when under high temperature, would have a higher shrinkage rate. Some could reach to 0.1–0.3% under 1300–1400 °C. That is how lead brings destruction to the hearth bottom.

Breakage caused by lead occurs from time to time. Lead is very close to the cooling device of the bottom, thus leading to stress deformation or even breakage for pipes carried with the cooling material.

8.3.4 *Inspection of the Cooling Device at the Hearth Bottom*

It is important for safety production of blast furnace to have a reasonable structural design. Since the cooling device at the hearth bottom was invented, the frequency of breakage decreases. The reason for the breakage of bottom in Fig. 8.26 is voidance of cooling device owning to the lack of maintaining works. Figure 8.28 shows another example that inspection and measurement works of cooling water at the bottom are undertaken frequently. The bottom cooling device needs to be inspected constantly, and with the measured temperature distribution of the water-cooled pipe at the bottom, operator can be well informed about the corrosion condition of the hearth bottom and can check the working condition of the device as well.

8.4 Ceramic Cup Structure

The longevity design of blast furnace follows two methods: thermal conductivity and thermal insulation method. For thermal conductivity method, it uses lining material with high thermal conductivity. With the help of cooling system, it lowers the temperature of the furnace cylinder hot surface and forms the slag condensation protective layer, thus slowing down corrosion speed of lining. The insulation method, on the contrary, takes advantage of the lining featured low heat conduction and anti-corrosion to reduce the heat loss in the furnace. So, it is also called refractory method.

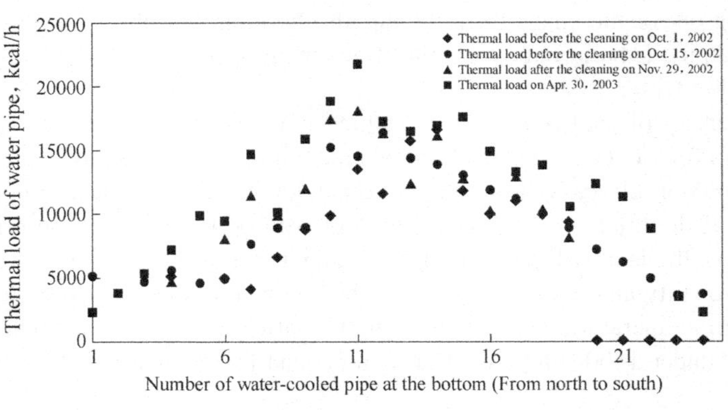

Fig. 8.28 Thermal road distribution of the cooling condition of hearth bottom of Capital Steel No. 1 blast furnace

In the 1980s, French proposed 'ceramic cup structure,' which actually is the continuation and development of the thermal insulation method. The lining design of hearth bottom and hearth combines low thermal conductivity materials with high or medium thermal conductivity materials. With this kind of design, people could map a special isotherm distribution graph. So that the hot metal solidification line is supposed to be close to the hot surface [19, 20]. Both designs have been translated into practice-successful masterpiece, and the life span of some blast furnaces is even over 20 years or longer.

The application of ceramic cup structure blast furnace is relatively successful in Europe. However, in China, it rises dispute widely: Some experts file an opposition to the ceramic cup structure design. They proposed that 'the ceramic cup furnace structure is very unreasonable in terms of theory and practise, it will seriously affect the extension of the life of the furnace.' At present, many blast furnaces only have a short life span (only 4–5 years) and some even less than three years. Unreasonable design of ceramic cup structure is the most important reason [21]".

8.4.1 The Practice of Ceramic Cups Structure

The first blast furnace applied with ceramic cups is the German company Thyssen Krupp (Thyssen Stahl AG) Ruhr Plant (Ruhrort) No. 6 blast furnace. It was put into operation in April 1984 [22]. Thyssen Krupp company Hamborn Factory No. 9 Furnace (1987), Schwergen Plant No. 1 blast furnace (1989) and No. 2 blast furnace (1993) have switched to ceramic cup furnace structure successively, and remarkable achievements have been achieved [4, 23].

Following are three examples of successful blast furnaces [24, 25].

Hamborn No. 9 blast furnace always puts up a good performance in production. In 2006, 17 years of the application of ceramic cup lining, the annual indicators are as follows: coke ratio: 232 kg/t; coal ratio: 174 kg/t; 84 kg/t nut coke of 10-–35 mm; fuel ratio: 490 kg/t; oxygen enrichment rate: 26.9%; blast furnace utilization coefficient: 2.54 t/(m^3 d). (Working volume utilization factor 3.04 t/(m^3 d).) (Table 8.9).

Figure 8.29 is calculated result of the corrosion of ceramic cup of Hamborn No. 9 blast furnace in different periods [26]. From Fig. 8.29, it can be found that about nine years after operation, the local corrosion of the side wall of hearth was close to the carbon brick lining. The thickness of ceramic cup furnace wall at that time was 400 mm. Ceramic pad was still in a safety condition.

Figure 8.30 shows the calculated result of the corrosion condition of Schwergen Plant No. 2 blast furnace after a 33.98 million tons' iron production [27]. According to this figure, it is obvious that the ceramic cup pad is in a safe condition.

Capital Steel No. 1 blast furnace is the first ceramic cup structure furnace to introduce French technology in China. It blew on in August 1994. The bottom has four layers of carbon brick with a 230 mm domestic aluminum silicate brick and two layers of 400 mm French aluminum silicate brick on the top. The hearth and bottom structure are shown in Fig. 8.31 [28, 29].

Table 8.9 First-generation indicators of three ceramic cup structure blast furnace

Factory	Serial number	Volume, m^3	Year of operation	The life of the first generation blast furnace, year	Volume of the first-generation blast furnace, t/m^3
Hamborn Factory	9	2132	December 1987	25	14,313.7
Schwergen Plant	2	5513	November 1993	26	18,136
Capital Steel	1	2536	August 1994	16.4 (Relocated and stopped in production)	13,328

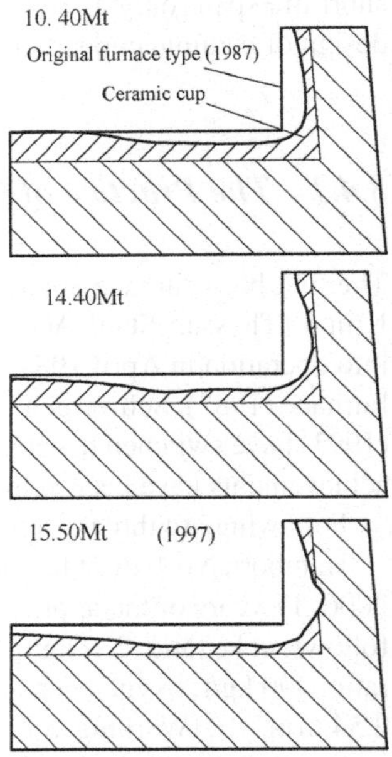

Fig. 8.29 Corrosion of ceramic cup of Hamborn No. 9 blast furnace

After eight years of production, the temperature of furnace bottom and hearth has been in a relatively stable condition. After 2003, the bottom temperature of the furnace increased faster with the center of the maximum temperature has reached an average of 710 °C. Figure 8.32 is the temperature change of the bottom.

In Fig. 8.31, the upper part of the bottom of the furnace is two layers of French ceramic cup brick, and the lower part is a domestic pad. Then, a domestic four-layer carbon brick and the next part are 24 water-cooled pipes at the bottom. The position of T3 thermocouple is in the middle of the bottom of the 3 and 4 furnace layer of carbon brick radial direction. The temperature rises up extremely fast from average 710 °C in the April to 876 °C on May 14, indicating that the bottom of the ceramic cup is corroded and destroyed after nine years' work. But there is no risk of breakage:

Fig. 8.30 Calculated result of temperature distribution and the corrosion condition of hearth bottom and hearth

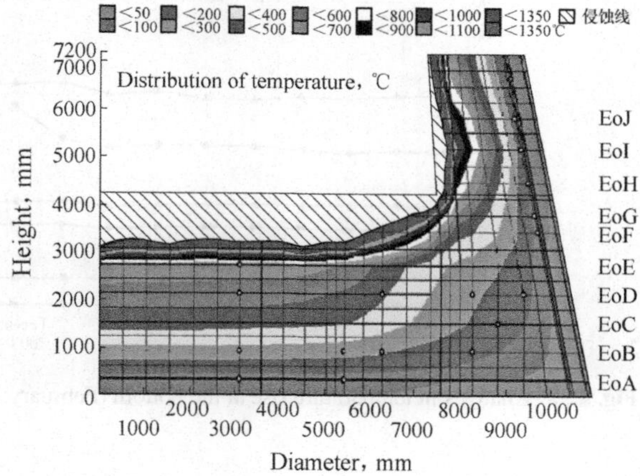

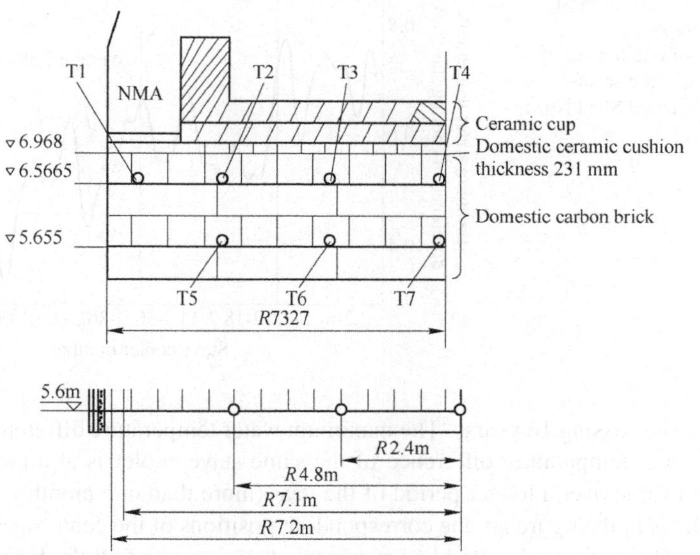

Fig. 8.31 Hearth and bottom of Capital Steel No. 1 blast furnace

The position of T3 thermocouple is in the middle of the lower part of the carbon brick furnace's fourth layer in radius direction. It is far from the stave cooler and the bottom cooling water pipe, resulting in a low risk of breakage; T1, which is close to the stave cooler, has a low temperature of only 85 °C and is in a very stable condition.

By December 18, 2010, according to the company's plan, all steel factories in Beijing would be closed. No. 1 blast furnace is closed. At that time, it has a stable level of production, and the furnace body is still in a favorable condition. In Fig. 8.33, it can be seen that the maximum water temperature difference of each stave cooler (two section A total of 60 cooling wall) in 2 sections of the furnace is less than

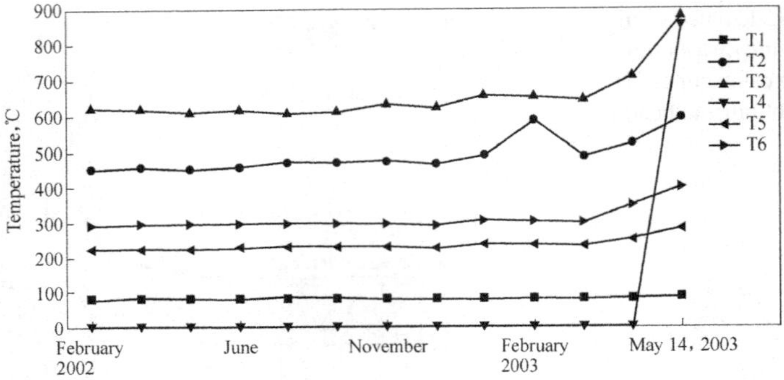

Fig. 8.32 Changes in temperature rise at the bottom (February 2002–May 2003)

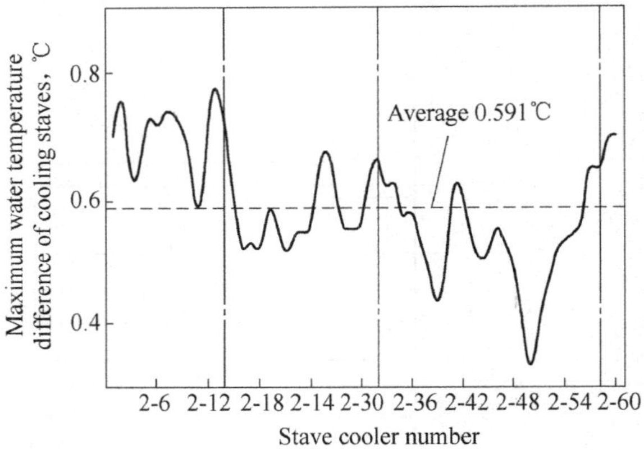

Fig. 8.33 Circumferential distribution of the highest value of temperature difference of cooling staves historically in the second section of Capital Steel No. 1 blast furnace

0.8 °C for the passing 16 years. "The maximum water temperature difference means that the water temperature difference of the same stave cooler is at a more stable maximum value over a longer period of the year (more than one month). The three straight lines in the figure are the corresponding positions of the centerlines of three tap holes. The horizontal dotted line represents the average of all the highest water temperature differences, which is 0.591 °C. As it can be seen from the figure, the maximum water temperature difference of the stave cooler corresponding to tap hole is higher than that of other stave coolers". [25]

The first generation of No. 1 blast furnace production is 16.4 years (it was closed due to relocation plan), and the cumulative iron production is 33.32 million tons with a generation of 13,328 t/m^3. Indicators are as shown in Table 8.10.

Table 8.10 Annual index of blowing on and blowing off

Year	1994	1996	1998	2000	2002	2004	2006	2008	2010
Index, t/(m^3 d)	1.832	2.05	2.191	2.254	2.327	2.164	2.370	2.423	2.353
Coke rate, kg/t	510.9	442.5	397.7	374.9	351.0	449.0	329.4	335.4	339.6
Coal ratio, kg/t	48.2	95.8	108.3	122.1	147.1	58.9	140.9	135.7	135.1
Temperature of the blast, °C	910	989	1046	1103	1119	1012	1149	1157	1139

8.4.2 Arguments About Ceramic Cup Structure

In Europe, ceramic cup structure is generalized. In China, although it is also promoted, but there are strong objections. The first reason is that ceramic cups have a short life span. Some experts believe that its life expectancy would only last for 2–3 years as described in the literature [30]. Indeed, some ceramic cup structure blast furnaces are short-lived.

8.4.2.1 Self-baked Carbon Brick-Ceramic Cup

Self-baked carbon brick is a creation of Zhang Yunzhong, a Chinese expert, and other researchers. Since 1974, it has been applied to more than 200 blast furnaces, contributing a lot to the construction of blast furnaces in China.

Self-baked carbon brick requires sufficient intensity: It must be baked with the temperature of more than 700 °C, and it also needs enough heat supply to ensure that the efficiency of binder; on the contrary, the thermal insulation of the ceramic cup can neither provide sufficient heat to the self-baked carbon brick nor create the required high temperature roasting conditions. Obviously, the combination of ceramic cup and self-baked carbon brick is a shortcoming in the design process. The larger the blast furnace is, the more prominent the shortcomings would be; due to a short diameter of the hearth of small blast furnace, the thickness of the bottom of the furnace is thinner. Therefore, the shortcoming is not obvious.

Figure 8.34 shows the result of cutting out different sectors from hearth and bottom of self-baked carbon brick-ceramic and rotating a pice of hearth back 90 degrees. The isotherm are connected with thin lines. From the figure, it is obvious that the temperature of hearth and hot surface of the furnace bottom (inside) are roughly the same, the same to the temperature of cold wall (stave cooler or cooling water pipe). Thickness (distance) between adjacent isotherms is different due to the different materials in middle part. The heat surface temperature in the figure refers to the temperature in the hearth; the cold side of the hearth refers to the inner side of the stave cooler of the hearth; the cold face at the bottom is the inner side of the cooling water pipe.

The inner lining of the picture is aluminum silicate brick, and the outer layer of the hearth and the bottom are charcoal bricks.

Now, suppose that in Fig. 8.35, the hot surface temperature of the furnace is

Fig. 8.34 Enlarged view of parts of the bottom and hearth

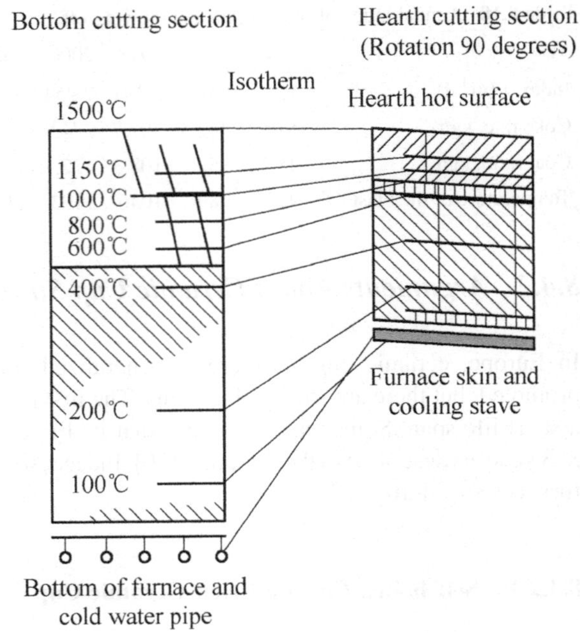

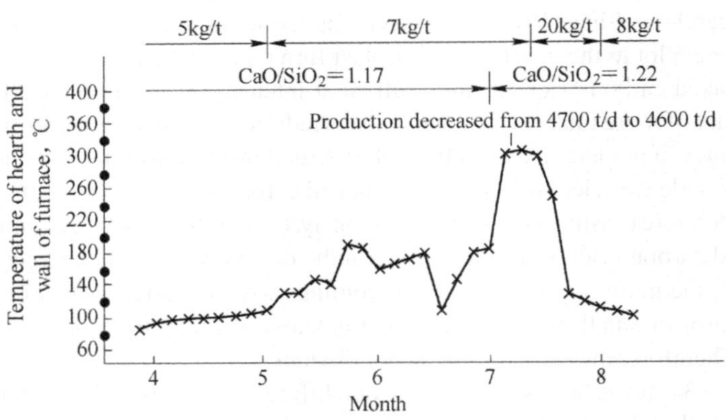

Fig. 8.35 Fettling operation of Fukuyama No. 1 blast furnace in 1969

1500 °C, and the cold surface is the 100 °C. And then, imagine that the inner layer is aluminum silicate brick, the exterior of the furnace and the bottom of the furnace are self-baked carbon brick. The isotherm in Fig. 8.35 still exists, but the distance between the isotherms varies according to the thermal conductivity. Compared with carbon brick, the thermal conductivity of self-baked carbon brick is quite different [21, 31].

From Table 8.11, it can be seen that self-baked carbon brick is characterized by a proportional relation: the lower the temperature, the lower the thermal conductivity.

Table 8.11 Heat conductivity index of fire clay brick

Temperature, °C	25	300	600	800	900	1000	1200
Self-baked carbon brick	3.05	5.24			9.00		
Heat pressing carbon brick			18.4	18.8		19.3	19.7
Normal microprorous carbon			10.4	10.4		10.5	10.9
Half graphitizing carbon brick			42	38		32	
Microporous carbon with low content of iron			120			70	
Aluminum silicate brick						1.8	

So, the thermal conductivity would be lower if in the lower part of the furnace. And if it in the higher part of cooling pipe, thermal conductivity would be close to 3, and it will be reduced nearly three times under a high temperature. In the lower part of the ceramic pad of the furnace, self-baked carbon brick is placed. This kind of arrangement structure is not conducive to heat export; but if we use special baked carbon brick, the thermal conductivity difference is small or with the temperature decrease, the thermal conductivity increases, which is favorable for heat exporting.

This is the basic reason for the short life of domestic "self-baked charcoal brick-ceramic cup".

8.4.2.2 Unreasonable Ceramic Cup Structure Leads to the Sort Life of Ceramic Cup Furnace

Corrosion in lining is inevitable. However, breakage can be avoided. Blast furnace campaign is a system engineering. Therefore, any "weakness" will lead to the short-time operation of blast furnace. Some short-time operations are caused by the mismatching of the cooling system and ceramic cups. For the first time, Tang Qinghua correctly pointed out the reason for the short life of An Steel No. 3 blast furnace and Bao Steel No. 4 ceramic cup structure blast furnace lies on the defects of the cooling system. Table 8.12 is the comparison between blast furnace stave cooler and cooling ratio surface [32].

"The cooling ratio surface area of Bao Steel No. 3 blast furnace is more than twice times that of the Bao Steel first generation of No. 4 blast furnace and the first generation of An Steel No. 3 blast furnace, which indicates that the over 0.6 cooling ratio surface area of hearth iron-casting stave cooler is not sufficient. It also confirmed by the fact that Bao Steel No. 4 blast furnace first generation has only been in operation for nine years" [32].

8.4.2.3 Unreasonable Design in the Structure of Ceramic Cup Hearth

The main defects of ceramic cup structure pointed out in the literature [27] are "artificial annular cracks between ceramic cups and carbon bricks". "After the temperature

Table 8.12 Stave cooler and cooling ration surface of An Steel No. 3 and Bao Steel No. 3 and No. 4 blast furnace

Blast furnace	Type of stave cooler and numbers of pipes	Diameter and distance of pipe	Cooling rate surface area
An Steel No. 3 blast furnace, first generation	Vertical type, four inlet pipes and four outlet pipes; width, 1040 mm	50 mm × 260 mm	0.604
Bao Steel No. 3 blast furnace, first generation	Horizontal type, ten inlet pipes and ten outlet pipes; height, 1450 mm	60.3 mm × 145 mm	1.31
Bao Steel No. 3 blast furnace, second generation	Horizontal type, ten inlet pipes and ten outlet pipes; height, 1600 mm	70 mm × 160 mm	1.374
Bao Steel No. 4 blast furnace, first generation	Vertical type, four inlet pipes and four outlet pipes; width, 920 mm	50 mm × 230 mm	0.684
Bao Steel No. 4 blast furnace, second generation	Horizontal type, ten inlet pipes and ten outlet pipes; height, 1600 mm	70 mm × 160 mm	1.374

of ceramic cup rising, the volume expands, leading to an outward movement of the carbon brick. Triangle seam if formed".

For any blast furnace hearth, unless built with the same fire clay brick, gaps and extractions will exist inevitably due to the mismatch of the expansion coefficient of two or more refractory bricks. Excellent ceramic cup structure does not have gaps owning to the matched masonry materials, and it would enjoy a long-time span because of a high intensity. In the summary report of ten years' ceramic cup structure application, the advantages of closely matched ceramic cup structure are underlined in particular [22].

"There is a 60 mm gap between the carbon material and the ceramic wall. To fill these gaps, the AMC 66 K Carbon of Savoie is used, which is helpful to absorb the radial expansion produced by the ceramic wall after blowing on, so that the carbon brick and ceramic wall are in close contact. The expansion seam is designed at the top and bottom of the ceramic wall, and the seam is filled with ceramic fibers to absorb the longitudinal expansion of the ceramic wall" [30]. Obviously, a reasonable design of ceramic cup structure is supposed to fully consider the impact of expansion and take corresponding measures.

8.4.3 *Advantage of Ceramic Cup Bottom and Hearth Structure*

In a thematic report at China's 2004 Annual National Iron-making Conference, Professor Yang Tianjun underlined that "the energy consumption of iron-making systems accounts for about 70% of the total energy consumption in the iron and steel industry and accounts for about 50% of the total energy consumption. Although the energy consumption of iron-making is decreasing year by year, it is still higher than about 20% of the international advanced level. In order to achieve the international level of comprehensive energy consumption of tons of steel, iron-making system, especially blast furnace iron-making, needs further development, and efforts are required in increasing energy conservation and expansion efforts". [33] Ceramic cup structure is an effective measure to reduce the energy consumption of blast furnace, and there is no alternative structure at present. According to the calculation of the French inventors, the heat dissipation loss of the furnace bottom and hearth is significantly reduced, and the hot metal temperature increased 10–20 °C [24] under the same conditions. These assertions have been proved by practice.

Capital Steel No. 1, 3 blast furnace shares same volume and similar original fuel. They have been worked for more than 16 years. The difference between the two blast furnaces is No. 1 blast furnace is ceramic cup structure, while No. 3 blast furnace is high aluminum brick–carbon brick furnace bottom structure. The heat flow intensity of them is shown in Table 8.13 [25]. The difference ratio of average heat flux intensity in two furnaces is 39.4%.

After comparing the sides wall temperature, we found that after 16 months, the cylinder brick lining temperature of the domestic carbon brick furnace is higher than that of the ceramic cup furnace with the same part hearth brick with the difference of 400–500 °C. The average heat flow intensity of the two-layer stave cooler of No. 3 blast furnace with domestic carbon brick is 8211 W/m^2, and the No. 1 blast furnace is 4498 W/m^2, only 55% of the former. Under the same conditions, ceramic cup hearth increases the temperature of hot metal about 15 °C [27].

The structure of blast furnace ceramic cup is an effective design in order to reduce energy consumption. Some of breakages in China's furnace with ceramic cup structure are caused by design and material defects. The word that "water is the best refractory" becomes a history; with the development of anti-corrosion and high temperature resistance of refractories, the thermal conductivity method will be combined with the resistant furnace lining to reduce heat loss and energy consumption, so as to realize the long life of blast furnace.

Table 8.13 Heat flow intensity ratio of stave cooler in No. 1 and No. 3 blast furnace hearth (Oct. 2015) (W/m^2)

Project	Maximum	Minimum	Average
No. 1 blast furnace	16,096	5505	7954
No. 3 blast furnace	18,825	5709	11,085
Difference	2729	204	3131

Bottom and hearth of ceramic cup furnace are still needed improved. At present, it should be promoted as an important measure of energy saving in blast furnace.

8.5 Fettling

Terms of Iron-making

Fettling: Generally, titanium mineral is used to fill furnace in order to fill up corroded lining. It is also known as "furnace protection". In this book, it is always called as fettling.

8.5.1 Evaluation of Methods in Preventing Breakage

If the blast furnace is threatened by breakage, the first remedial measure is to increase the cooling intensity of stave cooler, which is also the most common method. In 1984, the second-layer stave cooler of the Capital Steel No. 4 blast furnace was threatened by breakage. The cooling water pressure of stave cooler of the first, the third and the fourth layers of the hearth was reduced from 3.4 to 3.2 kg/cm^2, and the hydraulic pressure of the stave cooler of the furnace body was reduced from 2.3 to 2.2 kg/cm^2. It aimed to improve the water pressure of the second layer of stave cooler which was severely beset with breakage. Parts of stave cooler, in which the water temperature difference is too high, were shifted into 10–12 kg/cm^2 high-pressure water, aiming to dodge the danger of breakage temporarily.

At the same time, the tuyere, with the original length of 380 mm and a 5° tile down, was changed to the length of 400 mm. That is only a simple auxiliary measure for little help.

In July 1985, water temperature difference of No. 4 blast furnace's second layer of stave cooler exceeded the standard, of which eight pieces greater than 1°. At that time, it had been stopped for 12 h due to maintenance. Water temperature difference quickly reduced from 1° to 0.6–0.8°.

After blasting, operators controlled the air volume, reduced the smelting intensity to obtain temporary stability. While increasing the smelting intensity, water temperature difference, which exceed the limit of the stave cooler, was back to the original level, and the threat of BF burning through recurred.

In November, it went out of blast for 16 h, and remarkable and fast reduction was seen in temperature. Whereas, after putting into operation, breakage occurred again. The fundamental problems cannot be rooted out. The lining was seriously corroded, and a pause in blasting cannot be conductive to the loss of lining.

In the past, people try to block the tuyere of the upper part of breakage area. But that could only reduce production and the local temperature near the tuyere. It should not be denied that it is helpful, but it is bad for the operation and cannot solve the fundamental problem. In facing of a serious breakage, many factories try to block the tuyere. It is a good way to save time when urgent breakage occurs. Effective methods of filling the furnace should also be implemented to stop the breakage simultaneously.

It cannot relieve the threat of breakage if only relying on blocking tuyere. The life of blast furnace can be extended using titanium as the material filling furnace if used correctly.

8.5.2 History of Fettling with Burden Containing Titanium

Japan is short of iron ore resources. In coastal parts of the sea, Japan has sand iron, which contains Fe and Ti. As a raw material for iron-making, the Ti-containing material, during the production process, is found that it would deposit at the bottom and hearth, thus playing a protective role. Table 8.14 is a blast furnace of Sumitomo Corporation Kokura iron-making plant in Japan. The actual results of the use of sand iron in 1955 show that the furnace consumes with a maximum of sand iron 250 kg/ton of iron [12]. Later, due to depletion of sand iron, Japan began to import Ti-containing materials, which is used in filling furnace.

Japan Fukuyama Iron and Steel Co., Ltd. No. 1 blast furnace was out into operation in April 1969. Lining temperature increased gradually at about 1 m from the bottom of the hearth. In May, it rose from the 100 to 130 °C. At that time, in addition to reducing the production from 4700 to 4600 t/day, operator also enhanced the amount of TiO_2 in burden from 5 to 7.5 kg/t. But they still failed to stop the temperature rise. By July, the lining temperature rose to 300 °C. Then, the TiO_2 in the burden was increased to 20 kg/t, and the lining temperature dropped rapidly. It decreased back to the normal level of 100 °C within only ten days (Fig. 8.35) [34]. Figure 8.36 is the operation curve of No. 4 blast furnace in Wakayama-Ken, Japan. The original image is divided into four intervals, where three intervals are taken. The lower part is the hearth lining temperature of the furnace under the tap hole. When TiO_2 in the furnace is 12 kg/t, the hearth lining temperature did not decrease; when increased to 18–20 kg/t, the lining temperature decreased. The middle solid line of the graph indicates the amount of TiO_2 addition, while the dotted lines represent the amount of exclusion. The number (TiO_2 into the furnace, kg/t) and the interval (specific time interval) below the discharge are the date of the addition of different TiO_2. The upper part of the figure is the ratio of the total amount of TiO_2 added into the furnace

Table 8.14 Recovery rate of titanium Bao Steel

TiO_2, kg/t	[Si], %	[Ti], %	(TiO_2), %	Basicity of slag	η_n, %
9.78	0.57	0.177	2.28	1.18	30.2
15.23	0.51	0.245	3.59	1.18	26.86
19.53	0.60	0.295	4.17	1.17	22.66

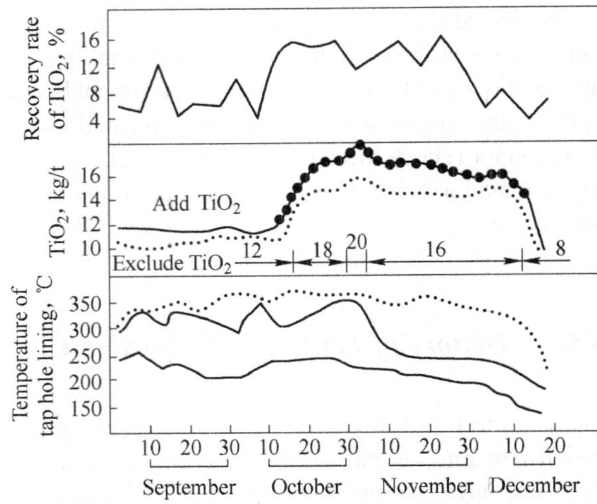

Fig. 8.36 Operation Curve of Wakayama No. 4 blast furnace fettling in 1979

and the amount of recovery (in the total amount of furnace minus remaining parts), which actually represents the recovery rate (between 6 and 17% [35]). Japanese experts found that Ti-containing materials are conductive to fettling and created a method of filling furnaces, which has made an important contribution to prolonging the campaign of blast furnaces.

The first plant to apply fettling technology in China is Liu Steel. Liu Steel was affected by the smelting of Panzhihua Iron and Steel Company and drawn on the successful experience of titanium ore filling technique in Japan. In January 1981, it started the test of adding vanadium and titanium ore in No. 2 blast furnace in. More details about operation are shown in Fig. 8.37. In Fig. 8.37, the lining thermometer is installed in the plane brick wall of tap hole with the depth of 250 mm. The

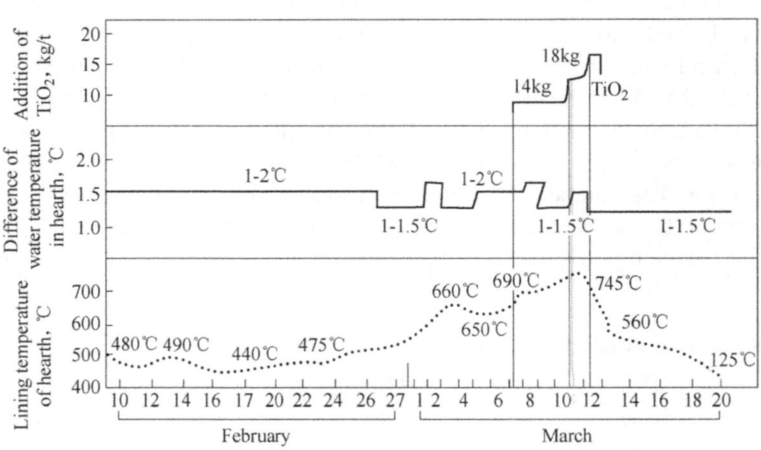

Fig. 8.37 Fettling operation curve of Liu Steel No. 2 blast furnace

normal temperature level is between 400 and 500 °C. When TiO_2 content reached 14 kg/t, although the furnace water temperature difference did not rise, the brick lining temperature continued to increase. On March 11, TiO_2 was added to 18 kg/t, and on March 12, it was added to 20 kg/t.

The temperature then quickly slipped, and the brick lining temperature dropped below 150 °C within one week. Liu Steel's success blazes a trail in the operation of the fettling in China, and Liu Steel's contributions cannot go unnoticed [36].

In September 1982, Xiang Steel No. 2 blast furnace made an experiment of fettling. Detailed analysis was made during the experiment. When the content of TiO_2 in the furnace was 12 kg/t, Ti contained in hot metal was about 0.15%. After a week, water temperature difference of stave cooler reduced to the normal level (stably below 2 °C), and blast furnace pulled through. Therefore, major capital was postponed until 1986, and the campaign of the furnace was extended to11 years. Figure 8.38 is the change of water temperature difference of stave cooler in the process of fettling of No. 2 blast furnace in Xiang Steel.

They had a meeting in 1984 and identify their work in fettling. At the meeting, they introduced successful experience of titanium ore fettling. It was since then this technique became available in China. This important technology made a significant contribution in prolonging the life of blast furnace in China [37].

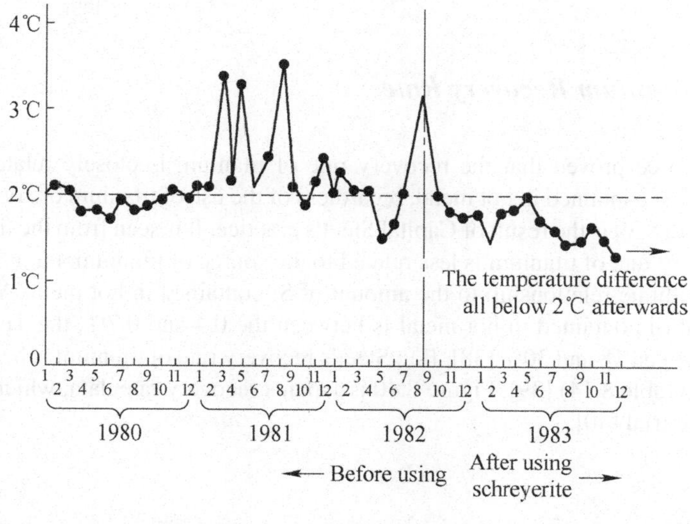

Fig. 8.38 Fettling of Xiang Steel No. 2 blast furnace and the change of water temperature difference of stave cooler

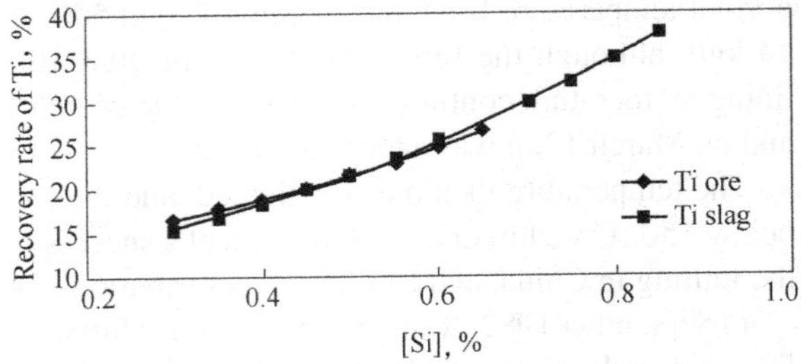

Fig. 8.39 Recovery rate of titanium in Capital Steel blast furnace

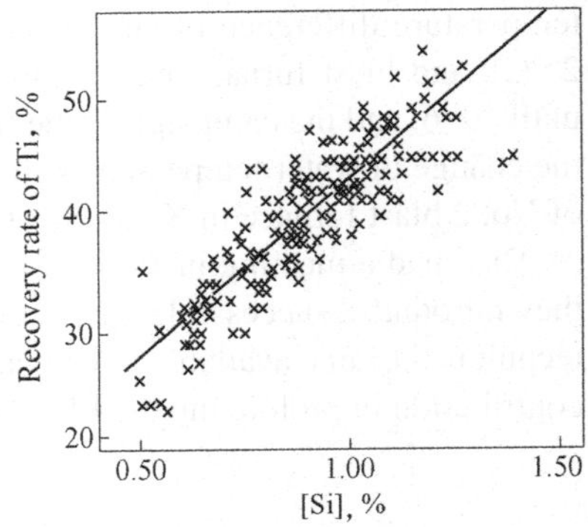

Fig. 8.40 Relation between recovery rate of titanium and [Si] in Japan

8.5.3 Titanium Recovery Rate

Years practice proved that the recovery rate of titanium is closely related to the amount of Si contained in hot metal, regardless of the use of titanium ore or titanium slag. Figure 8.39 is the result of Capital Steel's practice. It is seen from the figure that the recovery rate of titanium is less related to the source of titanium, but it is almost in a logarithmic relationship to the amount of Si contained in hot metal. When the amount of Si contained in hot metal is between the 0.4 and 0.7%, the Ti recovery rate is between 18 and 30% [38]. Bao Steel's recovery rate of titanium is between 22 and 30% (Table 8.14) [39]. Figure 8.40 is Japan's recovery rate data, which is close to Chinese trial [40].

8.5.4 Formation and Deposition of Ti-Containing Material in the Hearth

Many scholars have studied the precipitation and crystallization process of Ti-containing material in the blast furnace. These results have guided the work of fettling in our country, and remarkable results have been achieved. Figure 8.41 is the

Fig. 8.41 Relationship between the solubility of titanium in hot metal and temperature and nitrogen division pressure

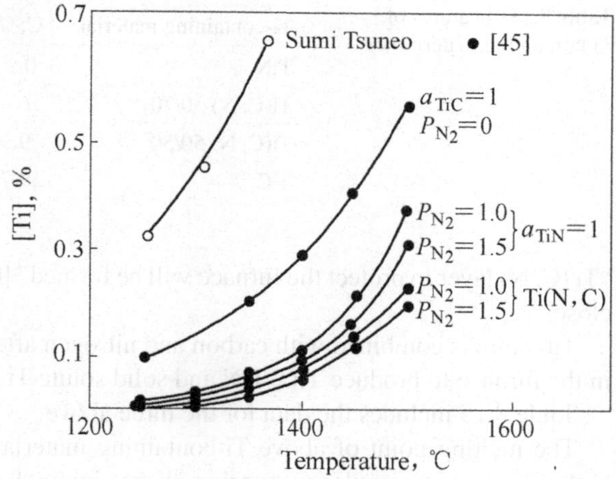

information about solubility of hot metal offered by Ren Yunfu and Jiang Lieying [41].

As can be seen from the figure, when the temperature of hot metal is lower than the temperature of the curve shown, there is metal titanium precipitation. Therefore, the lower limit of titanium content in hot metal is determined by the temperature of the furnace bottom. For example, when the bottom temperature is 1200 °C, the solubility of titanium in hot metal is only 0.012%. "The general control lower limit is 0.08%". [41] As you see from the diagram, the solubility and temperature of titanium are closely related: The lower the temperature, the more titanium is precipitated. When the hot metal is close to the stave cooler, the temperature is obviously the lowest, which happens to be conducive to titanium precipitation.

Figure 8.42 is the solubility of Ti with different conditions when Ti contains C 4% in hot metal. It is seen from the figure that the solubility of Ti does not exceed 0.15% [42] under a situation which is similar to that of the furnace. According to Professor Dong Yicheng and other studies, when the content of titanium in hot metal ≥ 0.1%,

Fig. 8.42 Solubility of Ti

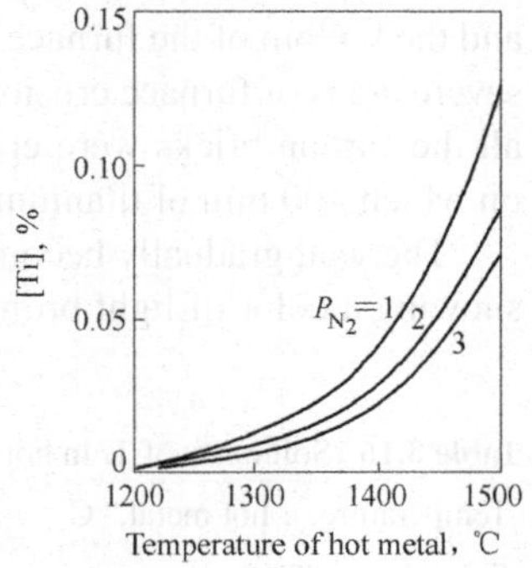

Table 8.15 Features of Ti-containing material

Ti-containing material	C, %	N, %	Melting point, °C
TiN	0.	20.7	2950
Ti(C, N) 30/70	6	15.1	
Ti(C, N) 50/50	9.6	11.2	
TiC	19.7	trace	6150

"Ti (C, N) layer to protect the furnace will be formed" [43]. Their inferences are very close.

Titanium is combined with carbon and nitrogen after reduction and precipitation in the furnace to produce TiC, TiN and solid solute Ti (C, N).

Table 8.15 includes the data for the three above.

The melting point of above Ti-containing material and solid solute is all very high. They are in granular suspension, dispersion in the hot metal, so that the molten iron becomes viscous. This Ti-containing material is the basic material for fettling. Professor Du Hegui and others concluded the solubility of Ti contained in hot metal in the blast furnace through thermodynamic calculations. The solubility is shown in Table 8.16 [44].

The results given by different authors are relatively close and can be consulted about fettling.

Breakage occurred in Capital Steel No. 4 blast furnace on March 5, 1985. After reparation, it was put into production for another year. Fettling mainly relied on Ti-containing materials. After blowing off in April 1987, it was found that "a large amount of titanium nitride Ti (C, N) and a small amount of graphite and α-Fe have deposited in the carbon brick of hearth, which was badly corroded. These sediments are fettling material with high melting point, hardness, density, magnetic and conductive properties. The deposition process of titanium carbide is that the TiO_2 is reduced to [Ti], and the [N] and [C] reactions dissolved in the hot metal form [TiN] and [TiC]," alternating between [TiN] and [TiC] when below the 1350 °C interface, forming the annular structure of the tree [45].

Figure 8.43 was drawn based on the results of the 1987 dismantling furnace. "A thick, bright layer of bronze minerals has been deposited at the junction of the hearth and the bottom of the furnace". "Titanium sediments are the thickest one in the most severe areas of furnace erosion". "In the hearth area on the west side of the tap hole, all the carbon bricks were eroded, leaving only about 30 mm of carbon pounding, on which 400 mm of titanium was deposited".

"The wall gradually becomes thinner upwards, and its thickness is about 20 mm, showing a color of light bronze". Extending along the bottom of the furnace to the

Table 8.16 Solubility of Ti in hot metal

Temperature of hot metal, °C	1350	1400	1450	1500
Solubility of Ti, %	0.212	0.299	1.414	0.567

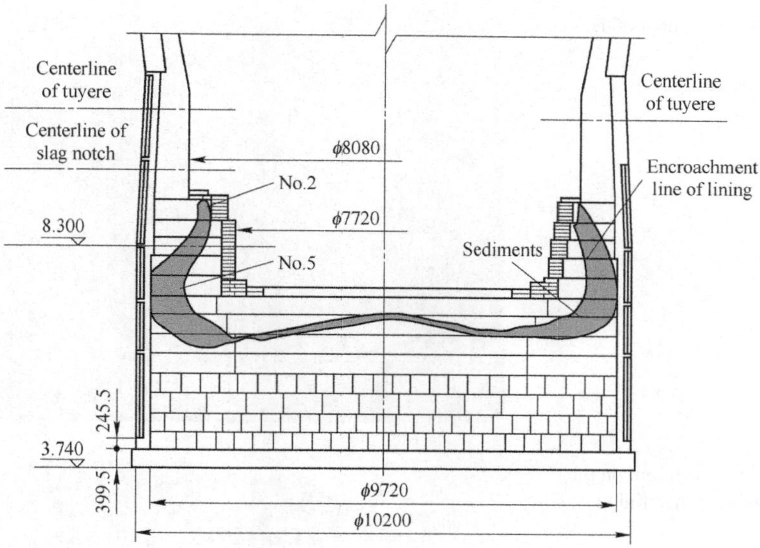

Fig. 8.43 Sediments in Capital Steel No. 4 blast furnace

high aluminum brick, it also shows a light bronze. The titanium sediment in the hearth of the east side of the tap hole is thinner than the west side but thicker than the sediment at the furnace block opposite the tap hole. The titanium sediment at 0.5 m under the two vents is mixed with more slag and coke. No titanium sediments were found on the surface of high aluminum brick at the bottom. Only a small amount of slag is found in the gap between certain carbon bricks and high aluminum bricks.

The actual corrosion line of Capital Steel No. 4 blast furnace after blowing off indicates that the study scholars are correct and reliable. First, the more the corrosion of serious places is, the thicker the sediments are. Besides, the more cooling intensity of the place is, the thicker the sediments are. This place is generally the closest to the stave cooler with the lowest temperature, and Ti-containing material is the easiest to be precipitated. It can be said that it is appropriate and reasonable to use burden containing titanium for fettling. Judging from the deposition thickness of Ti-containing material, safe production can be continued in the bottom of No. 4 blast furnace and the part of hearth.

The sedimentary condensate in Fig. 8.44, which is near the stave cooler, is the deposited titanium with the bright metallic sheen. The surface of the sediment consists of coagulum iron, slag and coke shredded grains, as shown in Fig. 8.45. The larger version of Ti(C, N) is shown in Fig. 8.46.

Table 8.17 is the composition of the sedimentary condensate at the position shown in Fig. 8.43 [45].

The data from abroad are similar to those of our country, and both clearly indicate the relationship between them. Figure 8.47 is the measured results of Kobe No. 3 blast furnace (1850 m^3). Table 8.18 is the titanium sediment composition of No. 3

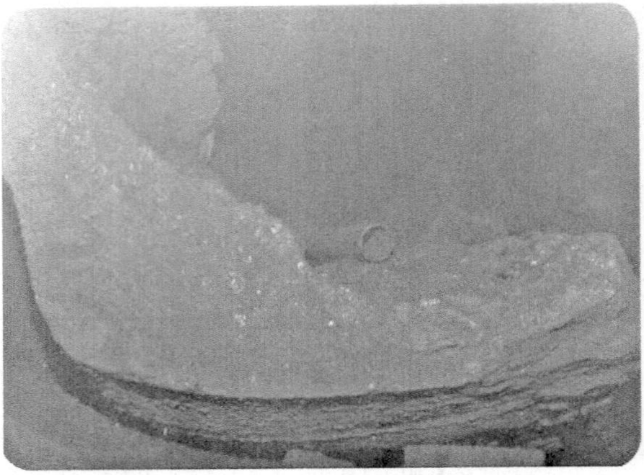

Fig. 8.44 Sediment of Ti

Fig. 8.45 Coagulum on the Ti-containing material or lining

Fig. 8.46 Ti (C, N) in 600 times Light-colored part: TiN; Dark-colored part: Carbon ferric nitride

Table 8.17 Composition of the sediments of No. 4 blast furnace

Part	Number	Layer	Ti	TFe	MFe	FeO	TC	Si	Ca	Al	Mg	S	K	Na
Thin layer of the Ti sediment	1–1	Outer	2.72	70.78	67.77	6.56	11.03	4.06	0.36	0.39	0.56	0.02	0.09	0.04
	1–2	Middle	18.00	57.24	53.27	5.79	5.70	4.25	0.50	0.90	0.04	0.008	0.07	0.02
	1–3	Inner	34.75	46.62	37.11	4.58	10.79	1.81	0.26	0.13	0.04	0.002	0.10	0.09
0.5 m below the tuyere	2	Average	33.40	32.53	34.54	4.31	9.01	4.57	0.55	0.82	1.18	0.77	1.00	0.12
Thin layer of the Ti sediment	5–1	Outer	17.48	32.11	22.76	6.30	11.93	10.92	1.43	0.56	0.20	0.02	0.17	0.09
	5–2	Middle	30.38	30.70	27.36	7.77	11.41	5.74	0.76	0.35	0.04	0.006	0.09	0.04
	5–3	Inner	24.70	46.44	36.63	3.33	3.94	0.92	0.34	0.16	0.04	0.006	0.07	0.02

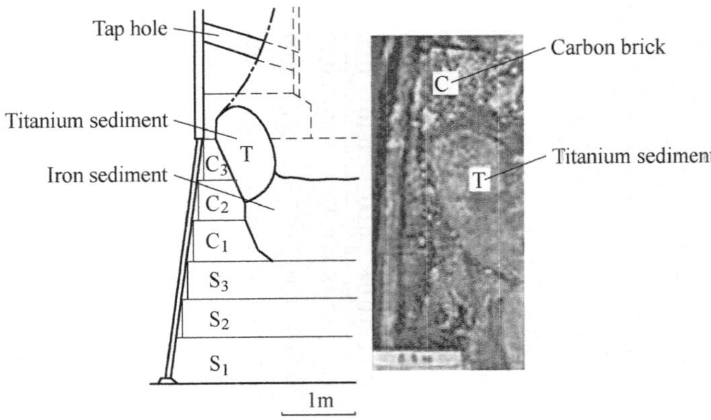

Fig. 8.47 Titanium sediment of Kobe No. 3 blast furnace

Table 8.18 Sediments in the junction of three carbon bricks

Position		C	Ti	Si	Mn	P	S
Junction of the first layer	Upside	3.63	14.70	–	–	–	–
	Downside	3.35	8.89	0.22	0.25	0.027	0.054
Junction of the second layer	Upside	1.77	0.06	–	–	0.016	–
	Downside	0.10	0.01	0.002	0.003	0.023	0.017
Junction of the third layer	Upside	0.014	0.004	0.01	0.003	0.70	0.020

blast furnace of Kobe Steel [46]. The figure shows that the three layers of bottom brick have been corroded, and the severest place of corrosion is the junction of the hearth and the bottom. Titanium sediments just deposit here. On top of the hearth are carbon bricks. The next three floors are clay bricks. In the photo on the right, C is the remaining carbon brick, and Ti is the titanium sediment. Table 8.18 is the sediment composition at the junction three layers of carbon bricks. The sedimentary state and chemical composition are consistent with the results of Capital Steel [46].

8.5.5 *Why Breakage Still Occur After Fettling?*

Breakage still occurs for some blast furnaces even after fettling.

There are two important conditions for fettling: the density of Ti in the hot metal and the lining temperature in contact with the hot metal. The density of Ti requires that there be enough titanium minerals in the furnace; the severely corroded lining should be built during fettling with a strong cooling strength, which is enough to reduce the temperature of the hot metal in contact with to less than 1200 °C in order to have a better precipitation and deposition of Ti in hot metal, timely. We have

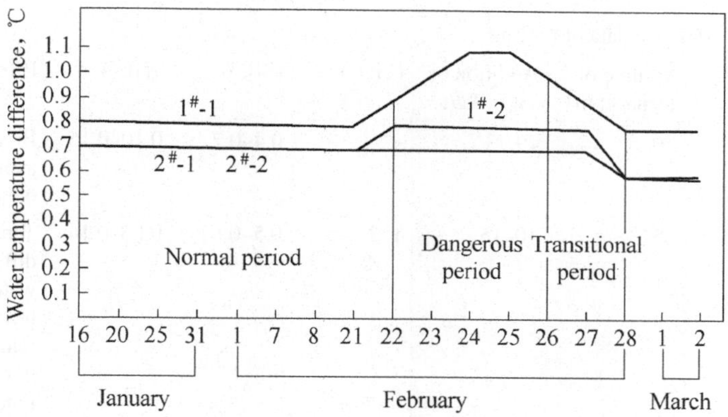

Fig. 8.48 Water temperature difference of stave cooler of Capital Steel No. 4 blast furnace

learned a lesson in the past: In 1995, due to a high frequency in cleaning the wall, Capital Steel No. 4 blast furnace (with volume of 2000 m^3), which has been put into production for two years, has been threatened with the danger of breakage as the temperature of the second layer of stave cooler being close to breakage. On February 22, Ti slag has been added as the materials of fettling. Ti content in hot metal was 0.06–0.08%. But the water temperature difference of the second layer of stave cooler continued to rise. The temperature difference of stave cooler $1^{\#}$-1 has reached the level of 16,500 kcal/(m^2 h), and the heat flux intensity was higher than the alarm limit stipulated by Capital Steel. On February 23, [Ti] content in hot metal was raised to the 0.12–0.16%. On February 24, at 11:00, heat flux jumped to 18,234 kcal/(m^2 h). That was unprecedented. Therefore, furnace immediately went out of blast, and four corresponding tuyeres were plugged. After the furnace was on blast, operator controlled the air volume to 4000 m^3/min. On February 25, Ti contents in the hot metal were about 0.12–0.16%. The temperature difference of stave cooler began to decline and fell back the normal level on February 28, and the tuyere gradually opened with a normal blast volume, as shown in Fig. 8.48.

The fact teaches us that deposition could only happen when the amount of Ti contained in hot metal is less than 0.07%, and the amount deposited in the furnace is very small (more than 0.08%).

Li [44], Song and Chen [47], Mo and Du [48], all of whom have compiled successful furnace experiences.

Professor Li Yongzhen once summarized seven examples (Table 8.19). The amount of TiO_2 in furnace is 10–15 kg/t of hot metal. Ti contained in hot metal is 0.06–0.25%, of which 0.06% refers to the lower limit of Ti fluctuation, while the actual fluctuation range is between 0.06 and 0.14%; Professor Song Jiancheng and Chen Peijian summed up 16 cases (Table 8.20). The Ti content in hot metal is between 0.09 and 0.36%, of which six cases, instead of fettling, are the normal production of vanadium titanium ore smelting. Mo Suizhi and Du Chunrong [49] summed up 14 cases (Table 8.21). The Ti content in hot metal is between 0.09 and 0.36%.

Table 8.19 Practice data of fettling

Factory	Volume of furnace, m^3	Addition of TiO_2, kg/t	(TiO_2), %	[Si], %	[Ti], %	Effect
Capital Steel	1036	9.73	<2	0.4–0.7	0.10–0.15	Breakage is avoided, high output
Wu Steel	1513	10–15	1–2	0.5–0.62	0.06–0.14	Temperature difference of water in hearth remains normal after nine days
Xiang Steel	750	10–15	1.15		0.154	Water temperature difference remains normal after seven days
Mei Steel	1000	about 12		0.6	about 0.15	Temperature difference of water in hearth remains normal after five days
Liu Steel	300	12–15	1.5–2.5	0.5–0.8	0.2–0.25	After three days, temperature difference of water in hearth reduces for a week
Tong Steel	300	10	1.75	0.8–1.0	0.1–0.15	Foundation temperature reduces from 605 to 585 °C
Ma Steel	300		2–4	0.43–0.45	0.137	Normal operation

Experiences show that the effect in fettling will be multiplied when Ti content in hot metal is between 0.08 and 0.12%, while if between 0.15 and 0.25%, it will be more effective. Higher content of Ti will lead to a good performance in fettling, but it will also result in hearth accumulation, and the operation will be damaged. Some blast furnace, due to a high amount of Ti in the furnace, will have "furnace hearth hot junction" [50], and iron jug and iron runner will be pasted to the furnace. Therefore, it cannot work orderly.

Table 8.20 Practice data of fettling

Factory	Number	Year	Ti load, kg/t	Composition of pig iron, %			Content of slag, %				
				Si	Ti	S	CaO	SiO_2	MgO	Al_2O_3	TiO_2
Pan Steel	1	1970		1.12	0.24	0.067	25.80	23.4	9.80	14.2	24.20
	1	1971		0.30	0.27	0.058	25.9	23.1	9.75	13.9	24.68
	1	1980		0.12	0.18	0.059	26.7	23.2	8.90	14.2	23.5
	3	1988		0.07	0.10	0.073	27.2	24.5	9.09	14.5	22.87
Cheng Steel	2	1989		0.35	0.33	0.048	32.2	24.2	4.84	15.71	17.89
	2	1990		0.27	0.36	0.059	31.7	22.6	4.31	15.12	19.60
Shui Steel	1	1978	138	0.32	0.25	0.038	35.7	30.7	9.45	11.10	13.40
Chong Steel		1978	108	0.41	0.25	0.034	37.7	31.8	8.45	16.38	9.37
Capital Steel	3	1986	22	0.40	0.09	0.040	38.6	36.8	12.3	12.30	1.24
Xiang Steel	2	1982	15	1.17	0.16	0.021					1.29
Shao Steel	2	1985	10	0.83	0.24	0.018	40.4	33.1	6.20	12.40	1.63
Baotuo Steel	3	1986	9	0.50	0.16	0.034	38.8	31.2	7.3	8.96	1.76
Tong Steel	2	1986	10	1.00	0.15	0.020					1.75
Shi Steel	1	1987	20	2.57	0.30	0.015	38.6	35.7	9.46	15.48	4.30
	1	1987	28	1.66	0.34	0.024	38.10	35.1	9.77	14.53	4.9
Wu Steel	3	1985	10	0.57	0.14	0.027	38.7	38.0	5.63	12.07	2.65

During fettling, it is necessary to keep titanium in balance regularly, and when the amount of Ti addition is greater than the amount of exclusion, the fettling is required; when they are equal, the situation is to be maintained; when the exclusion of Ti is greater than the addition, operators should be alarmed. Generally, when smelting iron for steelmaking, the amount of Ti in hot metal should be 0.02–0.04%. A high percentage of 0.8% generally is slightly above the balance. More than 0.8% will gear up fettling; if the amount of Ti in hot metal is less than 0.7%, the fettling would not be effective or it will only witness a little effect. Table 8.22 is the balancing process of Ti during fettling in An Steel. When the amount of Ti in hot metal is 0.075%, the Ti expelled from the blast furnace is larger than the Ti addition, indicating that Ti is of no help in fettling but dissolves part of the Ti-containing materials. Therefore, it is of great significance to keep Ti in balance during fettling [51].

Sumitomo Corporation Kokura Iron-making Plant produced sinteres ore with sand iron with the amount up to 250 kg/t. In January, 1955, TiO_2 residuals in the furnace

Table 8.21 Practice data of fettling

Factory	Number	Year	Ti load, kg/t	Composition of pig iron, %			Content of slag, %				
				Si	Ti	S	CaO	SiO_2	MgO	Al_2O_3	TiO_2
Pan Steel	3	1988		0.07	0.10	0.073	27.2	24.5	9.09	14.2	22.87
Cheng Steel	2	1990		0.27	0.36	0.059	31.7	22.6	4.31	15.71	19.60
Shui Steel	1	1978	138	0.32	0.25	0.038	35.7	30.7	9.45	11.10	13.40
Chong Steel		1978	108	0.41	0.25	0.034	37.7	31.8	8.45	16.38	9.37
Capital Steel	3	1986	22	0.40	0.09	0.040	38.6	36.8	12.30	12.30	1.24
Xiang Steel	2	1982	15	1.17	0.16	0.021					1.29
Shao Steel	2	1985	10	0.88	0.24	0.018	40.4	33.1	6.20	12.40	1.63
Baotou Steel	3	1986	9	0.50	0.16	0.034	38.8	31.2	7.30	8.96	1.76
Tong Steel	2	1986	10	1.00	0.15	0.020					1.75
Shi Steel	1	1987	20	2.57	0.30	0.015	38.6	35.7	9.46	15.48	4.30
	1	1987	28	1.66	0.34	0.024	38.10	35.1	9.77	14.53	4.90
Wu Steel	3	1983	10	0.37	0.14	0.027	38.7	38.0	5.63	12.07	2.63
Ben Steel	5	1985	15	0.69	0.15						2.00
An Steel	9	1987	10	0.65	0.10	0.025	41.1	38.0	7.70	8.01	1.23

Table 8.22 Ti balance during fettling

Overall addition of TiO_2, kg/t	11.0	11.0	11.0	11.0	11.0
[Si], %	1.910	1.305	0.900	0.560	0.095
CaO/SiO_2	1.09	1.11	1.11	1.10	1.11
[Ti], %	0.250	0.230	0.120	0.075	0.045
(TiO_2), %	1.12	1.30	1.72	2.12	2.47
TiO_2 brought by hot metal, kg/t	4.17	3.83	2.00	1.25	0.75
TiO_2 brought by slag, kg/t	5.60	6.50	8.60	10.60	12.35
Overall TiO_2 brought by metal, kg/t	9.77	10.33	10.60	11.85	13.10
TiO_2 residuals at the bottom of hearth, kg/t	1.23	0.67	0.40	−0.80	−2.10

on average were 21.58 kg/day and 6.95 kg/day in February. It is not clear just from Table 8.23 why there is such a huge difference in the amount of Ti left in the furnace. However, it is pointed out that Ti in iron is related to V, C and Zn. The slag rate also has an effect on the retention of Ti. Against the background that the research was still in its infancy at that time, relevant problems were not comprehensively studied.

8.5.6 Fettling Effect When Using Burden Containing Ti

It has a telling fettling effect using burden containing Ti. The effect of Ti contained in hot metal is shown in Table 8.24.

8.5.7 Advantage and Disadvantage of Fettling

Adding burden would pay heavy price as constant and longstanding materials are required at the risk that no iron would be produced. Besides, it also consumes coarse coke. In 1987, Shougang summed up and compared the smelting results of titanium ore and titanium slag which were added to the No. 2 and No. 3 blast furnaces, respectively. Practice has proved that the use of titanium ore for each increase of 0.01% Ti in hot metal, the consumption of coarse ratio will increase of 1.8–3 kg/t; however, with titanium slag, only 1 kg/t increase. Of course, due to the different composition of titanium ore and slag, the results also vary. Although it is certain that the use of titanium slag consumes less coarse coke, another consequence of the use of titanium slag is a significant increase in slag, which is detrimental to blast furnace reinforcement. Table 8.25 shows the composition of titanium materials.

Blast furnace longevity cannot be deepened on fettling, which will be a massive expense after accumulating over years. With modern science and technology, it is entirely possible to expand the life of blast furnace to 20 years without fettling. The economic benefits in this respect are very considerable.

8.6 Furnace Expansion and Crack of Furnace Shell

8.6.1 Crack of Furnace Shell

Crack of furnace shell is a common accident in blast furnace. Jiu Steel No. 2 blast furnace (1000 m^3) was put into operation since October 25, 2000. Five years later, the save cooler of the lower part of the furnace is seriously damaged, and spark and coke power were ejected out from the furnace in certain areas. By 2007, 18 blocks of stave cooler have been overall damaged with a local damage of 81 pieces. The

Table 8.23 Ti balance in blast furnace of Kokura

Date	TiO_2 brought by burden	Ti in iron, %	TiO_2 in iron, %	TiO_2 in slag, %	Amount of slag, kg	TiO_2 in slag, kg	TiO_2 output, kg	TiO_2 left in the furnace, kg	Basicity of slag	Slag rate, %
January, 1955										
1–3	21.28	0.31	5.15	4.09	441	18.04	23.19	−191	1.11	52.0
4–6	22.33	0.31	5.15	4.13	389	16.07	21.22	+111	1.30	50.3
7–9	24.45	0.29	4.82	4.04	433	17.49	22.31	+214	1.20	53.7
10–12	24.15	0.27	4.49	4.14	428	17.72	22.21	+194	1.11	51.6
13–15	25.33	0.30	4.98	4.72	472	22.28	27.26	−193	1.23	63.8
16–18	24.60	0.28	4.65	4.51	438	19.75	24.40	+20	1.19	68.3
19–21	25.23	0.30	4.98	4.56	402	18.33	23.31	+192	1.24	59.3
22–24	25.03	0.24	3.99	4.72	421	19.87	23.86	+117	1.16	52.1
25–27	25.67	0.27	4.49	4.46	429	19.13	23.62	+20	1.23	54.7
28–31	24.57	0.30	4.98	3.87	460	17.80	22.78	+179	1.19	61.1
February, 1955										
1–3	23.82	0.31	5.15	4.36	439	19.14	24.29	−47	1.09	60.1
4–6	22.74	0.28	4.65	4.66	419	19.53	24.18	−144	1.12	66.0
7–9	25.39	0.31	5.15	4.70	392	18.42	23.57	+182	1.26	56.6
10–12	23.59	0.28	4.65	4.62	392	18.11	22.76	+83	1.24	61.2
13–15	23.77	0.27	4.49	4.26	403	17.17	21.66	+211	1.25	52.8
16–18	22.34	0.28	4.65	4.08	438	17.87	22.52	−18	1.08	58.8
19–21	22.82	0.27	4.49	4.20	460	18.90	23.39	−57	1.17	64.1
22–24	22.93	0.29	4.82	4.15	405	16.81	21.63	+130	1.18	73.0

(continued)

Table 8.23 (continued)

Date	TiO_2 brought by burden	Ti in iron, %	TiO_2 in iron, %	TiO_2 in slag, %	Amount of slag, kg	TiO_2 in slag, kg	TiO_2 output, kg	TiO_2 left in the furnace, kg	Basicity of slag	Slag rate, %
25–28	21.69	0.25	4.15	4.23	449	18.99	23.14	−145	1.10	63.0

Table 8.24 TEffect of Ti contained in hot metal

[Ti] contained in hot metal, %	Fettling effect	Ti balancing result	Fettling effect
>0.25	Exceptional effect	$[Ti]_{addition}$ − $[Ti]_{discharge}$ > 0 (far outweigh)	Constant effect in fettling; the lining inside growing thick
0.15–0.25	Strong effect	$[Ti]_{addition}$ − $[Ti]_{discharge}$ > 0	Constant effect in fettling; the lining inside growing thick slowly
0.08–0.15	As low reaction or maintaining the status quo	$[Ti]_{addition}$ − $[Ti]_{discharge}$ ≥ 0	Effect can be seen in the thick part of lining or maintaining the status quo
0.06–0.08	Maintaining the status quo or dissolving slowly	$[Ti]_{addition}$ − $[Ti]_{discharge}$ ≤ 0	maintaining the status quo or the lining gradually becomes thin
<0.06	Being of no help	$[Ti]_{addition}$ − $[Ti]_{discharge}$ ≤ 0 (far outweigh)	Lining inside becomes thin in areas where fettling is finished, such as the high temperature area; place with low temperature still maintains the status quo

Table 8.25 Composition of titanium materials (%)

Composition	TFC	TiO_2	SiO_2	S	P	V_2O_3	Al_2O_3	MgO
Lamp ore of Panzhihua	30.89	10.50	20.00	0.605	0.0126	0.315	8.87	6.21
Iron fine powder of Panzhihua	51.56	12.73	4.64	0.53	0.045	0.564	4.69	3.91
Ti fine powder of Panzhihua	30.58	47.53	3.04	0.30	0.004	0.095	1.34	6.12
Slag of Panzhihua		22.62	25.29	0.39		0.32		8.09
Lamp ore of Cheng Steel	35.83	9.42	17.52	0.50				
Iron fine powder of Cheng Steel	61.51	6.75	2.93	0.051				
Ti fine powder of Cheng Steel	35.00	41.00	2.5	0.30	0.04	5.64		
Slag of Cheng Steel	77.43	15.90	23.85	0.73		0.29	15.06	5.25

corresponding furnace shells cracked seriously, mainly were No. 8, 9 and 10 sectors of stave cooler (Fig. 8.49) [52]. The cracking leads to a huge damage.

No. 1 blast furnace (300 m^3) of Nansteel once has a wide range of horizontal and longitudinal cracks in the furnace shell in the later stage of its operation and a transverse displacement between the stave cooler of the one or two layers of the furnace, with a 90–100 mm displacement and a longitudinal crack width up to 10–20 mm (Fig. 8.50). Although after many times welds, cracks still have a tendency to spread; there was a blue gas fire emitting at the cracks. Since May 1993, the bottom

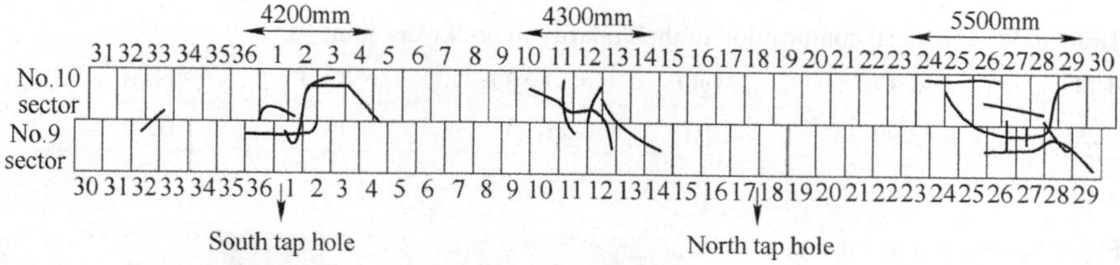

Fig. 8.49 Expanded draft of the cracking on furnace shell

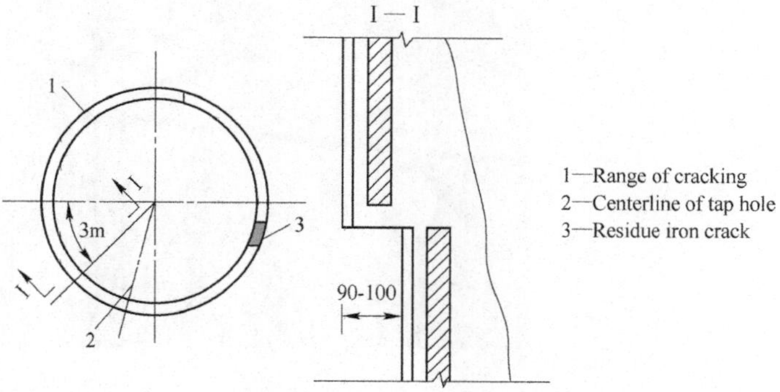

Fig. 8.50 Displacement of the first and second shell of furnace

temperature of the furnace has increased sharply, and the temperature of the bottom of the furnace has risen steeply from the original 644–754 °C. In view of the above, major capital of blast furnace was brought forward [53]. After disassembling the furnace, ring-shape cut-through cracks were found in the outermost brick of the two layers of the hearth from the lower slag notch to the near tap hole. It was at the outside brick of 200 mm, and the crack width changed from 10 mm in the lower slag notch to the tap hole of the 60 mm (Fig. 8.51). The bricks in the ring seam have clearly become grayish white and crisp and were in white leaf shape. After chemical

Fig. 8.51 Vertical ring-shape cut-through cracks

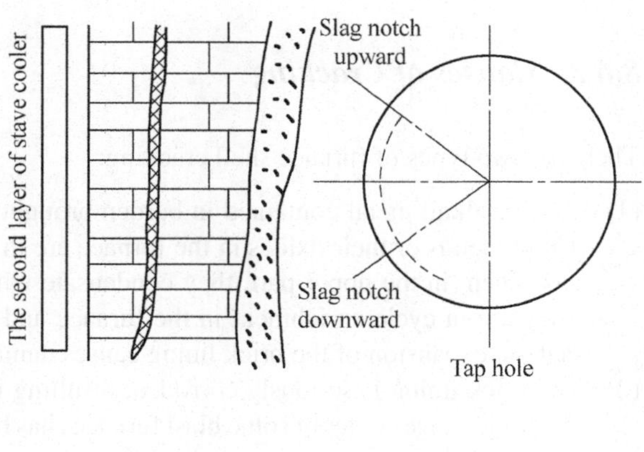

Table 8.26 Chemical composition of the impurity in brickwork joint (%)

CaO	Al_2O_3	MgO	Fe_2O_3	K_2O	Na_2O
1.31	32.99	0.10	0.41	25	0.80

Fig. 8.52 Shell cracking and welding up with strong rebar

analysis, it was found that K_2O, Na_2O content were 25% and 0.8%, respectively. Full analysis is shown in Table 8.26.

Xin Steel No. 6 blast furnace (1050 m^3) was put into operation in November 2005. On February 5, 2007, the first cracking occurred, and in the next two consecutive years, 46 times cracking happened. Major capital was required [54]. Figure 8.52 is the reinforcement of the weld after shell cracking of No. 6 furnace.

After sampling inspection, it was found that carbon brick ring seam generally contained a high level of ZnO or metal zinc, with the highest ZnO content of 64.79%. Carbon brick ash, compared with the original brick, increased greatly up to 81.94%, which is the result of the addition of zinc infiltration into the carbon brick after the formation of ZnO.

8.6.2 Causes of Cracking

There are two types of furnace shell cracking:

(1) Zinc or alkali metal contained in burden brought into the furnace is overdose. These metals or their oxides in the furnace are in gasification and rise with the gas. Then, in the upper part, they condensate with the furnace and then drop. They are in cycle enrichment in the furnace and are deep into brick lining so that the expansion of the brick lining causes damage [55].

(2) The brick lining is seriously corroded, resulting in high shell temperature and cracking. Stave cooler of some blast furnaces has been burned out, and if with no

remedial measures, or even cut down the supply of cooling water, under which the stave cooler was in the condition of water-break, the stave cooler will be quickly and completely burned. At this time, serious crack of shell and breakage is difficult to avoid.

8.6.3 *Measures to Avoid Cracking*

In the type 1 of cracking, the harmful components brought into the furnace by burden should be controlled, especially zinc, potassium and sodium. For most iron-making plants, due to the pressure of production costs, a complete and lasting use of the burden which would meet the needs of smelting is difficult to realize. First of all, circulation of harmful metals in the furnace should be blocked. Furnace dust, recovered by blast furnace, is the major assembling point for harmful metal. A more simple way is to select or collect the harmful metal compounds through "mineral processing" method. These materials will become useful chemical products, which will be sold to the relevant production plant. For example, as the ZnO in furnace dust, which is gathered and sold to the relevant chemical plant.

Now, a mature method of enrichment and selection is realized.

A scientific design of blast furnace, using sufficient qualified fire clay brick with high cooling intensity, can reduce the damage effect of harmful metals. The fundamental method is to control the quality of the burden according to the procurement requirements of qualified burden.

The type 2 of cracking should be avoided altogether.

The design of stave cooling has gained mature experiences [57]. Once occurring breakage, water should not be easily cut off. A more acceptable way is to change to a low pressure water in small amount, extend the life of the main body of stave cooler, while taking advantage of the opportunity to repair the stave cooler. If with no alternative but to cut off water supply, then do it and install cylindrical cooling rod, also known as cooling column, in order to improve the performance of stave cooler when water is cut off.

8.6.4 *Shell Cracking Lead by Expansion*

In 2013, a 450 m^3 blast furnace was put into operation after the major capital. After half a month, blast furnace utilization coefficient reached 3.1. Twenty days later, front part of blowpipe was found upturned, and air leakage was serious at tuyere. At the same time, ten water pipes, which were used to import water into stave cooler, were found that the hearth, furnace waist. Although a number of stop-air treatment were carried out, serious compression deformation in air supply support corrugated compensator was not ameliorated. It went out of blast 28 days after the operation.

A careful examination revealed that the bottom furnace shell was pulled up, about 150–160 mm from the base of the blast furnace, and it was expanded outwards. Thickness of straight section steel at the furnace shell bottom was 28 mm. A crack can be found (90 mm width in lower part, 1800 mm in length).

The research shows that the composition of heat-resistant concrete is magnesium brick waste lining dismantled from the construction of the base pier. Waste magnesium brick is in pulverization in the presence of water and swell, thus pushing up the blast furnace to move up. A little mistake will lead to irretrievable failure.

Once there is a large blast furnace, the front part of blowpipe was upturned because an irregular management and control of harmful metal contained in the burden added into the furnace. Although middle hearth was repeatedly changed, it still cannot avoid air leakage. Five years after the operation, it was found that the bottom plate of the blast furnace was in "pan-pot" shape. The circumference of it was about 160–170 mm in height and 1200–2000 mm in depth. The reason is that the metal zinc in the burden enters into the brick lining with the help of gas zinc oxide so that the volume expansion of the brick lining caused by the damage.

8.7 Grouting, Pressing and Spraying Lining

During the operation, the lining is corroded continually. The repair of the lining is a common method to prolong the campaign of the blast furnace, maintaining normal production of the regular means. Grouting, pressing and spraying lining are successful methods in repairing blast furnace. A number of specialized companies in China now are engaged in this aspect of the work. Only considerations are raised here.

8.7.1 Grouting

After production, it is found that the local furnace shell temperature of blast furnace is too high. First of all, it should be judged whether the reason lies on lining erosion or on the high temperature gas entering from the furnace shell and the stave cooler. If it is the second reason, operator should grout timely, clogging the channel. This type of grouting, also known as cool-side grouting, is safe and reliable. Attention should be paid to three points:

(1) Position selection of grouting hole: Some areas of shell have lots of pipelines but short of voids, which inhibited choices. Therefore, according to the estimated gas path near the hole, hole spacing should not exceed 500 mm in order to ensure that the filling mud can overlap each other.
(2) A moderate addition in single hole grouting: no more than 1 ton per hole. Sometimes many grouting holes have already been blocked and could not be

poured in. So, it should be poured from other hole. Under this situation, it is necessary to make a judgment that whether to pour into the mud filling channel, or make another "short circuit" to result in a large amount of mud loss. There are no very accurate instrument measurements at that time. If grouting holes are found near the grouting hole, operator should open the stop valve and to see if there is mud flowing out. Or to measure the temperature of the furnace skin and see whether the cooling area is enlarged. If the lining is more complete, or the filler between the stave cooler is reliable and sturdy, grouting can be tempted to continued, and if the end-stage blast furnace or lining has been seriously corroded, it should not be continued.

(3) Grouting pressure: Grouting pressure of cold side general is controlled less than1MPa. In 2013, grouting occurred in Xing Steel blast furnace (3200 m^3) between the shell and the stave cooler due to gas leaving out, with pressure of 2 MPa. The inflow and outflow water pipe of No. 10 stave cooler 4 of No. 10 Sect. 4 and 2 fixed bolt were crushed, and about 100 t/h grouting materials were added into the furnace. A few vents were changed of this blast furnace, however, the water vapor is spraying everywhere. The grouting pressure of this stave cooler was 2 MPa, with the actual total pressure of hundreds of tons, more than the tensile strength of water pipes and bolts and fracture occurred. Therefore, controlling the grouting point in the shell around 0.05–0.1 MPa is appropriate. Outside drum of Bao Steel No. 4 blast furnace, hearth shell was 100 mm, and it also caused by high cold side grouting pressure.

8.7.2 Hot Surface Grouting

Hot surface grouting refers to the process that the mud reaches the brick lining and stave cooler between or between the brick lining. If the gap between bricks is large or the corrosion of lining is serious, grouting should be performed. Explosion might happen due to grouting, through which the mud will reach red-hot coke and hydrogen gas, especially in the part of the hearth. Therefore, hot surface grouting is full of dangers. There have been many explosions due to hot surface grouting, leading to heavy losses. In the blast furnace where grouting explosion occurred, the furnace wall will be badly damaged, and the hot metal would flow out. The blast furnace needs to be gone out of blast; in some of the blast furnaces, explosion occurred, and some workers who were at that time near to the furnace were burned and killed. In February 2010, a blast furnace stopped the blast, many workers near the vents are working, and grouting was performed at the same time, resulting in explosion. Flames were ejected from the vents, and dozens of people were burned, of whom 13 were seriously injured.

In August 2010, a blast furnace was put into operation for seven months and then stopped for repair. "Grouting was operated in the lower right side of No. 1 tap hole, at the Sect. 2 No. 5 stave cooler. A sudden fire broke out in the furnace" [56]. One worker happened to pass in front of the tuyere and was killed by the fire ejected from

Fig. 8.53 Gaps formed from carbon brick after grouting explosion

the furnace. Repair began in three days later. After removing the stave cooler, it is found that there were two 30–70 mm gaps in the carbon brick inside, as shown in Fig. 8.53.

In 2010, breakage occurred in a 2500 m^3 blast furnace. "In the morning of August 20, grouting is operated normally. During the process, at 10:54, 7 tuyeres are black in front (16#-22#), while the blast pressure is increased from 355 to 400 kPa. In the tuyere 25#, tar substances outflow from large suites. After the blast releasing, it was found that the blocking materials are coke breeze mixed with coke powder, and 7 h later (at 20:08), breakage occurred [58].

Hot surface grouting should be carefully treated. If the lining is serious corrode, hot surface grouting is forbidden.

8.7.3 Hard Material Pressed into the Lining

This is first started by Bao Steel. As a regular maintenance furnace technology, it has been used in Bao Steel from No. 1 blast furnace. This kind of pressing lining mainly used in middle and low part of the cooling plate and the cooling rods inserted in maintenance part of the furnace. The effect is great. 'Patching mainly depends on hard material. Taking advantage of blast furnace regular wind suspension (usually 20 h) and forming a 17.4–19 MPa high-pressure pump near the furnace shell, as a result, reducing the furnace shell temperature.' [59] Fig. 8.54 is the position of the height and circumference along the blast furnace of Bao Steel No. 2 blast furnace, which was pressed with hard mud.

Many blast furnaces apply this method with the great performance. Attention must be paid to the pressure control of the pump, according to the situation of the blast furnace, to prevent excessive pressure, damage to the furnace body.

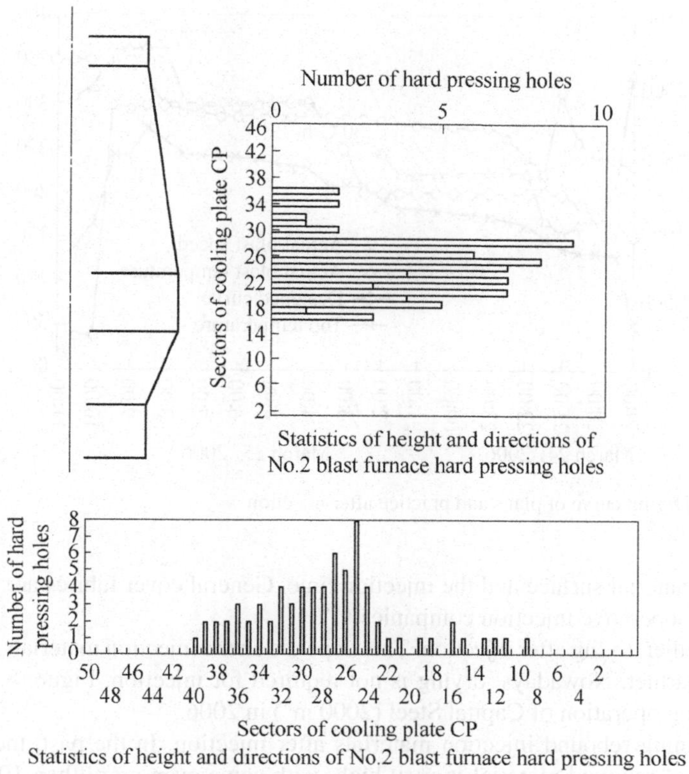

Fig. 8.54 Protection of No. 2 Bao Steel blast furnace

8.7.4 Injection Lining

Injection lining is prevalent in the country. The implementation of spraying is from the bosh to the body no matter in local area or all body. Many companies specialized in their own development of ingredients and specialized equipment, engaging in injection operations. Attention should be paid on injection operation in the following parts:

(1) Before spraying, furnace wall should be cleaned to remove the local adhesion connector. Slight bonding can help to clean edge airflow; if there is a nodule, the furnace should be effectively washed. If not suitable for washing, other measures should be taken to remove the furnace tumor. Injection paint should be injected on a relatively clean furnace wall.
(2) Dropping the material surface below the part where it needs to be injected. Drop surface operation is same with maintenance stop blast drop surface; the difference lies on that after the material surface drops in place, the fabric needs to be stamped, and the number of cover fabrics is determined by the depth of

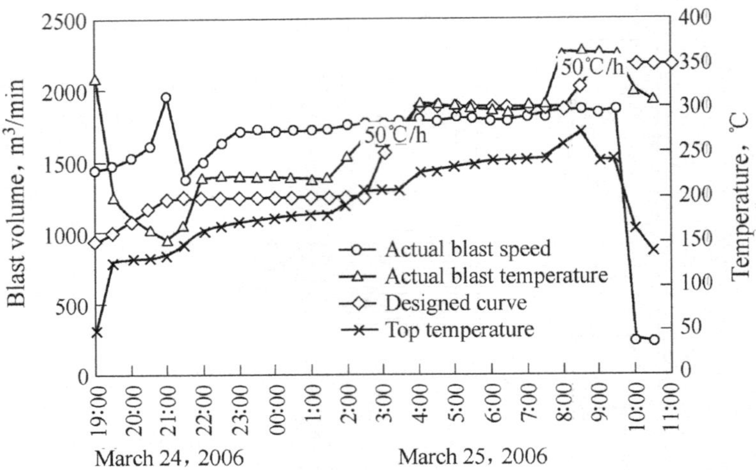

Fig. 8.55 Drying curve of plans and practice after injection

the material surface and the injection time. General cover fabrics are provided by cooperative injection companies.

(3) Whether drying after injection. That depends on the injection materials provided by partner. Nowadays, drying is not required for injection. Figure 8.55 is the drying operation of Capital Steel (2000 m^3) in 2006.

(4) Cleaning rebound injection materials after injection. In the past, the proportion of rebound material is very high, with some even more than 10%. Now, the implementation of "wet spraying" is realized, and the rebound material is generally about 5%, the future will be lower. Bounce material and cover fabric, conditional should be cleared out. In the future, the rebound material will be much less, and no need to clean.

8.8 Precaution of Furnace Damaged

8.8.1 Blast Furnace Structure Determines the Life of the Blast Furnace

Longevity is impossible if the design for furnace is wrong. Reasonable structure is an important factor. Since the invention of cooling bottom, furnace bottom breakage has rarely occurred, and that lies on the result of a reasonable structure. The first principle of blast furnace design is: from the stave cooler, the more inward (furnace bottom upward), the smaller the thermal conductivity should be. In the other words, the more outward (bottom of the furnace), the greater the heat conduction capacity should be. Blast furnaces are generally multi-layer structure: furnace shell steel plate, cooling device, cushioning filler, refractory brick lining, etc., and any part of the middle do

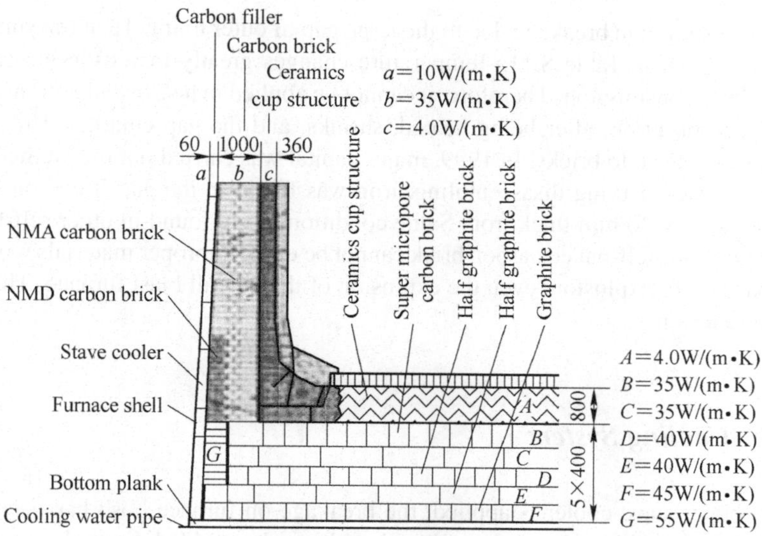

Fig. 8.56 Hearth structure from the perspective of heat transmission

not allow the existence of "thermal resistance". There is a "thermal resistance" in the middle, which leads to "heat accumulation" and the temperature will also rise here. That is the reason for the breakage of furnace body.

Figure 8.56 is a schematic diagram of the furnace cylinder structure. In general, the choice of refractory materials should be very prudent. Sometimes, inside padding wall of stave cooler will be ignored. Although it is very thin, but the requirements are very strict, that is, to have a high thermal conductivity, but also has a good "elasticity," which can absorb the deformation of the brick lining. Thermal conductivity is not a "thermal resistance layer"; a large shrinkage will lead to a gap between stave cooler and brick lining; if the shrinkage is of absence, "rigid body connection" will happen between brick lining and stave cooler. When displacement of the furnace shell or brick lining deformation happens, will produce stress, "crack" is possible.

8.8.2 Material of Furnace Lining

Working conditions of furnace hearth and bottom are bitter, and they should have the features such as anti-iron, slag erosion capacity. Practice has proved that high graphite brick lining is easy to be dissolved by hot metal, although they are very suitable for the furnace hearth work.

The material is analyzed separately and performs well, but it must be considered as a whole. The lining of different materials, in the whole, must adapt to each other. The most obvious is the ceramic cup furnace structure. Ceramic cup material expansion rate is higher than the outer layer of brick lining. It expands during production process, resulting in breakage due to the large gap of outer lining. Heat transmission index is shown in Table 8.11. Temperature changes greatly in self-drying carbon brick's heat transmission. Therefore, it cannot be applied in hearth and bottom. Self-drying carbon brick, after being heated, shrinks, and the gap enlarges. Hot metal will easily enter into bricks. In 1999, major capital was carried out on An Steel No. 4 blast furnace. During disassembling, iron was found in the gap if carbon brick, consisting of 5–30 mm thick iron. Same conditions were found in other self-baked brick stove. So, self-baked carbon brick cannot be used. Improper materials will lead to expansion or explosion, even the explosion of the overall blast furnace. This is a profound lesion.

8.8.3 Cooling System

Since copper stave cooler is applied, the breakage on furnace bosh has been very few. But the hearth was burnt through and could not be avoided. From the operation analysis, the structure, material and cooling are three main aspects. Some design experts put forward that China's blast furnace wall water consumption is higher than Western Europe and Japan. Why after breakage, people always tend to consider about the insufficient cooling strength and water. In fact, the cooling strength depends on whether the heat passed in the furnace to the furnace wall device can be taken out. Therefore, the cooling strength depends on the structure, material and construction quality. Some blast furnaces tend to ignore the heat transfer effect of the filling layer, although this layer is very thin, which rarely more than 80 mm. Some building furnace companies also do not pay attention to the construction quality of this thin layer, and they learn less about the filling layer voids, resulting in the formation of a "heat resistance layer". Some blast furnace construction quality is strict, but there are some improper selections of materials. It could also lead to the result that thermal conductivity is less than lining and easily causes "heat accumulation". As for such cases, operator can only increase the amount of water.

Water quality affects heat transfer capacity to play. If water temperature is too high or contains impurities, deposition and scaling are possible in the cooling device, reducing the cooling capacity. Regular detection of cooling device working state is necessary.

8.8.4 Construction Quality

In 1995, breakage occurred in a largest blast furnace hearth in a famous steel company at the time. Pulling down the cooling wall of the burning site, operator found that red coke can be seen from outside in the bricks close to the breakage. The brick was in a wedge-shaped, and the inside of the gap seems to be wider than the outside side. Company leaders attached great importance to it and asked for the chief engineer for help. According to him, it was not about construction quality. Here is the "brick seam"—the gap in the last brick of a circle of bricks, while others will not. This is clearly a hidden danger.

There are many reasons to burn through, not just masonry quality.

Furnace cylinder is the part of hot metal; construction quality cannot be ignored. Table 8.27 is a record of the construction quality inspection of Capital Steel hearth.

Breakage occurred sever times in Capital Steel. Therefore, great importance is been attached to the quality of the furnace. Operators with great experience are send

Table 8.27 Quality of Capital Steel No. 1 blast furnace ninth generation, compared with international standard

Name	GBJ8—64 stipulation, mm	Qualification rate of No. 1 furnace-constructing, %					
		≤0.4	≥0.5	≤0.75	≤1.0	≤1.5	≤2.5
I. High alumina brick or chamotte 1. Bottom: (1) vertical gap (2) Horizontal gap 2. Hearth 3. Bosh and belly 4. Body	≤0.5 ≤0.75 ≤0.5 ≤1.0 ≤1.5	84.0 81.6	100 85.4 100	100 78.1 92.05	99.9	100	
II. Carbon construction 5. Bottom (1) Oblique joint gap (2) Radioactive gap (3) Horizontal gap 6. Hearth: (1) Radioactive gap (2) Radioactive gap	≤1.5 ≤2.5 ≤2.5 ≤2.0 ≤2.0	<spread at whole bottom> <comprehensive bottom>		64.5	98.4 64.8 90.5 54.9 65.7 95.3		100 99.5 100 100 100
III. Flatness 7. High alumina brick (8 m with guiding rule) 8. Carbon (3 m with guiding rule) 9. Elevation difference in different points at the layer of bottom	≤3.0 ≤2.0 ≤5.0						≤1.5 100 ≤5.0 100

to furnace masonry, with plug, valet into the furnace inspection. They record the results of the check in detail.

From Table 8.27, it is seen that the bottom of the furnace, furnace cylinder brick seam is not more than 0.4 mm, accounting for more than 81%, vertical seam is 100% less than 0.5 mm.

8.8.5 Control the Amount of Harmful Elements into the Furnace

Control harmful elements, especially K, Na, Zn and Pb. The damage effect of lead on the bottom of the furnace is very prominent. When the lead vapor escapes from the hot metal outlet, it damages the health of the pre-furnace workers. Zinc, lead limit of the factory regulations are different and are generally 0.15 kg/t. High-lead ore should be smelted in the discharge plant of the blast furnace. Alkali metal is higher than the limit, and it should take alkali discharge operation.

8.8.6 Fettling Operation

When breakage appears in the furnace hearth or furnace bottom, titanium containing materials should be timely added. As long as the fettling operation is timely and correct, breakage can be avoided.

8.8.7 Interim Measures to Secure Time

Stop blast at the tuyere and stop blowing on are effective but reduce the level of production. It should not be applied unless necessary. In addition, these measures can be temporary emergency, in exchange for furnace time. However, operation will be powerless if breakage occurs. Therefore, in a critical moment, stopping the blast and reducing production are the last resort and should be implemented immediately to avoid breakage.

Increasing water pressure is an effective method to increase the amount of water. Figure 8.57 is a schematic diagram of the changes in water pressure and cooling water flow made by Capital Steel [38]. The relationship between cooling and hydraulic pressure depends on the characteristics of the pump, the structure of the cooler and the composition of the cooling water pipe network.

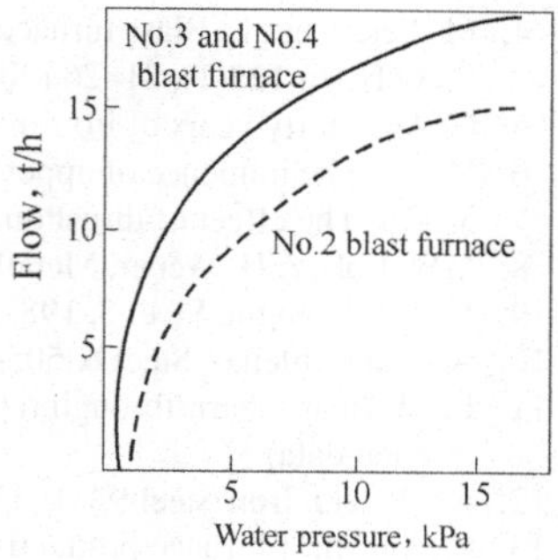

Fig. 8.57 Relationship between cooling water flow and water pressure

8.8.8 Reasonable Bake

Blast furnace design expert Zou Zhongping mentioned the impact of the oven on the life of the blast furnace, and it is very reasonable. He pointed out that the bake process should ensure the cold wall temperature to a certain level and to ensure that bonding material consolidation between the furnace shell and cooling wall and between the cooling wall and the brick lining. Otherwise, it would be easy to be blown by the blast or gas, resulting in a large thermal resistance of the "air gap". He pointed out that during Bao Steel No. 1 blast furnace baking, Japanese experts proposed that the furnace shell temperature should reach 65 °C. This temperature will ensure the furnace shell and stave cooler away from damage and can promote the consolidation of the unshaped bonding material here.

8.8.9 Improving and Tightening Testing System

Improving the detection means and timely understanding of the actual situation of the lining is conductive to take steps, including the end of the blast furnace cooling system inspection project. For the place which may be burned, attention should be attached, and operation should be under surveillance. At the same time, the inspection cycle should be short. For major danger points, a cordon should be put in place, alarm if so timely, and report to the relevant leaders.

References

1. Q. Tang, Re-understanding of some problems of prolonging the life of blast furnace hearth. Iron-making **5**, 7–11 (2014)
2. S. Zhang, Z. Yu et al., *BF Longevity Technology of Wu Steel* (Metallurgical Industry Press, Beijing, 2009), p. 137
3. Y. Ryoji, et al., Ten years life of No. 5 BF in Fukuyama, in *Iron-making Proceedings* (1985), pp. 149–163

4. M. Peters et al., Blast furnace relining strategics for campaigns lives of more than 20 years. Stahl Eisen **123**(1), 21–26 (2003)
5. C. An, Thirty years of iron-making at Capital Steel. Capital Steel Technol. 472–487 (1983)
6. Y. Liu, The influence of upper thrust on operation in blast furnace. Iron Steel **12**, 13–16 (1988)
7. Y. Liu, The effect of thrust on the hearth of blast furnace. Iron Steel **12**, 1–4 (1995)
8. V.W. Lobyz, H. Weber, Metall. Trans. **2**, 14–18 (1956)
9. N.N. Babaregin, Steel **3**, 198–200 (1961)
10. A.I. Sagoruleiko, Steel **6**, 505–512 (1951)
11. L. Li, Hearth burn-through repair and production of No. 4 blast furnace. (1985) (Capital Steel internal data)
12. X. He, etc. Iron Steel **9**, 31–33 (1956)
13. C. An, Blast furnace production. 479–487 (1983). (Capital Steel internal data)
14. M. Xu, Q. Liu, *100 Cases of Blast Furnace Accident Handling[M]* (Metallurgical Industry Press, Beijing, 1986), pp. 229–278
15. V. Paschkis. Iron and Steel Engineer, 1954(2): 53–66
16. V. Paschkis, Iron Steel Eng. **6**, 116–132 (1956)
17. Y. Liu, Treatment and maintenance of burning-out of BF bottom. Iron-Making **4**, 19–21 (1999)
18. G. Duan, Bottom burning accident analysis of 100 m^3 BF at Tang Steel. Iron-Making **2**, 47 (1994)
19. J.M. Bauer, J. Schoennahl, D. Kuster, Ceramic cup in the hearth of blast furnace, in *Iron-making Proceedings* (1988), pp. 27–36
20. J.M. Bauer, Technical trends of blast furnace refractory linings in Europe. Refractories **43**(10), 515–528 (1991)
21. M. Song, Z. Lu, Evaluating the harm of the structure of ceramic cup, in *Hearth of Blast Furnace Collected Works of the Annual Conference on Iron-making (Part II)* (2012), pp. 108–111
22. Translated by Q. Chen, Ten-year development of ceramic cups for lining hearth of blast furnace. Foreign Refractories **11**, 10–16 (1996)
23. H.J. Bachhofen, W. Kwalski, M. Peters, R. Peter, New experience with an improved ceramic cup lining in the blast furnace hearths of Thyssen Krupp Stahl AG, in *Iron-making Proceedings* (1999), pp. 681–690
24. Z. Xiang, X. Wang et al., *Blast Furnace Design—Theory and Practice of Iron-making Process Design* (Metallurgical Industry Press, Beijing, 2014), p. 563
25. F. Zhang et al., *Modern Blast Furnace Longevity Technology* (Metallurgical Industry Press, Beijing, 2012), pp. 504–524
26. W. Kowalshki, H.J. Bachhofen, H.P. Ruther, Investigation on tapping strategies at the blast furnace with special regard to the state of the hearth, in *Iron-Making Proceedings* (1998), pp. 595–606
27. M. Peters, P. Schmole, H.P. Ruther, Blast furnace relining strategies for campaigns of lives more than 20 years. Stahl Eisen **123**(1), 23–26 (2003)
28. F. Zhang, L. Liu, H. Xu, Design of ceramic cup for No. 1 BF of Capital Steel. Capital Steel Sci. Technol. **1**, 6–8 (1995)
29. B. Shan, Application analysis of ceramic cup hearth of Capital Steel No. 1 blast furnace. Iron-Making **1**, 9–13 (2000)
30. Y. Li, Structural and performance optimization of refractories for blast furnace hearth and hearth. Ph.D. Dissertation, Wuhan University of Science and Technology, 2014
31. Y. Hao, Q. Chen, Iron-Making **(Supplement)**, pp. 7–27
32. Q. Tang, Discussions on the structure of blast furnace hearth, in *Report at the Annual Meeting of Blast Furnace Iron-making* (2015)
33. T. Yang, H. Zuo, Ways of scientific development of blast furnace iron-making technology in China. Iron Steel **1**, 1–5 (2008)
34. M. Higuchi, in *Iron-Making Conference Proceedings* (1978), pp. 492–503
35. K. Chun, et al., Iron Steel **S**, 90 (1979)
36. Y. Han, Iron-Making **5**, 18–21 (1975)

37. Iron-making Branch of Xiang Steel. Information of the Conference on Appraisal of Vanadium-Titanium Magnetite Used in BF Furnace Protection, 1984; Q. Meng, K. Liu et al., Iron-Making **2**, 1–4 (1992)
38. S. Su, Iron-Making **5**, 21–23 (1987)
39. X. Wen, Baosteel's practice of protecting furnace with titanium-bearing raw materials, in *Theory and Practice of Protecting Furnace with Titanium-bearing Minerals*, ed. by J. Song (Metallurgical Industry Press, Beijing, 1994), pp. 98–102
40. A. Shigemi, *Iron-making Handbook* (Diren Library, Tokyo, 1979), p. 245
41. Y. Ren, L. Jiang, Low study on mineral composition and formation mechanism of titanium deposits in blast furnace. Iron Steel **5**, 1–5 (1988)
42. I. Toshihino et al., Titanium solidification in hearth of blast furnace. Stahl Eisen **94**(2), 47–53 (1974)
43. Y. Dong, S. Yu, Behavior of titanium in blast furnace and its effect on hearth and bottom life. Iron Steel **2**, 3–8 (1988)
44. Y. Li, Principle of smelting vanadium titanomagnetite by blast furnace, in *Principle of Smelting Vanadium Titanomagnetite in Blast Furnace*, ed. by H. Du et al. (Science Press, Beijing, 1996), p. 63
45. Z. Pei, Material composition and characteristics of titanium deposits in blast furnace hearth. Iron and Steel **1**, 13–18 (1990)
46. G. Narita, Iron Steel **5**, 65–74 (1976)
47. P. Chen, J. Song, Graphics of furnace protection and operation for titanium-bearing materials, in *Technology of Furnace Protection for Titanium-Bearing Materials of Blast Furnace*, ed. by J. Song (Metallurgical Industry Press, Beijing, 1994), pp. 65–70
48. J. Song, Y. Song, The present situation and Prospect of using titanium-containing materials to protect blast furnaces in China, in *Technology of Furnace Protection for Titanium-Bearing Materials of Blast Furnace*, ed. by J. Song (Metallurgical Industry Press, Beijing, 1994), pp. 1–8
49. C. Du, Z. Mo, Furnace protection technology for titanium-bearing materials in blast furnace]. Iron-Making **1**, 1–3 (1992)
50. T. Lin, Treatment of titanium hot junction in No. 1 blast furnace of Xiang Steel. Hunan Metall. **6**, 21–24 (1998)
51. Z. Miao, A. Li et al., Practice of vanadium-titanium mine furnace protection in An Steel, in *Technology of Furnace Protection for Titanium-Bearing Materials of Blast Furnace*, ed. by J. Song (Metallurgical Industry Press, Beijing, 1994), pp. 114–120
52. J. Li, et al., Reasons and countermeasures for furnace skin cracking of Jiu Steel No. 2 BF. Iron-Making **4**, 22–25 (2007)
53. J. Zhou, et al., in *Papers of the Annual Meeting of Small and Medium Blast Furnace Iron-making*. Harbin, 2010, pp. 147–149
54. B. Zhuang, C. Wang, J. Guo, Reasons and countermeasures for shell cracking of No. 1 BF in Nan Steel. Iron-Making **5**, 40–43 (1994)
55. T. Tang, Measures for prolonging the service life of blast furnaces, in *Blast Furnace Design: Iron-making Process Design—Theory and Practice*, vol. 2, ed. by Z. Xiang, X. Wang (Metallurgical Industry Press, Beijing, 2014), pp. 565–577
56. Q. Tang, Hearth and bottom burning accident, in *Blast Furnace Abnormality and Accident Handling*, ed. by S. Zhang (Beijing: Metallurgical Industry Press, 2012), pp. 93–124
57. Q. Wu, Blast furnace body, load, in *Blast Furnace Design: Iron-making Process Design—Theory and Practice*, vol. 2, ed. by Z. Xiang, X. Wang (Metallurgical Industry Press, Beijing, 2014), pp. 359–440
58. Q. Liu, H. Chen, H. Wei, Shagang No. 1# 2500 m^3 BF hearth burnout, repair and re-production, 2013 (internal information)
59. X. Wen, W. Li et al., Iron-making, in *New Century Vocational Training Series for Iron and Steel Industry*, ed. by L. Xu (2000), pp. 99–100

Chapter 9
Scaffolding

Abstract The damage caused by accretion to the blast furnace is very serious. It is because that the accretion on higher part of stack is difficult to remove by smelting and generally it is necessary to stop the smelting to remove it by explosion, which needs to open holes on the sealed furnace body in general situation and sometimes needs to remove the cooling staves and lining. This is not only the loss of production, but also a serious damage to furnace body. Although the accretion in the middle and lower parts of the stack can be flushed away, the loss and destructiveness are also amazing. There was a 300 m^3 small blast furnace, for the treatment of removing accretion, more than 1000 t of coke and 200 t fluorite were consumed in 20 days before and after and the output loss was more than 2000 t [1]. There was also a blast furnace, due to the late discovery of the accretion, the blast furnace was abnormal for a long time, and the loss was very much. Therefore, the prevention of accretion is really the primary task of blast furnace operators.

9.1 Symptom of Scaffolding

Scaffolding has been troubling China's iron-making workers for many years and almost all iron-making plants have had a painful experience of scaffolding. However, scaffolding could be prevented actually and some factories have already eliminated scaffolding, so the important experience depends on prevention: Once there is a sign of scaffolding, the operator should deal with it immediately in order to eliminate the latent accretion in the bud. Therefore, the key is to promptly determine the signs of latent accretion.

9.1.1 Heat Insulation at Scaffolding Parts

There is heat insulation of the furnace wall because of scaffolding, which leads to the reduction of heat transferred from the inner furnace to the furnace wall and results in changes in many measurement parameters:

Y. Liu, *The Operation of Contemporary Blast Furnaces*,
https://doi.org/10.1007/978-981-15-7074-2_9

(1) Decline of heat flow intensity. The intensity of heat flow through the furnace wall decreases obviously at the scaffolding site of blast furnace. It is more feasible to judge the thickness of the furnace wall according to the change of the thermal load of the staves. It is more complicated to measure the thermal load of staves, because it is necessary to measure the volume of cooling water and water temperature, especially the closed-circuit soft-water-cooling system connected upper and lower parts are difficult to measure. However, it is still effective to determine the scaffolding of the furnace wall by the change in the thermal load of the staves. In 1989, Capital Steel No. 1 blast furnace (2536 m^3) was compared with the thermal load changes of the staves. Later, the operators stopped the furnace and observed. It was confirmed that the scaffolding area of the furnace wall was basically consistent with the judgment [2].

(2) Lower temperature of furnace wall than normal level. The temperature of the furnace wall varies according to the installation position of the thermocouple and the depth of the embedded brick lined. The temperature value displayed in the normal furnace type is also different. Since November 7, 1980, the temperature of the lower part of the Capital Steel No. 2 blast furnace had generally declined and was lower than the normal level (200 °C) two days later, thus, the operators could judge there was scaffolding in the lower part of the furnace (Table 9.1).

(3) Water temperature difference lower than the normal. The temperature difference between the inlet and outlet of the staves could also reflect the condition of scaffolding, but the temperature change of the thermocouple in the furnace wall was slower because the staves were close to the shell. In the 1980s, Capital Steel often measured and counted the water temperature difference of the water tank to monitor the furnace type. Table 9.2 shows the variation of the furnace wall in Capital Steel No. 3 blast furnace before and after getting scaffolding in 1972.

It can be seen from the table that the number of staves with high water temperature difference began to decrease on March 9th. On March 10th, only one piece was left when the temperature was higher than 2 °C and 12 pieces were increased when the temperature was less than 1 °C. It was also the same for the temperature differences between the inlet and outlet water of staves at the bosh. And on March 12th, the blast furnace was seriously disordered. Experience has shown that once the water

Table 9.1 Temperature changes in the lower part of Capital Steel No. 2 blast furnace (November, 1980)

Date	Time	Temperature, °C			
		East	South	West	North
8	14:40	140	300	258	390
9	2:50	120	210	140	260
	10:45	110	210	220	170
	19:00	100	160	130	110
10	5:10	110	200	180	130

Table 9.2 Number of stave blocks at Capital Steel No. 3 blast furnace with temperature differences between inlet and outlet water

Date	Temperature differences of stave cooler at belly, °C				Temperature differences of stave cooler at bosh, °C					Furnace wall changes
	<1.0	1–1.5	1.6–2.0	2.0–2.5	1.1–1.5	1.6–2.0	2.1–2.5	>2.5	>2.1	
March 8th	2	25	14	6	6	16	14	6	20	Transition period
March 9th	1	25	19	3	8	15	16	1	17	Transition period
March 10th	12	23	12	1	8	15	16	6	22	Transition period
March 11th	11	29	5	3	5	17	14	4	18	Transition period
March 12th	11	22	14	1	11	27	2	0	2	Period of getting thick
March 13th	9	20	15	4	8	16	12	4	16	Period of getting thick
March 14th	13	26	8	1	14	23	2	1	3	Period of getting thick
March 15th	12	31	5	0	21	15	4	0	4	Period of getting thick
Apr. 1st	0	18	18	12	5	18	11	6	17	Normal period
Apr. 2nd	0	6	14	28	5	21	7	7	14	Normal period
Apr. 3rd	0	2	18	28	0	14	17	9	26	Normal period

temperature difference of the staves reflects the change, the scaffolding of the furnace wall is already serious, and it would be difficult to remove, except flushing the wall. Therefore, it was necessary to constantly grasp the temperature change of the furnace wall because the temperature change of the furnace wall was more sensitive and timely. It was the same to monitor the changes of furnace wall with heat load of water tank. The water temperature differences between the inlet and outlet of the water tank were multiplied by the flow rate of the cooling water, and the heat taken away by the cooling water of the water tank was obtained. If the amount of water was constant, the temperature differences between the inlet and outlet water could represent the heat load.

(4) Obvious decline of temperature. At the initial stage of nodulation, the temperature of the shell reduced not significantly; when the area of the scaffolding

Table 9.3 Surface temperature of the upper bodies of old No. 2 blast furnace

Gradient of vertical temperature, °C	Average transverse temperature, °C	Temperature of each measuring point, °C																	
+7.8	68.5	76	77	69	58	61	60	68	70	67	72	84	87	87	97	93	84	80	80
–0.7	76.3	101	95	79	70	61	66	72	57	59	66	87	93	107	106	96	90	92	99
–1.6	75.6	106	98	91	90	59	63	57	58	60	63	82	92	105	105	94	94	93	91
+4.1	74.0	98	100	93	92	61	60	58	59	60	62	82	98	107	93	90	92	82	93
–9.55	78.1	94	97	97	97	88	92	67	61	65	62	65	63	76	103	119	111	104	101
	68.6	80	92	88	105	94	67	52	52	57	62	62	62	56	58	99	100	85	100
Average vertical temperature, °C		92.5	53.2	89.3	85.3	70.5	68.0	62.3	59.5	61.3	64.5	77.0	82.5	89.7	93.7	98.5	95.2	89.3	94.0

Note: 1. Measuring time: September 27, 1977

2. Measuring tool: WREA-981M Surface Temperature Measurement Meter; Atmospheric Temperature: 25 °C

3. Wind level 3; Wind direction: north to south

4. Measuring height ranges from gas sampling hole of throat to the first layer of beam water tank. The longitudinal space of the measuring points is 1200, 1000, 1200, 1200,12000 mm from top to bottom. The lateral distribution of the measuring points is from the No. 1 tuyere to the No. 10 tuyere

expanded or the thickness increased, the temperature of the shell reduced obviously. For the blast furnace at the end of its campaign, most of the thermocouples on the furnace wall had been burned out, and the water tank was also quite damaged. The normal means of monitoring the furnace wall had been unable to fully play a role. Under such conditions, it was sometimes necessary to use the shell temperature to judge the accretion at furnace wall. The temperature of the shell was slow to reflect, if it was not the accretion at furnace wall, it would be generally not easy to detect. Table 9.3 shows the results of measuring the shell temperature during the scaffolding at furnace wall of the Capital Steel old No. 2 blast furnace. The temperature in the area of the wireframe in the table was 10–30 °C lower than the temperature in the adjacent zone, which was the scaffolding range of the furnace wall.

9.1.2 Scaffolding Makes the Distribution of Burden Changing

9.1.2.1 Accretion Causes Burden Surface Deflection

Accretion at furnace wall affects the distribution of burden. Hadak and others had used the model to study the influence of accretion thickness of furnace wall on the burden distribution [3]. Figure 9.1 shows the experimental result. Figures A and B represent the normal furnace type and the furnace type with accretion, respectively. Figure 9.1a shows the trajectory and duration of the burden descending from the throat. Accretion hinders the decline of burden, changes the burden trajectory and extrudes the burden moving toward the center, thus, the movement trajectory and duration of each part have changed significantly. Figure 9.1b is a comparison of the velocity of the burden movement. The ratio of the descending velocity of the burden of throat w_0 to the velocity at each point w, w/w_0 was used to draw the line of is velocity distribution.

It can be seen from the figure that the moving speed of burden at the upper part of the accretion, but the moving speed is fast at the lower part as opposed to the upper part. Due to the velocity difference, the distribution density of the burden changes, and the accretion was extruded by the upper burden, of which the density was larger and the lower part was looser. As a result, the burden distribution was seriously affected. In the upper part of the accretion, the velocity of burden near the furnace wall was relatively slower than the burden velocity in other areas, thus, causing the burden surface to tilt.

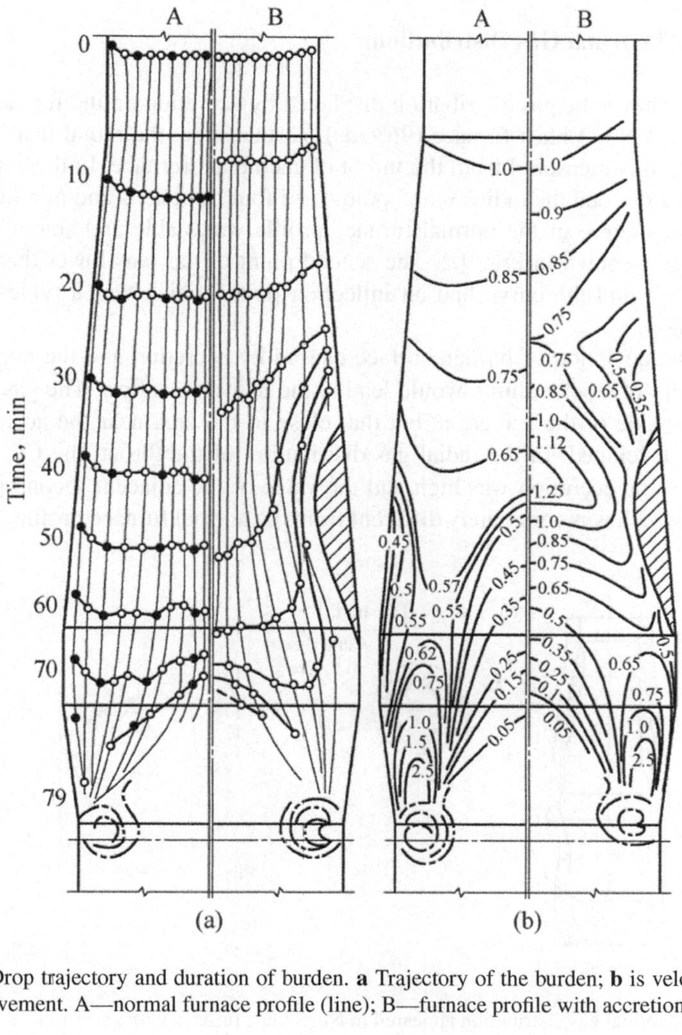

Fig. 9.1 Drop trajectory and duration of burden. **a** Trajectory of the burden; **b** is velocity of the burden movement. A—normal furnace profile (line); B—furnace profile with accretion

The above-research results have been confirmed by practice. Scaffolding often causes the deviation of burden surface and causes a larger gap in the depth of the stock rod. The more serious the scaffolding, the larger the differences of stock rods. The difference of stock rod of some reached 1 m because of the accretion at furnace wall. Of course, the segregation of burden does not necessarily because of accretion. Sometimes, the channeling or the defect of burden distribution would also result in the segregation of burden. It should be judged comprehensively with other parameters.

The influence of annular accretion on burden surface is sometimes not obvious. To a large extent, it depends on the uniformity of the distribution of accretion. If the accretions in all directions are not much different, the stock rod does not tilt obviously.

9.1.2.2 Abnormal Gas Distribution

Figure 9.2 shows the gas distribution displayed by accretion on the furnace wall of Capital Steel No. 3 blast furnace (963 m^3). At that time, the radial distribution of CO_2 in gas was measured from the throat of each blast furnace. In the experiment, the operator divided the radius into 5 points and four directions and measured every 4 h. The gas curve of the normal furnace profile was stable and smooth, and the scaffolding is shown in Fig. 9.2. The second point of gas was lower than the first point (edge), and the curve had an inflection point, which was a typical sign of scaffolding.

Both the deviation of burden surface caused by accretion and the non-uniform permeability of stock column would lead to the uneven gas flow. The gas flow was less on the side of the accretion, but that close to the area near the accretion was relatively developed. In the radial gas distribution of the throat, the CO_2 value at the edge of the accretion was high and the value of the adjacent second point was very low, which was completely different from the normal furnace profile. However,

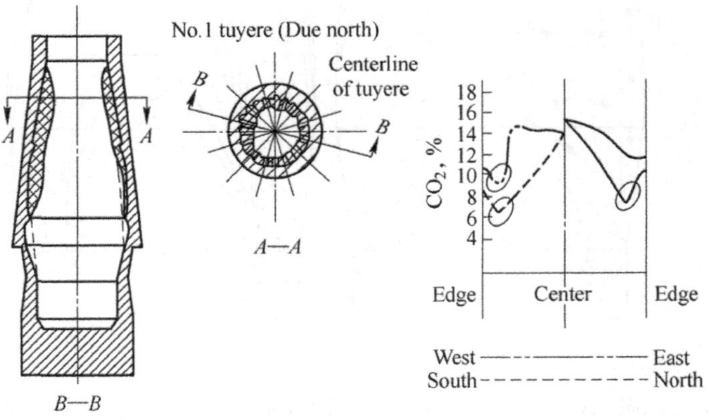

Fig. 9.2 Abnormal gas distribution appeared in No. 3 blast furnace with accretion

the radial temperature measurement (cross-beam temperature measurement) of the throat was totally opposite to the gas curve: The edge temperature measurement at the upper part of the accretion was lower because of less gas passing through, while the temperature of the second point near the accretion was higher because more gas passes through it.

When there is accretion in the upper part of furnace, the ore rolled to the opposite side after charging, so the opposite side was quick to discharge, and the CO_2 content in the gas was relatively high. However, the scaffolding side was opposite: The top temperature was low and the CO_2 content was also low.

The gas distribution in the annular accretion was heavy at the periphery and light at the center, and the gas distribution was seriously uneven due to local scaffolding. For example, if the accretion forms on the upper part of the blast furnace, the gas curve was irregular, and the value of CO_2 of edge at the first point was higher than that at the second point in serious cases, the phenomenon of barb occurs. If the accretion formed at the lower part, the gas curve would appear to be automatically aggravated. Under this situation, although the charging system had not changed, in serious cases, changing the charging system had no effect, and the gas distribution remained the same.

The radial temperature distribution of the throat was opposite to the gas CO_2 value. The temperature at the low point of CO_2 corresponded to the high point. The temperature at the second point was higher than that at the first point at the same position as measured.

9.1.2.3 Frequent Channeling, Constant Slip and Bridging in Serious Cases

The gas distribution was abnormal and sometimes the local gas volume was too large, thus, the channeling was inevitable. Serious channeling would inevitably lead to slip.

Accretion blocks the drop of burden and was easily to cause bridging. Slipping and bridging were the associated phenomena of severe scaffolding. Figure 9.3 shows a typical record of the stock rod with accretion. This blast furnace was 963 m^3, and it could be seen from the records of the two stock rods in the north and south within 24 h that there were many collapsed materials, the blanking was not smooth, there were moments of difficulty to operate instantaneously and the stock rod displayed in a horizontal direction. The more serious the scaffolding, the harder the furnace operation.

9.1.2.4 The Speed of Burden Was Unstable; the Hourly Speed of Burden Varied Greatly

Accompanied by channeling and bridging, the burden velocity fluctuates greatly. It can also be seen from the record of burden speed that the interval of each batch of stock rod was different: Sometimes, the time of a batch of stock rod was very

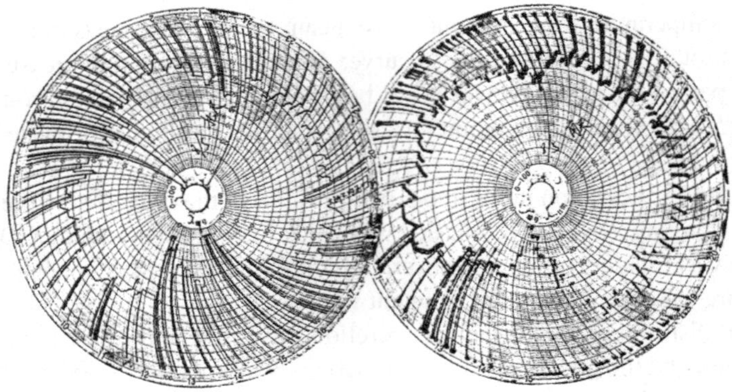

Fig. 9.3 Record of stock rod with accretion

long; sometimes, because of the slipping, the charging speed was very fast, and the instability of actual burden speed was the inevitable result of scaffolding.

9.1.2.5 Low Gas Utilization Ratio

Scaffolding causes the distribution of charge and gas to be abnormal, and the gas utilization ratio is generally low.

9.1.3 Effect of Scaffolding on Hearth

Because the accretion blocked and squeezed burden distribution, burden descending and gas rising are affected. At tuyeres area on the side of the accretion location, there is some burden, which has not been fully heated and reduced, often entering into the hearth and resulting in fluctuation of furnace temperature.

(1) Because of insufficient heating and reduction of local burden entering into hearth, the furnace temperature drops, thus, it is difficult to control the temperature stability of blast furnace with accretion. In order to prevent the furnace from being cold, the furnace temperature control level should be appropriately increased during the scaffolding period.
(2) When "cold coke descending" appears in the front of tuyeres, there is always black coke, which is not fully heated moving up and down before the tuyere below the scaffolding site, that is, "cold coke descending" occurs from time to time.
(3) Slag blooding and filling in tuyere. The furnace temperature is slightly lower, sometimes the blooding of tuyere or even slag filling in tuyere.

9.1.4 Operational Characteristics After Scaffolding

Scaffolding occupies a part of rising space of blast furnace gas. Although the operating blast volume is lower than the normal level, the channeling is easy to appear when the blast is added. The blast pressure is high and the permeability index is low. When the accretion is severely formed, the lower blast volume is also difficult to stabilize, and stiff (tight) furnace and bridging would happen often. (Figure 9.4)

9.1.4.1 Higher Temperature at the Top of the Furnace

Despite the low blast volume caused by scaffolding, the furnace top temperature is still high due to the low utilization ratio of gas and frequent channeling.

When the annular accretion is formed, the top temperature sat four directions are close each other, which is generally high and the temperature difference is small. Reflected in the radial temperature distribution of the throat (cross-beam temperature measurement), the peripheral temperature is low and the center is high. For local accretion, the temperature of top varies greatly in the four directions, and the top temperature on the opposite side of scaffolding is higher and fluctuates greatly.

9.1.4.2 Dust Quantity Increased

The section of blast furnace at scaffolding site decreases and the gas flow rate increases. Because the blast furnace is not operating smoothly, the quantity of blast furnace dust blown out increases obviously. The burden strength is not enough with more fines under this condition, thus, the blast furnace dust would increase several times compared with the normal condition.

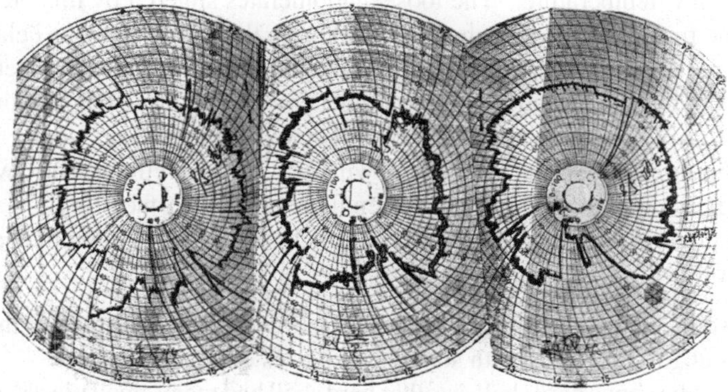

Fig. 9.4 Operation record of scaffolding in 24 h

9.1.5 *Characteristics of Scaffolding*

To sum up, the characteristics of different scaffolding sites and shapes of accretion are listed in Table 9.4.

9.2 Organization Structure and Formation of Accretion

There are many specific reasons for the formation of scaffolding. The stoppage of the soft-melting burden contacting the furnace wall and re-solidification of the soft-melting burden are essential prerequisites for accretion in blast furnace. If the soft-melting burden re-solidified for some reason, as long as those soft-melting continues to descend with the stock column, it would not stick at the furnace wall. Similarly, if the soft-melting burden contacting the furnace wall stops moving but not solidified, naturally it would not stick to the furnace wall. Scaffolding is likely to appear only when the above two conditions occur simultaneously. Of course, if the two conditions appear at the same time, it is not certain to form accretion; but these two conditions are necessary for scaffolding.

9.2.1 *Organization Structure of Accretion*

From a large number of investigations, it can be seen that the general accretion can be divided into three parts: the root, the middle and the upper part. Figure 9.5 is a general outline of the shape. The specific shape and thickness of the accretion vary with the blast furnace and scaffolding time.

The upper part of the accretion: The upper part is the burden, which melts less because of low temperature. "The loose agglomerates sintered by fine ore particles close to the furnace wall were about 150–200 mm thick with obvious delamination along the radius of the blast furnace. A thin layer of slag was clamped between the two layers, which indicated that these binders were not formed at one time. They were not firmly bonded to the lining, and (sometimes) had been clearly separated from the lining due to shrinkage after blowing out. The accretion surface was bound by small pieces of burden, which had been fluxed on the surface" [4].

The middle part of the accretion: "The structure is relatively complex and solid, with the following characteristics:

(1) The solid shell formed by liquid condensation close to the lining of the furnace is about 150 mm thick with some large pieces of coke.
(2) Next to the above hard shell, another hard shell had many small pieces of sintered ore and coke adhered to the inner surface. On the inner surface of the furnace, it is composed of slag, metallic iron particles and fine precipitates with thin layers alternating with each other with glabrous surface" [4].

Table 9.4 Characteristics of scaffolding

Blast furnace parameters	Scaffolding site		Shapes of accretion	
	Upper part	Lower part	Annular accretion	Local accretion
Top temperature	Lower at scaffolding side	Higher at scaffolding side	Temperature in four directions is close; generally high	Sharp difference in four directions; great fluctuation, generally > 100 °C
Radial temperature of throat	The peripheral temperature on the scaffolding side is low		Higher center temperature	
Gas distribution of throat	Irregular; sometimes edge with fold points	The periphery is automatically aggravated and the charging is changed with little effect	Heavy at periphery; light in center	Abnormal gas distribution at scaffolding side; low CO_2 volume
Gas utilization ratio	Low	Low; fluctuate sometimes	Low	
Stock rod	The depth of stock line is uneven and fluctuates greatly			Scaffolding at upper part; High burden surface and shallow stock rod
Velocity of burden	Unstable speed of burden; slip often			Low speed of burden at scaffolding side
State of tuyere	Tuyere below scaffolding site is bright	Tuyere below scaffolding site is dark and cool	Working state of tuyeres are close	Great differences between working state of tuyeres; clear distinction between dark and bright
		Constant "cold coke descending"; easy to blooding and fill in slag		
Channeling			Much channeling in center	Much channeling at periphery
Slip	Slip more	Slip less	Gas permeability drops sharply and the blast volume decreases a lot after slipping	Gas permeability drops less after slipping

(continued)

Table 9.4 (continued)

Blast furnace parameters	Scaffolding site		Shapes of accretion	
	Upper part	Lower part	Annular accretion	Local accretion
Bridging	Bridging sometimes	Difficult to deal with after bridging; Stock rod is deep after dealing with bridging	Easy to bridge after slipping	
Common characteristics				
Furnace wall temperature	The temperature of the scaffolding site is lower than the normal level and lower than the area nearby without accretion. The temperature at the scaffolding site changes slowly			
Temperature at backward of stave	At the beginning of scaffolding, the water temperature difference of the corresponding part gradually decreased, and the scaffolding area was lower than the normal level and adjacent area			
Temperature of shell	At the initial stage of scaffolding, the decrease was not obvious; as the scaffolding area enlarged or the thickness increased, the temperature of the shell was significantly reduced			
Heat flow intensity of furnace	The above reflections are the results of the decrease of heat flow due to scaffolding			
Dust quantity	Significantly increased, in severe cases increased to dozens of kilograms per ton of iron			
Blast volume	The blast volume is lower than the normal level. It is difficult to add blast, and it is easy to produce channel, which is difficult to run smoothly and easy to bridge			
Blast pressure	Compared with the normal level, the blast pressure is high; and sometimes the blast pressure fluctuates greatly due to the serious channeling			
Permeability	Because of the frequent channeling, the fluctuation is large; the overall level is low, and the pressure relationship is tense			
Si content in hot metal	Fluctuate greatly; hard to be stable			

Figure 9.6 is a picture of a middle part of accretion [5]. From the picture, it can be seen that the coke and other massive burden bound together by a thin liquid-phase condensate.

Root of the accretion: "The root contains more iron and was hard. The outer part of the accretion was 200–400 mm thick incrustation, and the inner part of the incrustation was covered with coke, mineral and slag. The lower part became a cohesive whole, while the upper part was loose. The shell had more than one layer and sometimes up to 3–4 layers" [6].

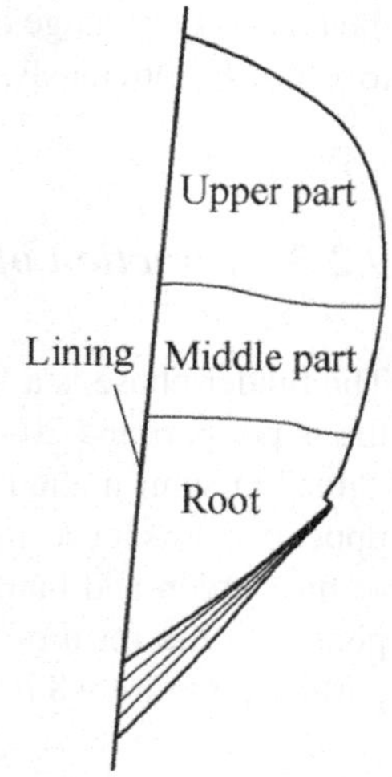

Fig. 9.5 Schematic diagram of accretion structure

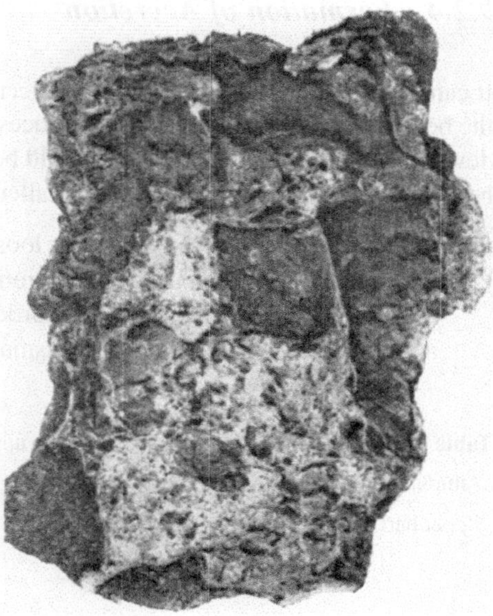

Fig. 9.6 Photograph of the middle part of accretion

9.2.2 *Physical Properties of Accretion*

The composition of the upper and middle parts of the accretion was similar and the melting points of the binding phase were similar. The difference was that the middle part was denser than the upper part, and the "binder phase" was wider and tighter than the upper part. Table 9.5 shows the temperature measurements in the upper and middle sections of the furnace. Due to the different sampling components, the temperature of semi-melting and full-melting was also different. Generally speaking, the temperature range of liquefaction begins at 1000–1100 °C from individual point to 1400 °C. Obviously, this temperature level could disintegrate the accretion.

9.2.3 Function of Binder Phase in Accretion

The binder phase is a flaking and slag binder, which was about 20–30 mm thick in the upper part and 20–60 mm thick in the middle part. It bound them together like "glue" in semi-melted solid burden or solid burden that had not been melted. The upper part was loose and the middle part was tight. Figure 9.7 is a schematic diagram of the burden and binder phase at the scaffolding site. The figure shows a "binder phase" in the form of a sheet, which sticks together with various types of burden particles. Figure 9.8 is a micrograph of the binder phase [7].

9.2.4 Formation of Accretion

It can be seen from the structure of the accretion that the accretion is a thin layer at the beginning and does not affect the process of the furnace operation. After gradual development, the smelting process would be seriously affected only when the layers become thicker, the accretion become taller and longer.

(1) The upper part of the accretion has loose tissue and thin liquid-phase bond coat.
(2) Although the middle part of the accretion has a compact structure, it still keeps the original quality of the furnace burden. Most of the ore or coke remains lump, although the particle size becomes smaller. Since then, as the furnace accretions

Table 9.5 Temperature of binding phase in the upper and middle part of accretion (°C)

Sampling part	Number	T_1	T_2	T_3
Upper part of accretion	1	1090	1290	>1400
	2	1100	>1400	
	3	1140	1390	1400
	4	1100	1200	1250
Middle part of accretion	1	1090	1360	>1400
	2	1040	1230	1290
	3	1080	1400	1420
	4	1100	1190	1215

Note The liquid-phase appeared in the upper and middle parts of the furnace, and the temperature T_1 was close to 1000–1140 °C; T_2 was the temperature at which the binder phase half melted by, 1200–1400 °C; T_3 was the total melting temperature of the binder phase

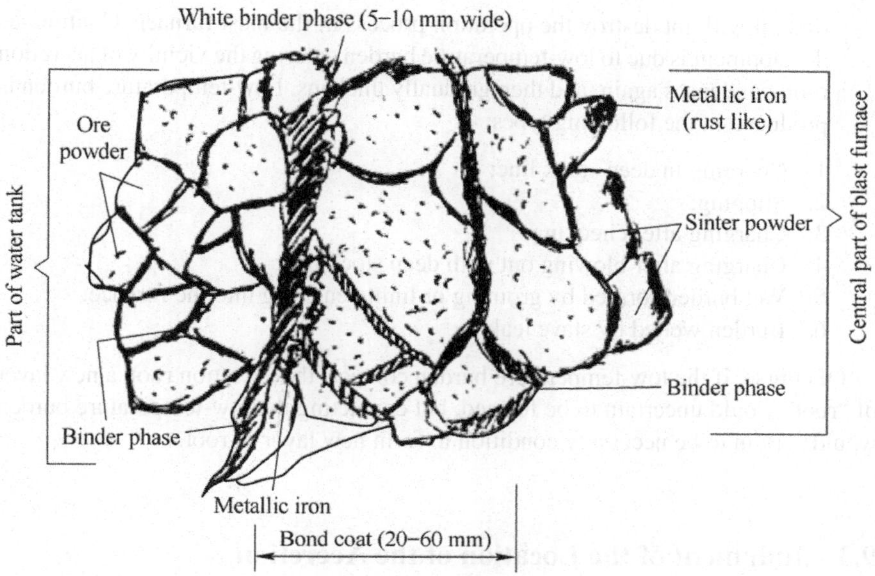

Fig. 9.7 Schematic diagram of the burden and binder phase

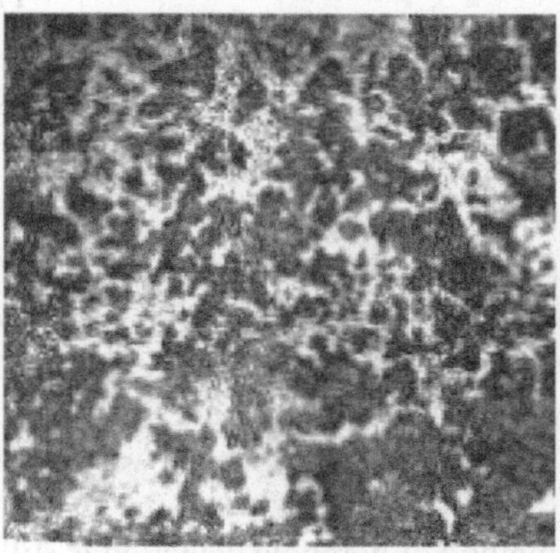

Fig. 9.8 Micrograph of the binder phase of accretion (× 100)

continue to develop and gradually become thicker and taller, it has become a serious obstacle to the production of blast furnaces.

(3) Formation of the roots. The root of the accretion is strong and the iron content is generally high. Because of insufficient carburization, the carbon content is low and the melting point is about 1400 °C, which makes it difficult to melt. It is a thin layer at its initial stage and later due to unrestricted development layer by layer, a thicker support point is formed. The scaffolding began and gradually developed. The formation of the roots, first of all, is the melting burden close to the furnace wall without moving and sticks to the furnace wall. If it is restricted at this

time, it will not destroy the operation process of the blast furnace. Continuous development is due to low-temperature burden entering the vicinity of accretion root, condenses again, and then gradually thickens. Low-temperature burden is produced in the following types:

1. Charging in deep stock line;
2. Slipping;
3. Charging after checking;
4. Charging after blowing out with deep stock line;
5. Wet burden formed by grouting or lining entering into the furnace;
6. Burden wetted by stave leaking.

Of course, if the low-temperature burden contacts the accretion root, a new layer of "root" would uncertain to be formed, but contacting the low-temperature burden would certain to be necessary condition to form new layer of root.

9.3 Judgment of the Location of the Accretion

In the symptoms of scaffolding in Sect. 9.1, the commonly used methods of judging the location and shape of accretions have been explained. Here are some supplements.

9.3.1 Borehole Method

A number of rows of probing holes are reserved at the belly and shaft of the blast furnace. Specialists regularly probe holes to determine whether or not the accretions are formed. Figure 9.9 shows the form of the early hole-drilling record of Capital Steel. There were five rows. Each row had eight holes. Figure 9.10 is an accretion schematic diagram of No. 3 BF of Capital Steel 1962, which was determined by borehole detection.

In modern large blast furnaces, the method of borehole detection is still the basic method to observe the change of wall thickness. The methods of retaining holes in furnace shell and lining are the same as those in the 1950s. Only the drilling tools are not steel brazing, but special drilling machine rigs. Bao Steel No. 1 BF introduced this device from Japan.

Although the method is simple and accurate, it is no longer suitable for direct use in blast furnaces operated under high pressure. New direct measurement methods need to be developed.

9.3.2 Method for Heat Transfer Calculating

It is also feasible to determine the location of the accretion by means of heat transfer calculating, but the heat flow intensity of the relevant parts of the furnace body must be known. Some blast furnaces are equipped with two thermocouples at the same location, and their temperature difference is used to calculate the heat flow intensity. Although there might have a big error in calculating the location of accretion with the formula, it is still a simple and easy way to judge the location of accretion at blast furnaces with known heat flow intensity.

9.3.3 Observation Method By Stock Line Lowering Down

This is also commonly used to determine the location of accretion. The location and distribution of accretions can be observed directly by taking opportunity of maintenance to lower down the stock line. Although this method is simple, it can be generally replaced by measuring furnace shell temperature because of much time-consuming and large loss for stock line lowering down.

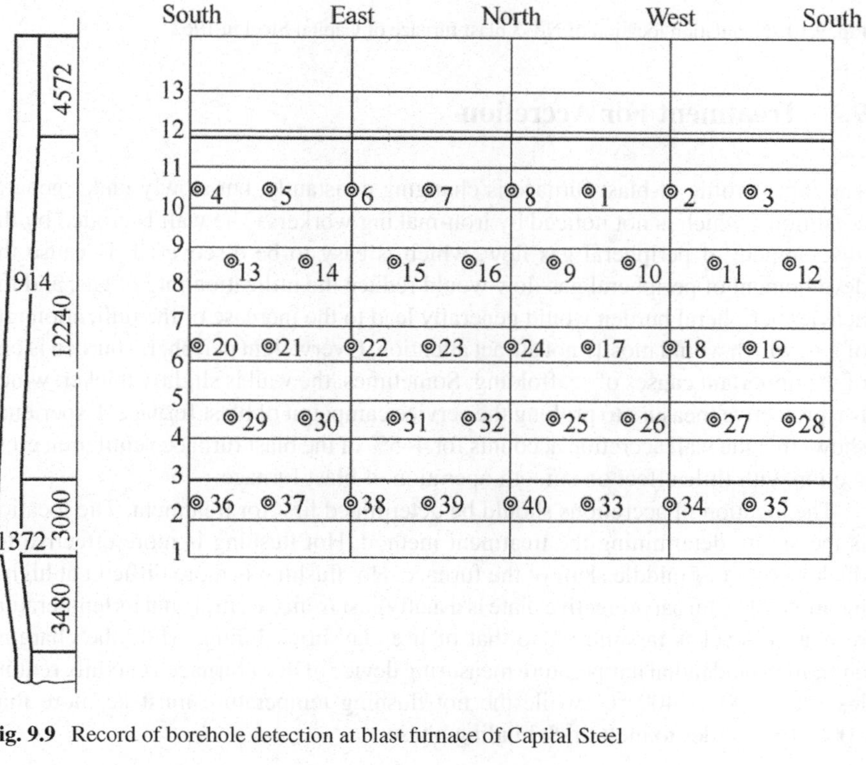

Fig. 9.9 Record of borehole detection at blast furnace of Capital Steel

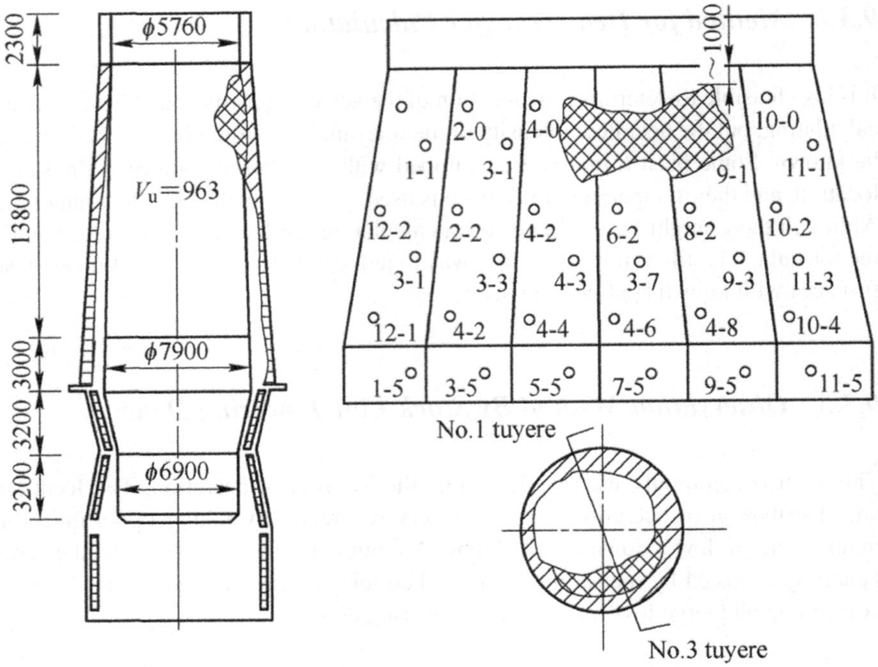

Fig. 9.10 Nodulation location of No. 3 blast furnace of Capital Steel in 1962

9.4 Treatment For Accretion

The inner profile of blast furnace is changing constantly, but slowly under general conditions, which is not noticed by iron-making workers. The wall is eroded by the development of peripheral gas flow, which is easy to be discovered. Because the development of peripheral gas flow would reduce the utilization rate of gas, and the heavier peripheral burden would generally lead to the increase of the utilization rate of gas, which would mostly not attract attention. Overweight peripheral burden is one of the important causes of scaffolding. Sometimes, the wall is slightly thicker, which is a protective measure to prolong the service campaign of blast furnace. Experience shows that the wall accretion accounts for 4–5% of the blast furnace ventilation cross section with little effect on smooth operation of blast furnace.

The location of accretions should be determined first for treatment. The location is the key to determining the treatment method. Hot flushing is more effective for the lower part of middle shaft of the furnace. Hot flushing is more difficult at higher locations. The throat protective plate is usually cast from cast iron, and its temperature resistance level is far inferior to that of fire clay brick lining, while the charging equipment and radial temperature measuring device of blast furnace generally require less than 300 or 400 °C, while the hot flushing temperature must be more than 1000 °C, in order to melt thick binding phase.

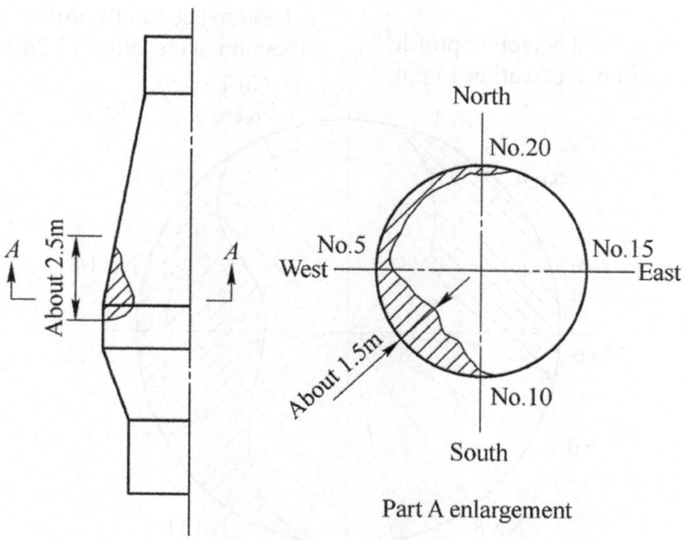

Fig. 9.11 Blast furnace accretion shape of No. 8 blast furnace in Xuan Steel (observation on March 25, 1991)

9.4.1 Blast Furnace Accretions Are Movable, Can Grow up and also Can Break Away

9.4.1.1 Some Accretions Can Break Away Automatically

The 1260 m^3 blast furnace in Xuan Steel had scaffolding at the middle of shaft (Fig. 9.11). At that time, the hearth accumulation and the blast furnace blast volume was small. Because the hearth accumulation and the shaft scaffolding, the furnace condition was often serious. At that time, it was decided that the hearth treatment process should strive to keep the blast furnace operate smoothly and let the accretions break away automatically [8]. Later, after the treatment of hearth accumulation and the blast furnace operating smoothly, the accretions broke away automatically.

9.4.1.2 Accretions Growing

A 1036 m^3 blast furnace in Capital Steel was blowing on in May 1989, but it was not smooth because of its heavy peripheral burden. Observation at lower stock line in July discovered accretion at shaft (Fig. 9.12).

Due to the eagerness to deal with accretions, insufficient attention was paid to keeping the furnace running smoothly. Channels and slip were continuous. Although the furnace had been flushed several times, the effect was very poor. Finally, the furnace was blowing out to blast off the accretions.

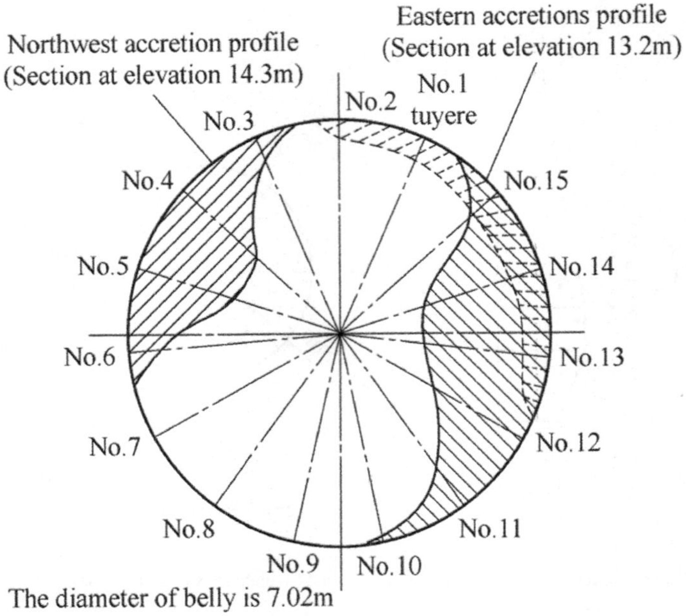

Fig. 9.12 Section of blast furnace with accretions. Dotted line: observation in July; solid line: observation in October

9.4.2 Flushing Furnace

In fact, the melting point of the cohesive phase (accretion) is lower especially the upper and middle parts of accretion are not firm. To adjust the distribution of gas flow and develop peripheral gas flow could make most of the cohesive phase melting and scatter the accretions. Hot flushing is most commonly used to remove accretions.

9.4.2.1 Temperature Distribution of Peripheral Burden in Furnace

In order to melt and loosen the thick binding phase, it is necessary to make the burden temperature close to the furnace wall reaching 1000–1100 °C. The temperature distribution in the furnace is affected by many factors. From the blast furnace dissection results, the main factors are gas distribution and burden characteristics. Table 9.6 shows the measured results of partial dissection of blast furnaces. The temperature of the peripheral burden is around 1000–1100 °C.

From Fig. 9.13, it can be seen that according to analysis of the temperature distribution of peripheral burden, it would be easier to treat accretions by hot flushing below the middle of the shaft, but more difficult at upper part of shaft.

Table 9.6 Location of peripheral burden

Furnace number	Volume, m^3	Gas distribution type	Peripheral burden location	References
Experimental BF of Capital Steel	23	Peripheral gas flow developing type	Middle part of shaft	[9]
Lai Steel No. 3 BF	125	Peripheral gas flow developing type	Middle part of shaft	[10]
Nizaki No. 1 BF	721	Peripheral gas developing type	Middle part of shaft	[11]
Kukioka No. 4 BF	1297	Doublet type	Middle part of shaft	[12]
Hirohata No. 1 BF	1407	Central gas flow development type	Belly	[12]

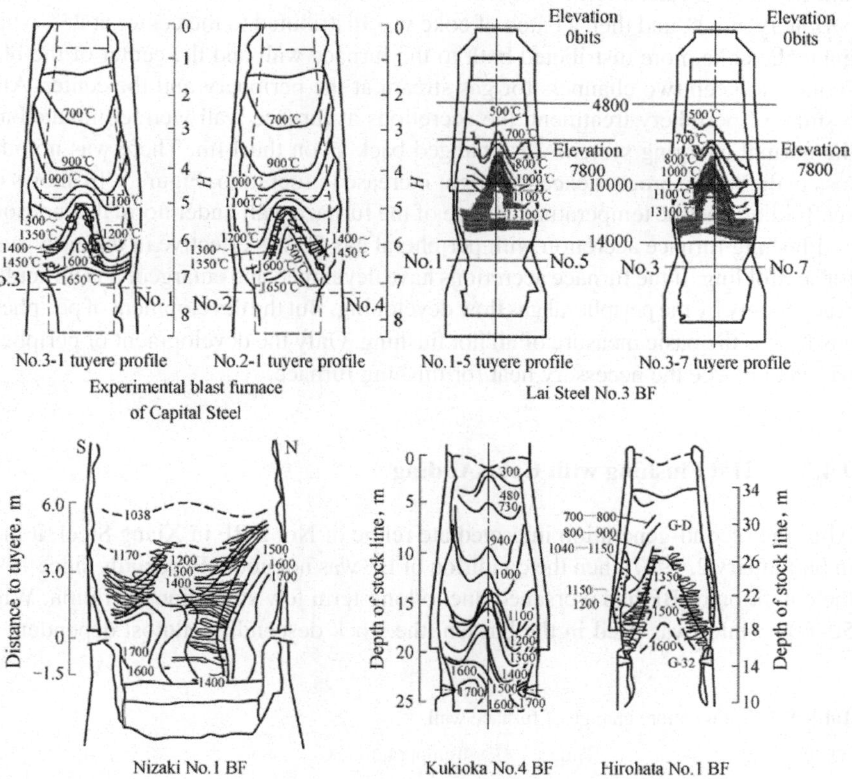

Fig. 9.13 Peripheral burden location map

9.4.2.2 To Develop Peripheral Gas for Furnace Purging

It is simple and effective to treat the accretions at lower part of the furnace shaft with developing peripheral gas flow at the initial stage of scaffolding. On November 7, 1980, the temperature of the lower shaft wall of No. 2 BF at Capital Steel generally decreased, and on November 9, the temperature at four directions generally fell below the normal level (200 °C) (Table 9.7).

Table 9.7 shows that the reason, which made the temperature of shaft lower than the normal level, was the blowing off for lack of sinter, but the temperature of the furnace wall did not rise after re-blast. On the 11th, the leakage of water from No. 50 vaporization cooling pipe was found. The furnace wall did scaffold. On the 12th, the angle of chute distributor was adjusted at 8:10, but the angle of sinter distributor ($\alpha_K = 30°$) did not change. The angle of coke distribution was changed from 25° to 32° and 23° half of each alternatively that meant a batch of coke was distributed to the periphery at 32°, and then a batch of coke was distributed to the center at 23°, which made the coke more distributed both to the furnace wall and the center of the blast furnace to keep two channels for gas stream at the periphery and the center. After 5 shifts of periphery treatment, the accretions at furnace wall were eliminated and the normal charging system was changed back as on the 14th. There was no other loss in the whole process except a slight increase in fuel ratio. Figure 9.14 shows the scaffolding and the temperature change of the furnace wall under normal conditions.

Flushing furnace accretion with peripheral gas flow is effective in the early stage for scaffolding. If the furnace accretions have developed and enlarged, it is difficult to recover only by the peripheral gas flow developing. But the development of peripheral gas flow is the basic measure of all hot flushing. Only the development of peripheral gas can provide the necessary heat for flushing furnace.

9.4.2.3 Hot Flushing with Coke Adding

After the second-generation intermediate reline of No. 2 BF in Xiang Steel, it blew in on January 2, 1997, then the condition of BF was not good. Especially since 1998, the condition of BF had worsened due to long-term low stock line operation. When 50–60% sinter was used in the burden, the stock descending almost dependent on

Table 9.7 Temperature change of furnace wall

Date	Time	Temperature, °C			
		East	South	West	North
December 8, 1980	14:40	140	300	258	390
December 9,1980	2:50	120	210	140	260
	10:45	110	210	220	170
	19:00	100	160	130	110
December 10,1980	5:10	110	200	180	130

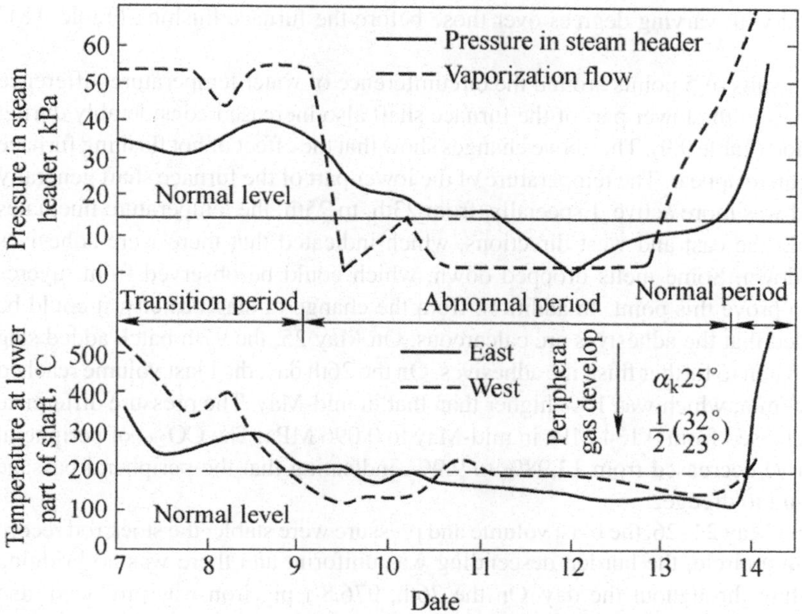

Fig. 9.14 Furnace wall nodulation and temperature change under normal conditions

slip and check. The number of bridging increases month by month, the temperature of lower part of the furnace shaft was low and stagnant, and the average temperature difference of water in staves of 7 sections was only 1.52 °C. The accretions at the lower part of the furnace shaft were determined by analysis and study, then furnace purging was conducted, but the effect was not good in a period, then switch to hot flushing with additional coke on the base of old experience.

The ore batch weight is 10.8 t, with 400 kg manganese ore added in each batch. "Coke blank was added from the 9th batch of coke, in the order of 6C + 5A + 3C + 2 (4A + 3C) + 5A + 3C + 4A + 3C + 2 (10A + C), and then as normal burden (A—normal burden, C-4 cars of coke). In order to burn and melt the binder off, 207 t coke were added throughout the day. [Si] increased from 0.87% on May 22 to 2.43%. The manganese content in the slag also increased from 0.5% on May 22 to 0.93%. In order to dredge the central and peripheral gas flow, the charging sequence was changed to 3A + B: A is α_{ore} 30° and α_{coke} 28° in the 54 batches on the 23rd; B is ore angle without change, α_{coke} 32°, and coke was added to the periphery. The coke and ore stock lines were adjusted to 2.5 and 2.0 m, respectively. On May 23, 107 t coke blank was added in fractions."

"Due to the basicity of slag at the first tapping was too high, the fluidity of slag and iron was poor. When the average [Si] of the whole day is maintained at the level of the previous day, the basicity of slag jumped from 1.12 on the 23rd day to 1.28 with the maximum 1.38. From the temperature, variation of the lower part of the furnace shaft at the east, south and west directions, all of them have different degrees increased with varying degrees over those before the furnace flushing (Table 9.8)" [13].

The results of 5 points around the circumference of water temperature difference of 7 staves in the lower part of the furnace shaft also increased considerably during this period (Table 9.9). The above changes show that the effect of hot flushing furnace had begun to appear. The temperature of the lower part of the furnace shaft generally rose and was more active. Especially, from 23th, to 25th, the temperature fluctuates greatly in the east and west directions, which indicated that there were adhesives falling down. Some melts dropped down, which could be observed from tuyeres that also prove this point. In addition, from the change of slag basicity, it could be concluded that the adhesives are calcareous. On May 25, the 97th batch added slag 300 kg/batch to further flush the adhesives. On the 26th day, the blast volume reached 1017 m^3/min, which was 14% higher than that in mid-May. The pressure difference also decreased from 0.134 MPa in mid-May to 0.096 MPa. The $CO_{2\,\%}$ of peripheral CO_2 curve decreased from 13.98% to 9.9%, indicating that the peripheral passage had begun to dredge.

"From May 24–26, the blast volume and pressure were stable, the stock rod record was with a circle, the burden descending was uniform, and there was no bridging and sliding throughout the day. On the 26th, 976.8 t pig iron was produced, and the charge coke ratio was 719 kg. The furnace condition was getting better and the flushing effect was achieved. The coke load increased from 1.53–2.36 on May 25th,

Table 9.8 Temperature change in the bottom of No. 2 blast furnace of Xiang Steel before and after flushing

Date	East	South	West	North
May 18th	270	50	350	350
May 20th	305	85	310	340
May 22th	320	110	600	330
May 23th	670	100	404	280
May 24th	730	140	240	420
May 25th	400	190	620	250
May 26th	420	250	600	300

Table 9.9 Temperature variation of lower part of furnace before and after flushing of No. 2 blast furnace in Xiang Steel Group

Date	No. 1	No. 8	No. 16	No. 24	No. 32
May 21st	2.7	2.2	0.5	1.7	0.5
May 23rd	2.0	0.5	1.0	3.0	1.0
May 24th	3.5	3.0	2.0	2.5	2.5
May 25th	3.5	2.5	1.5	1.5	1.5
May 28th	4	2	5	2	4

and the additional slag was stopped on May 26th. [Si] decreased from over 2.0% to about 1.5%" [13].

The hot flushing was successful. The furnace temperature control of hot flushing mainly depends on the interval and proportion of normal burden and coke blank. When the number of normal batches between coke blanks is more, the furnace temperature would raise less; on the contrary, the increase would be more. The quantity of coke blank is the key to flushing furnace, which decides the effect of flushing furnace.

9.4.2.4 Hot Flushing Furnace with Full Coke Blank

The charging system for hot flushing, which is divided into groups consists of normal burden plus coke blank. The normal burden is used to balance the furnace temperature and absorb heat to prevent overheating the furnace. Actually, it is the coke blank to play the role of flushing. As mentioned above, during the coke hot flushing furnace of No. 2 BF in Xiang Steel, the normal burden was removed off that was the coke blank hot flushing furnace:

$$6C + 5A + 3C + 2 \times (4A + 3C) + 5A + 3C + 4A + 3C + 2 \times (10A + C) \rightarrow 6C + 3C + 2 \times 3C + 3C + 3C + 2C = 22C$$

It is equivalent to 22 batches of coke blank.

For loosing and melting binding phase of accretion, adding fluorites would be better certainly.

9.4.2.5 Hot Flushing Furnace with Additional Fluorite

"Calcium fluoride is the main fluorinated mineral for furnace flushing. The smelting and tests of fluorine-bearing ores in China have fully proved the good fluidity of the slag (Table 9.10). The melting temperature of fluorine-containing slag is relatively low. The experimental results indicated that the melting temperature of fluorine-containing slag is 1150–1200 °C, the basicity of slag is increased from 0.6 to 3.0 with difference of melting temperature 200 °C" [14].

Table 9.10 CaF_2 content and slag viscosity (Pa s)

Experimental temperature, °C	CaF_2						Slag composition, %			
	0%	5%	10%	15%	20%	25%	CaO	MgO	Al_2O_3	$CaF_2 + SiO_2$
1400	5.55	0.214	0.199	0.109	0.085	0.065	55	2	13	30
1200	4.3	1.0				0.3	45	2	13	40

Fluorite dilutes slag viscosity, but also corrodes lining. The quantity of fluorite should be controlled. Generally, CaF_2 does not exceed 4%.

9.4.3 Treatment on the Scaffolding with Lowing Stock Line

Cheng Steel—"Since 1993, scaffolding had occurred for 12 times in the four furnaces of the plant. To eliminate those accretions, blowing off has lasted for 240 h 27 min, which consumed as much as 2271 t of additional coke and lost 18,582 t of hot metal. The scaffolding during the year of 1998–1999 is shown in Table 9.11."

Blowing off and exploration would greatly damage on brick lining. "We found that, the gas would have high flow rate and high temperature in the process of lowering stock line, which contributed a lot to the treatment of lower accretions." Then, the method of dealing with middle and lower accretions with "deep lowering stock line burning" method summed up that means descending burden level to the underneath of the scaffold, then putting proper amount of coke into the furnace according to the scaffold volume. After stopping charging, blow into control top temperature with water spraying and scouring accretions by gas flow with high temperature and high flow rate. The advantages of this method are as follows:

(1) Before the arrival of coke blank, furnace temperature would not be too high, so hot metal and slag could tap freely.
(2) After the arrival of coke blank, the gas, ascending from tuyere to top, could keep high temperature until the gas meets the spray water on the top because there is no burden. Therefore, a high temperature area is formed between the tuyere region and the water spraying area on the furnace top.

Table 9.11 Scaffolding of blast furnaces of Cheng Steel in 1998–1999

Scaffolding time	BF number	Scaffolding position	Accretion shape	Treatment method	Blowing off time
February, 1998	No. 4	The area from lower shaft to belly	Crescent	Thermal washing	11 h 38 min
February, 1998	No. 5	Middle and lower shaft	Strip	Exploring	9 h
March, 1998	No. 2	Upper and middle shaft to belly	Annular	Exploring	8 h 5 min
March, 1998	No. 4	Upper and middle shaft	Strip	Exploring	6 h 45 min
May, 1995	No. 4	Lower shaft to belly	Crescent	Thermal flushing with descending stock line	4 h 30 min

In March 1999, it was scaffolding at No. 4 blast furnace of Cheng Steel. The stock line lowered down to tuyere belt, then blowing off. The scaffold was exposed, locating at lower part of shaft with a crescent-shaped and a volume of about 30 m^3 above the tuyere. "According to the scaffolding location and the scaffold volume, 80 t blank coke was put into the furnace. Then, an aqueduct was fixed at the top sampling hole. After that, started to blow in at 13:00 after debugging and testing, and turned the blast temperature up to the maximum and adjusted the blast volume to 600 m^3/min which was equivalent to 70% of normal volume."

In order to "developing peripheral gas flow to scouring accretions, at the beginning of blowing in, due to the high level and low temperature of coke stock column, top temperature increased slowly. Top temperature could be controlled under 400 °C with 10t/h of water spraying. A little green burden could be seen at several tuyeres at 15:00. First tap of hot metal 3.68 t with slag 8 t was at 17:00. Since then, top temperature gradually rose, spraying water increased to 15 t/h accordingly. However, there was still a little green burden in front of tuyeres. Second tap of hot metal was discharged at 19:00 with a hot metal 0.5 t and slag 2 t. Since 21:00, there was a gradually increasing green burden in front of tuyere and a rapidly rising in top temperature. At 22:00, the water sprayed from the top had been increased to 25 t/h. Blocks of cohesive materials dropped from several tuyeres until 23:10. There was almost empty in front of tuyeres. Tapping began at 23:20, lasting for 30 min due to the low blast pressure, with an amount of hot metal of 27.31 t and slag of 8 t. After tapping and blowing out, workers observed from manhole at throat and found that accretion had been completely cleaned. Then, blowing in and whole process of cleaning accretion lasted for 10 h50 min. The treatment effect was also ideal. Two problems need to be paid attention to in the treatment of accretions with "lowering stock line thermal flushing method":

(1) The cooling intensity of furnace body, where accretions were formed, should be controlled to facilitate the adhesives breaking away.
(2) The cooling equipment should be focused on its inspection in the process of burning and flushing. The furnace body has great heat flow intensity during the treatment of accretions. On May 4, 1999, a dangerous accident of rubefaction at furnace shell was happened in the treatment of accretions" [15].

In No. 2 iron-making plant of Ji Steel indicated that the portent of scaffolding [16] appeared on the walls of No. 1 blast furnace (128 m^3) in April, 2003:

(1) Bridging appeared and checking became difficult.
(2) The stock line on east side was 400 mm deeper than that on west side. The gauge rod on the west side often slid.
(3) About staves equipping at belly and shaft, the cooling water temperature difference decreased 8–10 °C at the northwest. The temperature of four spots of the BF shaft was all decreased, among which the northwest dropped 100 °C.
(4) The curve of radial CO_2 distribution in the BF throat gas was irregular. The CO_2 content in northwest gas was 18–20% and that in southeast was only 4–6%.
(5) The BF blast pressure fluctuated, the blast volume was relatively low, and the hot metal output was significantly reduced.

There were totally eight tuyeres of No. 1 blast furnace. The lengths of No. 3 tuyere and No. 4 tuyere were 10 mm longer that others and the diameters of them were 5 mm smaller than others. Due to the poor quality of coke, the scaffold was about 400 mm above (northeast direction) No. 3 and No. 4 tuyere. Then, we shortened the lengths of the No. 3 tuyere and No. 4 tuyere by 30 mm and enlarged their diameters by 5 mm, and strengthened gas flow flushing in this direction.

For removing the accretions, "we increased the amount of coke for hot flushing. Burdens in flushing furnace were composed by: 16 batches of coke and 10 batches of normal burden were put into Group 1; eight batches of coke and 10 batches of normal burden were put into Group 2; four batches of coke and 10 batches of normal burden were put into Group 3. 300 kg fluorite was added at every 4 batches of coke. After this treatment, although the scaffolding of the furnace wall was not fundamentally solved, the smooth running of the furnace was improved."

"We spent three days to take the time of planning maintain and use the method of lowing stock line to eliminate accretions in BF. The specific methods of the treatment are as follows:

(1) To stop charging in advance and blowing down.
(2) The temperature of BF top rose rapidly due to stopping feeding. Therefore, fifteen batches of coke were planned to put into the furnace to control top temperature. Wetted the coke with water before putting in the furnace. To prevent producing detonations in the furnace by the wet coke and damaging equipment, we opened the relief valve on the furnace top. This step also prevented the furnace from dropping in temperature and the difficulty of opening slag notch and tap hole caused by large pieces of adhesives fell from scaffolding position into the hearth after lowering stock line.
(3) Coke charging time was decided by top temperature. When the furnace top temperature was as high as 500 °C, put one batch of coke into furnace. It would spend one hour to charge 15 batches of coke. The spray pipe was equipped for cooling after blowing out till the top temperature raised to 500 °C. The pipe was equipped in the sampling hole. However, the northeast scaffolding position was not equipped, which avoided the water cooling and hardening of scaffolding thus caused the difficulty of scaffolds removing off.
(4) When the work was ready, we re-blast and sped up lowering stock line with full blast. By adjusting the water quantity, the top temperature was controlled below 500 °C. In 60 min, the relief valve squealed for the first time, and the scaffold began to fall off. Every 10–15 min, the accretion fell one part. There were totally 11 times before and after. With the lowering stock line, the blast volume also decreased with the later blast pressure in control of 0.04 MPa.

We observed from the inside of BF that accretions were almost eliminated in which a few of them did not affect furnace condition."

It is a creation for eliminating scaffolds with lowering stock line, which significantly improved the effect of flushing furnace.

9.4.4 *Accretion Blasting*

The hot flushing is less effective on the treatment of scaffolds on upper part of BF, while blasting is usually used for.

9.4.4.1 Opening and Blasting

In the past, people used to cut apart furnace shell and dismantle stave sat scaffolding position for removing accretions by blasting. But it takes too much time and energy. In March 1974, abnormality occurred in the process of gas sampling in throat of No. 3 BF in Capital Steel. In the northeast and northwest, CO_2 content at Point 2 was lower than that at Point 1, as shown in Fig. 9.15. The abnormality time in each month has been listed in Table 9.12. We can see from Table 9.12, since March to September, scaffolding had already been in the upper part in the northwest direction, but did not grow significantly.

Typical gas curves in Table 9.12 are shown in Fig. 9.16.

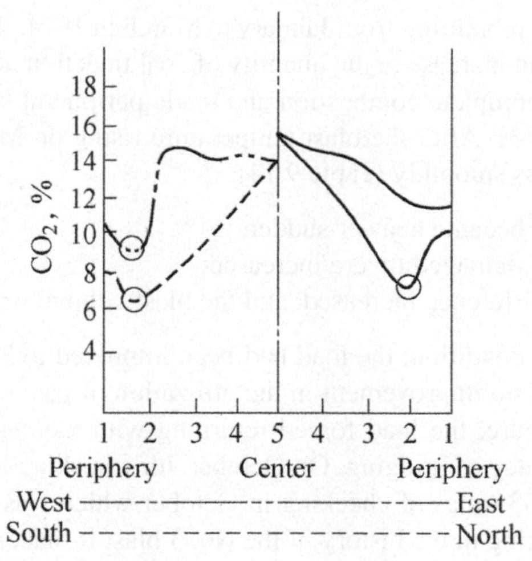

Fig. 9.15 Legend of sampling gas at throat

Table 9.12 Abnormality frequency of gas sampling process (times per month)

Project	Sampling direction	Month in 1974–1975												
		1	2	3	4	5	6	7	8	9	10	11	12	1
Abnormality frequency	Northeast	8	7	4	3	4		4		3	5	19	14	14
	Northwest	1	1	26	29	32	29	18	27	25	37	48	59	51

Note It was failed to collect samples of northeast in June and August due to the sampling tube motor was damaged

Fig. 9.16 Typical gas curves

Table 9.13 Low blast temperature period

Time	Blast temperature, °C	Coal ratio, kg/t	Oil ratio, kg/t	Total injection, kg/t
November, 1973	1042	83	28	111
December, 1973	1011	93	35	128
January, 1974	850	100	43	143
February, 1974	795	96	39	135
March, 1974	840	101	32	133

Since a hot stove rebuilding from January to March in 1974, the blast temperature was very low with an increase in the quantity of coal injection and fuel oil injection, which increased incomplete combustion and made peripheral burden heavier automatically (Table 9.13). After the blast temperature rising on March 28th, the blast furnace operated less smoothly (Table 9.14):

(1) The peripheral became heavier suddenly;
(2) The number of damaged tuyere increased;
(3) The pressure difference increased, and the blast volume was not accepted.

In the unsmooth condition, the load had been improved to 3.88 on October 4th. However, there was no improvement in the utilization of gas, which led to the lack of furnace temperature, the load forced returning with increase of coke charging and increase frequency of bridging. On October 7th, checking had been taken for 7 times. There were 53 times of checking in October which was the month with the most times of checking in the history of the No. 3 blast furnace. The scaffolds grew rapidly during this period. In December, the scaffold was enlarged continuously, and the first, second and third points of the gas curve generally decreased in east and north, especially in west and south due to temperature difference of water rose in No. 2 stave at the hearth, three tuyeres above it had to be narrowed accordingly, tuyere area from 0.391 to 0.352 m^2, and the periphery was to be heavier certainly. The blast furnace had been flushed several times during this period but was less effective, so it was necessary to use blasting to eliminate accretions.

The location and size of furnace accretions were determined by the measurement of the shell temperature and observing from shaft hole, as shown in Fig. 9.17. On the cross section of the most severe scaffold, the area accounted for about 8% of the BF cross section, part of which may had already fallen off before blowing down.

Table 9.14 High blast temperature period

Time	Blast volume, m^3/h	Blast pressure, kPa	Top pressure, kPa	Pressure difference, kPa	Blast temperature, °C	Gas distribution		Damaged tuyeres
						Side	Center	
March 1–10, 1974	2189	1.82	0.7	1.12	769	9.1	10.7	–
March 21–26, 1974	2028	1.82	0.7	1.12	820	9.7	9.3	3
March 29, 1974	1854	1.87	0.7	1.17	1034	11.4	10.8	1
March 30, 1974	1893	1.86	0.7	1.16	1033	10.6	13.4	1
March 31, 1974	1946	1.84	0.68	1.16	1022	11.4	11.9	1
April 1–10, 1974	1952	1.86	0.69	1.17	1039	11.0	12.5	13

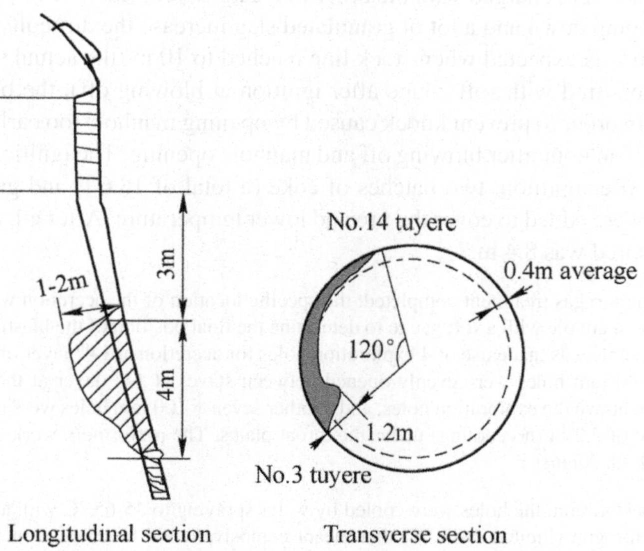

Fig. 9.17 Position and size of scaffolds

On January 28, 1975, the furnace was blowing off at 8:00 and it was observed from holes that scaffolding was very severe and weight about 17–18 t. A total of 16 explosions were fired with 9 kg dynamite. Blowing in was started at 23:00 the same day.

9.4.4.2 Eliminating Scaffolds with the Drilling and Blasting

"On June 5, 2008, blasting was operated in the middle part of the shaft. When blowing out and lowering the stock line to 8 m, a sever accretion was found on the upper part of shaft. Irregular adhesions were found at 0.5 m below throat plates in due east and due south with maximum thickness 1.5 m. The lowest part was at 3.5 m from the bottom of throat plates with uneven surface. On July 14th, when blowing out and lowing stock line for the next time, it was found that the scaffolding had further developed to a ring. Then, the peripheral gas flow was developed strongly and greatly reducing the cooling water of the staves of shaft to eliminate the scaffolds, but the effect was very poor" [17].

Blowing out began on August 6, 2008 lasting for 2590 min (about 43 h), then we started to eliminate accretions with explosion.

"At 0:30 August 6, 2008, the coke load was adjusted to 2.72. At 4:30 began to lower stock line, while a total of 2 batches burdens were charged into BF. The last two batches were charged with the order of 6 cars coke, 6 cars ore to prevent big fire after damping down and a lot of granulated slag increase the difficulty for recovery. At 6:56, it was expected when stock line reached to 10 m (the actual stock line was 9.2 m, measured with soft gauge after ignition at blowing off), the blast would be stopped. In order to prevent knock caused by opening manhole too early, the ignition was in half an hour after blowing off and manhole opening. The ignition process was smooth. After ignition, two batches of coke (a total of 18.6 t) and granulated slag (12.76 t) were added to cover the fire and lower temperature. After all, the final stock line measured was 8.4 m."

> After the top gas treatment completed, the specific location of the accretion was measured from the manhole with a soft gauge to determine the final position of the blasting hole. The blasting hole was made use of 4 exploration holes for accretion at 14th layer of the furnace. Eight 100 mm holes were evenly opened between staves at 14th layer at the position of 300 mm above the exploration holes, and another seven ϕ 100 mm holes were opened at the position of 2.2 m (no cooling) under the throat plates. The performing work completed at about 9:30, August 7.
>
> After perforation, the holes were cooled by water spraying to 45-65 °C with a temperature measuring gun (limited < 80 °C) that meant explosive could be charged in. Professional blasting personnel inserted 500 g of explosives into each of the 19 holes (12 holes at 14th layer and 7 holes at non-cooling zone) to conduct blasting. At 10:28, August 7th, the blast was finished. Through the observation from the manhole of throat, the accretion at upper shaft was removed successfully, the furnace wall was smooth and the blasting was successful.

Compared with the blasting by removing water tank and cutting shell, the method of drilling and blasting is simple and easy to operate, saving time and energy.

9.4.4.3 Slide Blasting

The "slide blasting" can be effectively taken into effort in the following conditions:The accretion is thin or formed before long, its surface contacts the rising gas only for a short time with thin ironclads. A 6 mm oxygen pipe or steel rod could be put into the hole as the "channel" of the explosive, with its lower end placed at accretions. The explosive is hung in the "channel" together with the fuse, which can eliminate scaffolds by blasting directly without opening holes.

A blast furnace (1260 m^3) in Han Steel used to successfully blast off accretions at bosh and belly with the method of blasting by a sliding rod. It spent 608 min to lower stock line above the tuyere level. They were conveying explosives through a manhole with a sliding rod to the accretions at belly and bosh. The blasting operation was carried out for a total of 17 times, with the use of 24 kg of explosive, lasting for 3h18min (14:48–18:06) [18].

9.4.4.4 Experiences in Baotou Steel

Due to the particularity of Baiyun'ebo ores, the alkali metal load is 7.5 kg/t, whose destructive effect on the strength of coke and sinter is aggravated due to high fluorine content. In 20 years from 1959 to 1979, the blast furnace was in the state of scaffolding in 70% of the time [19]. The experience of eliminating scaffolds with blasting method was summarized by Baotou Steel, which is very rich and precious:

"Only when accretion's location is accurately judged and confirmed at the upper part of the furnace shaft, the scaffolds removing could be suitable for blasting. The principles are as follows:

(1) The blast furnace temperature should be maintained at a higher level before the explosion. Enough coke blank should be added before blowing out.
(2) When blowing out, the stock line should be released below the root and accurately measured. When the top temperature is high, sinter can be added to cover the surface for cooling.
(3) Burning with oxygen to unclog all holes, then measuring depth, observing from the top to draw a plan and section of scaffolds and to approximate accretions volume.
(4) Lower part should be blasted first and upper part next. The weak part first and then the thick and dense part.
(5) Type II rock explosive is usually used in the blasting of accretions. First put the explosive into the thin-sheet explosive tube, and then insert the explosive tube into the depth of the blasting hole. The explosive part in the tube is strictly prohibited to be close to the brick lining in the furnace.
(6) Each explosive hole can charge 1–3 kg explosives, and 15–20 kg explosives for extra-thick accretions.
(7) Check the blasting effect and damage on shell and staves after blasting. According to the distribution of residual scaffolds, the next turn of blasting

would be determined. Special attention should be paid to the rise of burden surface. Blasting holes below burdens are strictly prohibited to be used in blasting."

9.4.4.5 Lessons on Blasting Accretions

Roots must be also blasted off, because remaining root would be possible to grow in the process of production and to lead to scaffolding again.

After lowering stock line, charging enough coke in accordance with scaffolds volume and blowing out, the accretions are supposed to fall off on the coke surface. After blasting, enough coke should be continuously charged on accretions according to actual quantity. In past years, No. 2 blast furnace (516 m^3) in Capital Steel started blowing out and blasting scaffolds on March 27, 1979 and was dismantled in October. The accretion had a volume of 50 m^3 with weight of 60 t. Coke blank of 28.7 t were put into the burdens below the scaffold and another coke blank of 25.8 t were put above the accretion. It was observed after operation that accretions were not fully melted into the hearth and blocking there. The later added coke burned around the accretions, which worsened the heat transfer process and led to tight furnace. The burden before blowing off should meet actual requirement of accretions, so the coke should be added below accretions that could accelerate the combustion of coke after blowing in and to upwardly heat and melt accretions, on the other hand, it deferred the entrance of accretions into the hearth, which facilitated the recovery of furnace condition.

9.5 Prevention of Accretion

The prevention of accretion should be considered from two preconditions for the formation of accretions. As long as one of them is eliminated, the accretions would not be formed. Therefore, the following aspects should be paid attention:

(1) Using beneficiated burden is the basic measure of BF operation and the precondition of BF developing. It provides an important guarantee for the stability and smooth operation of BF operation. Here are explanations of a few points related to scaffolding.

   1. Too much fines in the burden and poor permeability of stock column often led to channeling and slipping. The peripheral channeling makes the temperature of peripheral burden and wall increase sharply. Slipping is often occurred with channeling and with the upper cold burdens suddenly slipped into high temperature zone. Slipping also often accompanied with uneven descending of burden with local stagnation from time to time, resulting in the cohesion of burden into high temperature zone.

The content of breeze in burdens, especially the ore fines less than 4 mm or with coke less than 8 mm, poses a great threat to the blast furnace. Screening burden is an important way to eliminate accretions.

2. Burdens with poor strength would continue be broken and pulverized in the falling process in BF, resulting in the production of fines. A little careless would affect the blast furnace operating smooth, resulting in scaffolding.
3. The stability of raw material composition is an important precondition for the stability of BF temperature. Stable temperature could maintain the stability of the composition and melting zone of hot metal and slag. Although the melting zone is also affected by operation, the stability of raw materials plays more important role anyway.
4. The alkali metal content ($K_2O + Na_2O$) should not exceed 3 kg/t of iron. If too much, low alkali content ores should be partially used or external desulfurization devices should be equipped for removing alkali metals with lowering slag basicity periodically. The burden containing [Zn] should also be controlled, otherwise, ore blending should be changed to Zn content less than 0.15 kg/t iron, because alkali metal and Zn has a destructive effect on burdens and the brick lining, even lead to scaffolding.

(2) Maintain reasonable gas distribution. Reasonable gas distribution is necessary to keep the smooth running of blast furnace. Excessive heavy periphery is the common reason of the unsmooth running. After No. 7 BF (2580 m^3) of AN Steel blowing on in October 1984, scaffolds were caused by too heavy peripheral burden. Similarly, After No. 1 BF in Capital Steel blowing on in 1989, they tried to increase peripheral burden from very beginning, but the accretions also appeared. In the initial stage, the wall was not fully heated but the blast furnace needs sufficient heat. Therefore, it is obviously inappropriate to use heavy peripheral burden in the initial stage of blowing on. In 1978, No. 3 BF of Wu Steel was also scaffolding due to excessive heavy peripheral burden [20]. Bao Steel had also appeared scaffolds on belly for many times as a result of excessive peripheral burden, leading to the falling of scaffolds and the tuyere buckling bend to inside of the BF.

Similarly, excessive heavy center burden due to peripheral burden too light also lead to BF unsmooth running. Therefore, it is very important to maintain reasonable gas distribution according to burden conditions and operation demands.

(3) Pay attention to the change on permeability of stock column, and always keep the blast volume accommodate with it. When the strength of burden decreased or the fines content increased, the blast furnace should reduce the blast in time and the load should be reduced when necessary, which is the basis of daily operation, to keep the furnace running smoothly and stable. After World War II, scaffolds generally occurred in Ukraine BF. Due to the recovery and rebuild after war, iron and steel were in urgent need. But raw material conditions were extremely poor at that time. BF workers intensified blowing to increase production, as a result, sever scaffolds happened. In 1953, the book "Scaffolding of Blast Furnace" [5]

was published in Moscow, in which summarized this experience. Many iron-smelting experts analyzed the actual results of 39 cases of scaffolding before 1951. One of the conclusions was that the permeability of stock column must be kept suit with the blast volume.

At the beginning of the 1950s, there was a prevalence of scaffolding in Chinese blast furnaces, due to similar factors in Ukraine. In 1955, at the Second National Blast Furnace Conference, Zhuang Zhen'e, a famous expert in China, analyzed of scaffolding cases of AN Steel over the years [6]. As Wang Zhixi, the deputy director of Iron and Steel Department, saying at the opening ceremony of the conference: "To pursue the output task, acting against the basic principle of that blast volume should suit with permeability of stock, and blindly over blowing in regardless of raw material conditions is wrong. These operations caused clipping, bridging and also caused scaffolding in severe case, which will result in a significant loss. BF is prone to irregular working with unqualified raw materials, but if the above operation guidelines can be strictly followed, the accident of scaffolding or bridge can be avoided. Recently, the scaffolding of some blast furnaces in various factories had mostly related to the violation of this basic guideline" [21].

(4) Activating hearth condition. The activity of hearth is a necessary condition for BF to run smoothly. The accumulation of hearth, resulting in the channeling and clipping continuously, provides the precondition for the scaffolding of BF. Therefore, the hearth accumulation should be prevented and treated in time.

Plugging tuyere is possible to destroy the circular uniformity and normal working of hearth. Generally speaking, tuyeres should not be plugged for a long time. If tuyeres must be plugged, the period should be limited in 20 days; if the tuyere need to be continuously plugged, it should be replaced by another one. Plugging small tuyeres would be better. If there is no small tuyere, temporarily putting a sleeve inside the tuyere, copper or brick could be available.

(5) Maintaining the smooth running of blast furnace:

1. When raw materials contain a lot of fines and be forced to increase blast volume, the channeling must happen. "Channeling" frequently without reducing the blast volume, but with high-pressure difference to hold out against the BF process in which the stock and blast volume do not adapt to each other, then the stable operation would be destroyed. This is the major cause of scaffolding.
2. Maintaining smooth running is necessary for BF intensify and also the basic method of eliminating scaffolding. If raw material is unqualified, we should control smelting strength to maintain BF smooth operation. If BF process is smooth operation, scaffolding would not happen.
3. When burden with high alkali metal is used, smooth running would be particularly important. To keep smooth running and to remove alkali in time is a good way to reduce that destructive role of alkali metal to BF process. Reducing the basicity of slag at regular intervals can reduce the circulation

and accumulation of alkali metal in furnace. The smooth running reduces the damage of alkali metal to burdens. When the basicity of slag is below 1.0–1.05, while the basicity increases or decreases by 0.1, the alkali content in slag correspondently decreases or increases by 0.200–0.300. It is effective to reduce the basicity of slag, to remove alkali in time with the external desulphurization of hot metal [20].

(6) Preventing frequent operation in the state of low stock line. It is dangerous that BF always operate in the state of low stock line. It is a bad habit that should be paid more attention. Low stock line, especially deep low stock line, is possible to make condense melted burden re-condensing. Low stock line is the most common precondition of scaffolding. Therefore, burden should be charged at proper stock line strictly. Low stock line is supposed to be recorded in duty records. Operation in the state of low stock line in terms of duty should be checked and found out the responsible person in charge of low stock line.

The blast should be stopped when the burdening is not available to prevent operation at low stock line. The blast volume should be reduced at deep stock line, so it is not allowed to operate at low stock line. Standard stock line operation is an important condition to guarantee top temperature stability. It is especially important to smelt Zn containing ores, because when the top temperature is too high, the Zn steam would be condensed at the off-take or furnace throat, sometimes the off-take is blocked and forced to stop the blast.

(7) No water leaking into the furnace. Leaking into the hearth may cause re-condensing of smelted burdens. Some plants cut water from leaky coolers to prevent scaffolding, which is an inappropriate method. Because the leaky cooler may be burned off in a short while, which may accelerate the damage of neighboring coolers. Therefore, low-pressure artesian flow can be joined with the damaged coolers, as shown in Fig. 9.18. Water pressure inside the cooler is lower than that in the furnace, so water would not leak into the furnace. To close low-pressure water supply pipe as blowing off and to open it before blowing in, which cannot only maintain a small supply of cooling water to protect the leaky coolers, but also prevents leaking into the furnace.

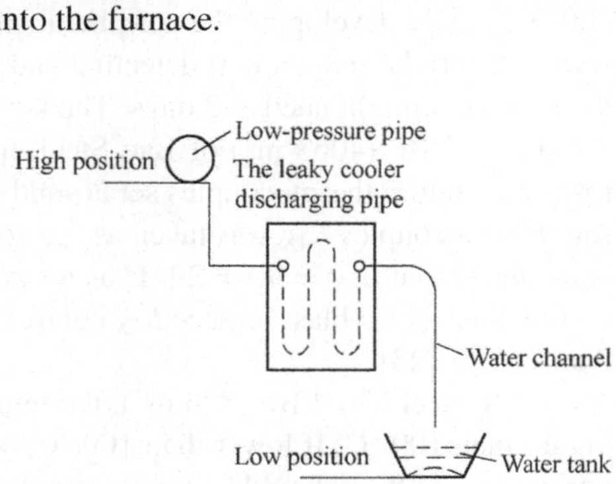

Fig. 9.18 System of low-pressure automatically flowing of the leaky cistern

(8) Handling the condition that blowing off for a long time. Although blowing off for a long time is common cause of two preconditions of scaffolding, it can be avoided completely if handled properly.

Before stopping the blast for a long time, if the furnace wall is not clean, the peripheral gas flow should be properly developed to clean the furnace wall. Before blowing off, enough coke should be put according to the time for blowing off before the coke reaches the tuyere. The purpose of adding coke is to add heat and improve the permeability of the stock column to ensure BF operating smoothly. After blast off, the cooling water pressure and cooling strength should be reduced. Generally speaking, the large blast furnace would reduce to 0.15 MPa or lower, and 0.088 MPa or lower for the small blast furnace. When analyzing the causes of scaffolding in No. 3 BF of Benxi Steel, it was emphasized that the high cooling intensity was the main reason. So, it was regulated that when BF blowing off more than 3 h or more, the water pressure should be controlled between 0.09 and 0.12 MPa, and add coke blank into central area in advance. [Si] should be controlled more than 0.8% before blast off [20].

Re-blowing after a long-term blowing out, peripheral burden and coke load should be light to ensure enough BF temperature and better permeability of stock column to melt the burden bonding on the wall and fall off.

(9) Controlling the temperature of belly and lower shaft. The important characteristic of the start of scaffolding is that the heat flow intensity at the position of scaffolding begins to decrease. The sensitive locations for scaffolding are belly and the lower shaft, so controlling the temperature of the brick lining of belly and lower shaft is an effective measure to control scaffolding.

Different blast furnaces have different structures, operating conditions and the interpolation depth of thermocouple is different; the belly temperature of normal type is also different. In the 1980s, the belly temperature of No. 2 BF in Capital Steel was higher than 200 °C. Under normal operating conditions, scaffolding begins to form once the temperature was below 200 °C. During the 1960s, the belly temperature of No. 1 BF in Capital Steel was higher than 300 °C for normal type. Once below 300 °C, timely developing the peripheral gas flow could make BF return to normal levels, generally in 8–24 h, if detecting and processing timely. If detecting was late, the treatment might need 1–2 days. The key is in time.

At No. 1 BF (4063 m^3) of Bao Steel, there were 8 thermocouples including 4 low-temperature thermocouples set around belly. The average temperature of these four thermocouples L_{4X} was taken as the normal control index of furnace profile. It is regulated that $L_{4X} = 90 \pm 20$ °C as normal level [22].

Ben Steel No. 3 blast furnace L_{4X} control target is 200 °C, scaffolding when lower than 200 °C [23].

In AN Steel No. 2 BF (826 m^3), the temperature of stave back at belly must be higher than 100 °C. If lower than 100 °C, scaffolding would occur on the wall, the peripheral gas flow should be timely developing [24].

(10) Stable operation. Stable operation, mainly refers to the control of furnace temperature stability. Specifically, the content of silicon in hot metal fluctuates little, and the general rule [Si] fluctuates in ± 0.15%. Secondly, the basicity of slag should be stable. The stable silicon content of hot metal and the basicity of slag are the foundation of BF process stably.

In operation, the basic system should be as stable as possible without frequent changes. The basic systems of blast furnace are blasting system and charging system. When basic systems are stable, the long-term operation of the blast furnace would be becoming better and better. Frequent changes would result in serious consequences. Once the furnace is upset and unstable without stable operating system to follow, the operation must be chaotic. Stability is not only the target of a blast furnace production, but also an important medium to achieve the goal.

As Xu Juliang saying: "At present, the scaffolding of blast furnace is less and less. As long as we pay attention to the high-grade burden, according to the corresponding requirements of high-grade burden for different furnace capacity, the occurrence of scaffolding could be avoided. For example, the coke strength of large blast furnace $M_{40} > 85\%$, $M_{10} < 5\%$, CSR > 65%, CRI < 25%; sinter strength > 80%, particle size less than 5 mm < 5%. The scaffolding of blast furnace could be prevented by careful work with high-grade burden, careful operation and management."

Expert in Iron-making

Xu Juliang (1928-): After studying in the former Soviet Union Magnitogorsk Iron and Steel Plant for half a year, he returned to No. 9 Blast Furnace of Anshan Iron and Steel Plant, which was the largest blast furnace in China at that time and the chief of the first high-pressure blast furnace in China. In 1973, he was appointed as the deputy director, director and chief engineer of the Iron and Steel Division of the Iron and Steel Department of the Metallurgy Ministry.

Combining field experience, he published the articles of BF operation to sum up both positive and negative experiences in order to reduce accidents. To establish a platform for discussion and exchange, he founded the magazine "Iron-making" in 1982, summarized the progress of iron-making technology continuously and proposed the development direction.

In 1984, he and Liu Qi co-edited the first book "100 Cases of Blast Furnace Accident Handling" (published in 1986) in the world, which was an important reference book for blast furnace operation.

References

1. J. Liu, D. Han, Analysis and treatment of partial accretion on the furnace body of No. 1 blast furnace of An Steel. Henan Metall. **6**, 35–36 (1999)
2. J. Shan, An attempt to determine the thickness of blast furnace walls through thermal road of stave cooler. Capital Steel Sci. Technol. **4**, 6–9 (1990)
3. L.Z. Hodak, Fragments of the mixture in the furnace. Steel **3**, 199–203 (1971)

4. S. Shi, The analysis of scaffolding of No. 3 blast furnace of Capital Steel in 1962, in *100 Cases of the Treatment of Blast Furnace Accident*, J. Xu, Q. Liu ed. by (Metallurgical Industry Press, Beijing, 1986), pp. 96–104
5. N.I. Krasavchev, in *Styles of Blast Furnace Belly*. (Moscow, 1953), pp. 38–39
6. Z. Zhuang, Scaffolding in blast furnace of An Steel, in *Heavy Industry Science and Technology Division Document. Assembly of National Conference of Blast Furnace Production Technology 1955*. (Heavy Industry Press, Beijing, 1956), pp. 316–414
7. Capital Steel Research Institute, Preliminary analysis of nodules in Capital Steel No. 1 blast furnace, (1989) (internal data)
8. C. Zhang, The treatment of unsmooth conditions in No. 8 blast furnace in Xuan Steel. Xuan Steel Technol. **3**, 17–21 (1991); Y. Liu, Hearth accumulation, in *The Malfunction of Blast Furnace and the Treatment of Hearth*, S. Zhang ed. by (Metallurgical Industry Press, Beijing, 2012), p. 85
9. J. Zhu, Anatomic study of experiments on blast furnace by Capital Steel. Iron Steel **11**, 1–8 (1982)
10. J. Zhang, D. Luo, H. Zeng et al., *Dissection Investigation of Blast Furnace* (Metallurgical Industry Press, Beijing, 2019), p. 235
11. N. Kiichi, et al., The survey result on the collapse of No. 1 blast furnace of Amagasaki Steel. Iron Steel **13**, 195–204 (1980)
12. K. Kenjiro, et al., The survey on the collapse of blast furnace. Iron-Making Res. **288**, 37–45 (1976); K. Kanbara, et al., Discussion of blast furnace and their internal state. Trans. ISIJ, 371–380 (1977)
13. Z. Yang, Discussion on hot washing furnace of No. 2 blast furnace in Xiang Steel. Hunan Metall. **6**, 25–28 (1998)
14. Xu Tongyan, The analysis of flushing furnace process. Steel Technol. **8**, 18–22 (1988)
15. L. Wang, Z. Zhang, S. Na, The treatment and analysis of blast furnace scaffolds of Cheng Steel. Cheng Steel Technol. **1**, 10–13 (2000)
16. Y. Huang, G. Xin, X. Liu et al., Eliminating scaffolds on the upper wall with the method of lowing charge level. Shandong Metall **5**, 8–9 (2004)
17. J. Kou, X. Bai, et al., The production practice of the exploring accretion of No. 2 blast furnace of Jiu Steel. Iron Steel **4**, 50–52 (2009)
18. G. Liu, The practice of eliminating thickness of 1260 m^3 blast furnace of Han Steel with the method of lowing charge level, in *CSM Annual Meeting Proceedings*, vol. 2. (Metallurgical Industry Press, Beijing, 2005), pp. 472–475
19. D. Lin, C. Li, H. Wu, et al., in *Key Problems and Technical Progress of the Picking and Iron-smelting Technology of Baiyun'ebo Special Ores*. (Metallurgical Industry Press, Beijing, 2007), pp. 384–396
20. J. Xu, Q. Liu et al., in *100 Cases of the Treatment of Blast Furnace Accident*. (Metallurgical Industry Press, Beijing, 1986), pp. 135–141, pp. 165–173, pp. 115–123
21. Z. Wang, Speech on the national blast furnace production technology conference. Heavy Industry Science and Technology Division, in *Document Assembly of 1955 National Conference of Blast Furnace Production Technology*. (Heavy Industry Press, Beijing, 1956), pp. 1–11
22. W. Li, Operation practice during starting period No. 1 blast furnace at Bao Steel. Iron Steel **10**, 6–11 (1991)
23. K. Feng, C. Fan, Operation and practice of blast furnace by the structure of intensifying middle part. in: *CSM Annual Meeting Proceedings*, vol. 2. (Anshan: CSM Iron-making Committee, 1991), pp. 33–37
24. Z. Xia, in *The Practice and Operation of Blast Furnace*, vol. 117. (Liaoning People's Publishing House, Liaoning, 1988), pp. 120, 132

Chapter 10
The Operation of BF Blowing on (in)

Abstract The main target of blowing on (BF on steam/BF starting up) is safe and smooth. Usually, some equipment faulty would happen in the blow-in process, but less human injury. A lot of BFs started up in China frequently to accumulate rich experience to operate smoothly and realize the designed standard very soon. It was indicated that irregular situation generally came from less preparing, not enough burden, wrong operating, especially unqualified united test run for facilities.

10.1 Preparation Before Blowing on

10.1.1 Personnel Providing and Training

BF is a complicated huge equipment of production including many big systems of facilities. Before production starting, personnel providing and training should be arranged in accordance to the stations of the production system to have an intimate knowledge and skill of the facility operation in the whole process of installing, fixing, and test run; otherwise, it would be very hard or impossible to master and know well.

10.1.2 Test Run and Acceptance Check for Facility

Firstly, the standard for acceptance check should be formulated to be the regulation to accept term by term. The acceptance personnel should involve relative expert in charge, operators and maintainers. Facilities united testing would be necessary after monomer equipment testing to resolve emerging problems leaving no further trouble, which used to harm further production of BFs.

Some facilities such as TRT, cooling and watering system do not belong to iron-making plant, so BF men in charge should contact and coordinate with them.

Y. Liu, *The Operation of Contemporary Blast Furnaces*,
https://doi.org/10.1007/978-981-15-7074-2_10

10.1.3 Preparation for Burden and Reserve Parts

Before BF blow on, all the spares, reserve parts and burden should prepare perfectly according to the requirement of production, especially the coke had better with good strength and larger size to ensure the permeability of stock and active hearth in BF blowing on.

Many BFs in China now consist of sinters, pellets and coke-chemicals, so management for BF blow on should be centralized and unified in iron-making plant to simplify management procedures and pledge BF blow on smoothly.

10.1.4 Investigate and Survey for Every Side of Conditions

There are a lot of needs of BF production such as water, power, cold blast, oxygen, steam, compressed air, nitrogen, gas and so on. Some of them belong to other departments, so there must be a coordination group for dynamic management, particularly the situation of steelmaking, where is the hot metal to go. Once a BF blow on very well, but successive faults in steelmaking that forced BF blow out to wait. This is terrific for BF just blow on resulting to make irregular performance.

10.2 BF Drying (Heating Up)

Furnace drying starts after test running of facilities according to the procedure of blow on. The main purpose of BF drying is to prevent stress, displacement, extrusion or pull-cut of refractory liner which occur at the crystallographic transformation and its volume changing during temperature rising process to form a stead body of refractory liner safely.

For new BF, drying the gas from other BF would be convenient, but instead coke oven gas could be used for single new BF. If both were absent, alternative drying facilities should be built to make use of coal, woody, heavy oil or other fuel depending on local situation.

10.2.1 Hot Stove Drying

Hot stove drying is arranged before BF drying to make use of the hot blast. The contemporary BF is usually with high blast temperature hot stove, which is lined with silicon brick at upper part needing a longer drying period. The producer of silicon brick would offer the curve of drying. Figure 10.1 shows the curve of a large BF's hot stove drying.

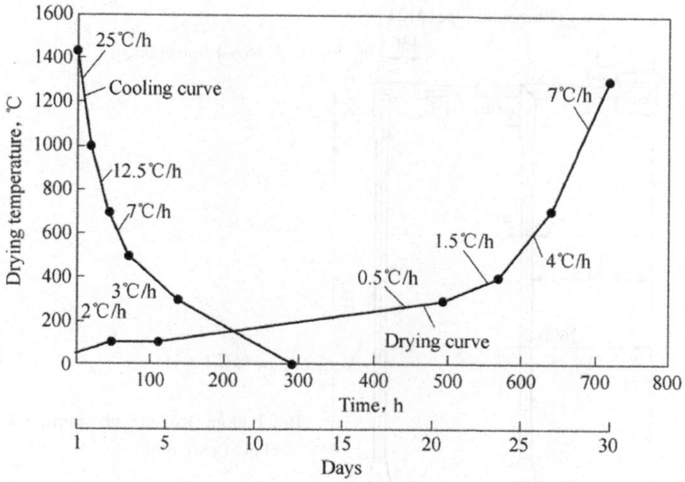

Fig. 10.1 Hot stove drying curve

(1) 40–100 °C, 2 °C/h, 30 h.
(2) 84 h for constant temperature.
(3) 100–300 °C, 0.5 °C/h, 400 h.
(4) 300–400 °C, 1.5 °C/h, 67 h.
(5) 400–700 °C, 4 °C/h, 75 h.
(6) 700–1200 °C, 7 °C/h, 72 h.

31 days in total.
The freezing water had to discharge from the bottom of the stove.

10.2.2 Preparation for BF Body Drying

The main preparation work includes:

(1) It is necessary to install drying guide tube at tuyeres according to the size of BF with one ring for small BF and two rings (outer and inner) for large one. The outer horizontal guide tube is located at 800 mm from the furnace wall. Some of guide tubes connected with bottom drying tubes, which located at 300–500 mm above the bottom and the outer ring connects with tape hole gas guide tube.

Figures 10.2, 10.3 and 10.4 show the tuyere drying guide tube's vertical and horizontal section and support layout drawing.

(2) It does not matter whether the section form of guide tube is round or square, just for the convenience of manufacture and install.
(3) The tape hole gas guide tube is along the angle of lined tap hole imbedding 0.5 m at the passway and keep a certain length inside the furnace according to

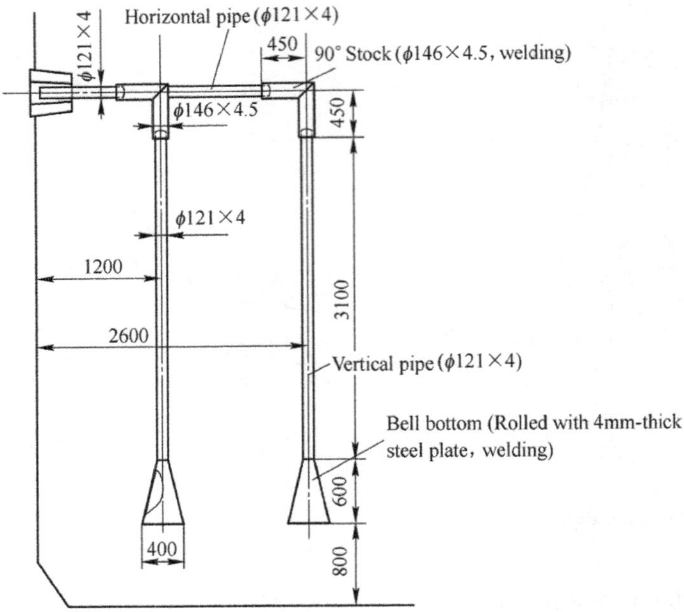

Fig. 10.2 Vertical view of drying pipe

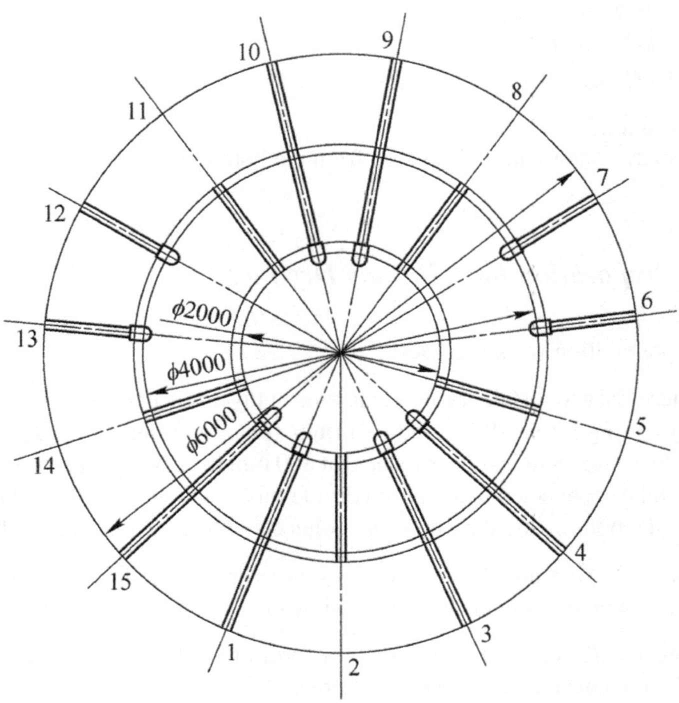

Fig. 10.3 Plan of drying pipe

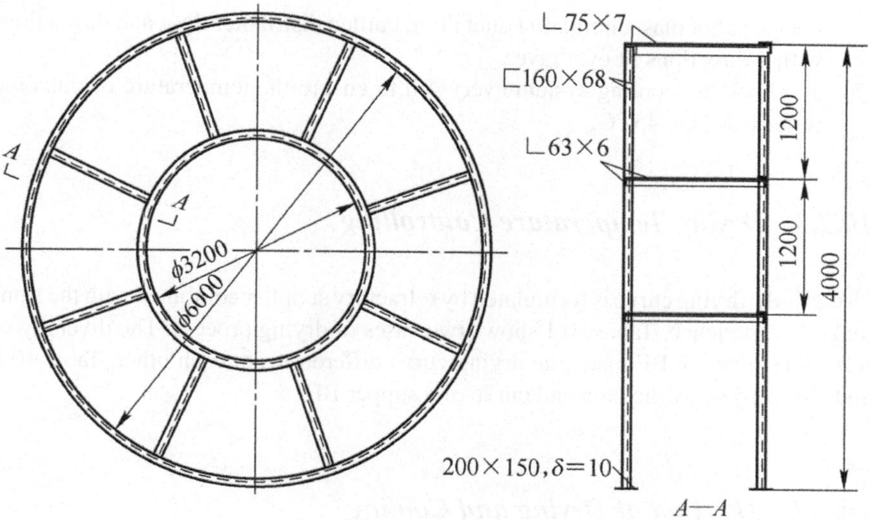

Fig. 10.4 Double scaffolds of drying pipe

the radius of hearth and supported by lined bricks and fixed with steel angle. The gap between tap hole guide tube and passway must be stuffed with dens mud and put a steel tube with 2.5 m long, 114 mm diameter inside another larger tube with 159 m diameter, then stuff the gap between the two tubes with mud, then pull out inner tube leaving a mud pass of diameter 114 mm to make up the brick and mud sheath of tape hole.

The connection of bottom drying tube and gas ejection guide tube of tap hole could improve efficiency of drying.

After drying guide tube fixing, personnel withdraw from site and fix the tuyeres and blowpipes to be going to blast for drying.

(4) Several thermocouples (0–600 °C) and their recording meters should be prepared for drying temperature recording. The thermocouples extend into hearth through tape hole, one of which was put on the center of bottom and another one at somewhere 300 mm from the wall. One more thermocouple for large BF was put on the half of bottom diameter. Every thermocouple was covered under a big block of refractory brick to ensure the temperature was of brick instead of blast. The external effective length was about 2 m.
(5) To seal tape hole with a piece of steel plate (10 mm thick) welded on the frame without leaking and built-in gas sampling facility, while the thermocouple wires should be protected. A tube (1 in) installed (weld) on the manhole cover (extend in 200 mm) at throat extend to cast house floor to draw down drying waste gas for its water testing with a valve to control.
(6) To install expansion testing facility for BF's body during the drying process. It was made of steel bars (3–5 mm diameter) with sharp head to be welded between the body and frame located at 2–3 layers of the different height along

uptake at hot blast circulating duct floor, burden distributer floor and throat floor with 4 directions of ever layer.

(7) To check the cooling system every 4 h to ensure the temperature of outcome water less than 45 °C.

10.2.3 Drying Temperature Controlling

Usually, BF drying curve is formulated by refractory supplier combined with the iron-maker's experience. Table 10.1 shows two cases of drying process. The diversity of refractory of some BF made the drying curve different from each other. Table 10.2 and Fig. 10.5 show the data and curve of a supper BF.

10.2.4 The End of Drying and Cooling

The water content of drying gas should be tested continuously during drying. The drying process could be ended when water content became less than atmosphere humidity in 8 h with bottom temperature higher than 450 °C. Figure 10.6 shows the water content changing during drying process.

The furnace cooling after drying should be operated according to designed temperature. The fluctuation of blast should be less than 20 °C, as the same of drying process. The temperature at hearth lower level should be lower than 300 °C for full coke blowing on or 60 °C for timber blowing on.

10.2.5 Rigorous to Operate and to Stop Accident

(1) The speed of drying temperature rising should be controlled strictly to prevent furnace body hurting or even breaking.
(2) To control top temperature <350 °C and bell-less sealing room temperature <70 °C.
(3) Must pay attention to the safety in the drying process. Although the BF not yet blow on, the fuel gas had been used for hot stove heating, the area around BF is still dangerous for gas poisoning and fire burning, so nobody could stay on every level floor of BF.
(4) The atmosphere and temperature in the furnace should be checked carefully to satisfy the safety standard that personnel could enter the furnace to install guide tube before drying and remove it after cooling.

Table 10.1 Drying temperature of two BFs

Plant	Furnace Number	Volume, m^3	Starting	Original temperature, °C	Heating rate	Constant temperature	Heating rate	Constant temperature	Cooling rate	Demand, °C	Drying time, h
					°C/h(h)	°C(h)	°C/h(h)	°C(h)	°C/h(h)		
An Steel	11	2580	May 24, 1990	150	20(9)	300(31)	20(10)	500(106)	30(13)	<200	209
Ben Steel	5	2000	July 24, 1990	180	24(5)	300(16)	20(5)	600(120)	25(12)	<300	168

Table 10.2 Drying temperature of a BF

Number	Hot blast temperature, °C	Heating rate,°C/h	Operating time, h	Accumulated time, h
1	Normal atmospheric temperature—150	10	24	24
2	150	Constant temperature	240	264
3	150–300	6.25	24	288
4	300	Constant temperature	48	336
5	300–550	12.5	20	356
6	550	Constant temperature	96	452
7	550–300	−12.5	20	472
9	300–50	−15	16	488
10	Cooling naturally		20	508

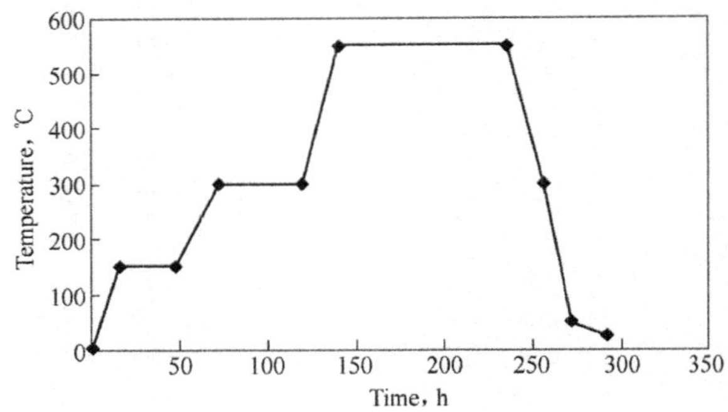

Fig. 10.5 Drying curve of a BF (Table 10.2)

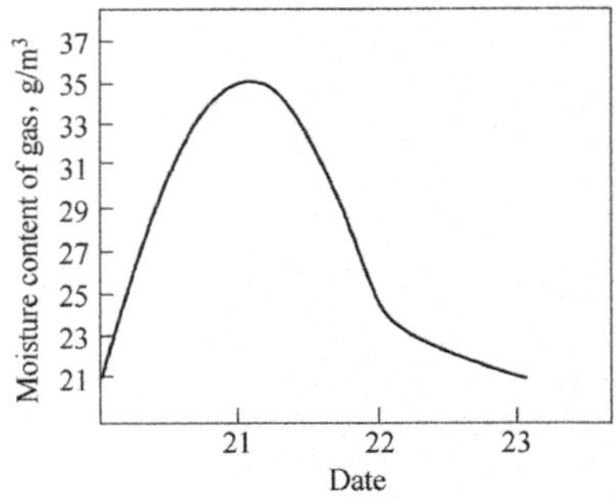

Fig. 10.6 Moisture content change when drying, Capital Steel No. 2 BF, 4th generation

10.3 The Smelting Index and Burdening Calculation for Blowing on

The recent idea of blowing on is quite different from 1960s while the first tapping of hot metal was at lower temperature (about 1400 °C), so to increase the basicity and volume of slag to prevent [S] getting higher. Table 10.3 shows an example from Capital Steel to compare the data of slag in different time that the basicity was down from 1.1–1.2 to 0.75–0.8 and the slag ratio from 0.924–1.2 to 0.413–0.651 t/t. It was a big progress to operate at lower basicity and ratio, not only save burden but also reduce the precious heat at high temperature in blowing on.

10.3.1 The Election for Total Coke Rate and Burden of Blowing on

(1) The precondition is to ensure enough temperature of hearth that indicated hot metal [Si] about 3.0–3.5% and total coke rate of full furnace 3.0 t/t.
(2) According to the burden condition, the basicity-and-slag ratio should be as low as possible.
(3) To charge the iron-containing burden (sinter, pellets) to the middle shaft (for small BF) or to lower shaft (for large BF) to guarantee iron ores to stay enough in the indirect reduction zone to be heated and reduced to obtain the good fluidity of hot metal and slag with enough hearth temperature.
(4) The excellent result would be obtained whether coke or timber was filled in the hearth if the above requirements had satisfied.

Table 10.3 Index of blowing on and location of empty coke

Plant	Volume, m^3	Open date	Charging				Normal burden	
			Coke rate, t/t	CaO/SiO_2	Mn, %	Slag volume, t/t	Coke rate, t/t	Basicity (CaO/SiO_2)
Capital Steel	576	May 30, 1965	2.35	1	0.7	1.2	0.95	1.1
Capital Steel	1327	December 15, 1979	3	1.2	0.7	0.924	0.9	1.1
Jingtang Steel	5705	May 18, 2009	3	0.8	0.7	0.413	0.6	1.13
Shouqian Steel	4078	January 8, 2010	3.03	0.75	0.7	0.651	0.76	1.06

10.3.2 The Method of Calculation (After Blowing on Program of No. 2 BF Jingtang Steel)

(1) The composition and bulk density of raw materials and fuel for furnace (Table 10.4).
(2) Table 10.5 shows comprise of coke blank, empty coke and normal burden.
(3) According to the demand of blowing on, to calculate the volume of every subdivided region of BF and its weight (or volume) of burden (or without region dividing at the same kind of burden in the furnace).

$$V_z = V_1 + V_2 + V_3 + V_4 + V_5 + V_6 + V_7 + V_8 + V_9$$

Table 10.4 Composition and bulk of raw materials and fuels for furnace

Type	TFe	SiO_2	CaO	Al_2O_3	MgO	MnO	S	CaF_2	Ash content	Bulk density
Unit	%	%	%	%	%	%	%	%	%	t/m^3
Sinter	56.92	5.28	10.26	1.66	1.70	0.170	0.023			1.8
Shouqin pellet	63.21	7.23	0.38	0.88	0.26	0.065	0.009			2.4
Limestone		0.64	51.04		4.04		0.081			1.6
Fluorite		16.88					0.012	78.39		1.6
Coke		6.07	0.36	4.20	0.18		0.64		12.00	0.53

Table 10.5 Comprise of coke blank, empty coke and normal burden

Type	Bed coke	Lower coke blank	Middle empty coke	Upper empty coke	Normal burden
Coke dry basis	36	36	36	36	36
Coke wet weight, t	36.1	36.1	36.1	36.1	36.1
Limestone, t			4.5	4.5	
Fluorite, t			0.6	0.6	0.6
Sinter, t					40
Shouqin pellet					50
Binary basicity R_2			1.05	1.05	1.04
Compressibility, %	15	15	14	13	13
Volume after compression, m^3	57.90	57.90	61.32	62.03	97.04

where V_z is the total volume of furnace; V_1 is volume of salamander; V_2 is volume from lower edge of hearth to 0.5 m below tuyere central line; V_3 is from tuyere central line to 0.5 m below it; V_4 is from tuyere central line to upper edge of hearth; V_5 is volume of bosh; V_6 is volume of barrel; V_7 is volume from upper edge of barrel to 0–0.5 m of shaft; V_8 is 0.5–6 m above the upper edge of barrel in shaft; V_9 is from 6.0 m above barrel to lower edge of throat.

$V_1 + V_2$ would be filled with coke (blowing on with coke) or timber basically (blowing on with timber).

(4) The number of burden batches calculating (bottom layer coke involved).

Coke blank:

According to the plan coke blank should be put to *X m* of barrel upper edge (0.5 m in the case of Jingtang Steel No. 2 BF) with 36 t on dry basis or 36.1 t wet per charge, volume of which was 57.90 m^3 (36.1/0.53 × (1–15%) = 57.90). The charge was weighted 30 (1767.948/57.90 = 30.5).

Empty coke:

The empty coke was charged in V_8 (1142.651 m^3) which consists of coke dry basis (36 t), limestone (4.5 t) and fluorspar (0.6 t), the compressed volume of which was 61.32 m^3 (36.1/0.53 + 4.5/1.6 + 0.6/1.6) × (1–14%) = 61.32). The number of burden batches was 19 (1142.651/61.32 = 18.6).

The total coke rate could be calculated with the following formula:

$$36 \times (2 + 30 + 19 + A + B)/58.174 = 3.0$$

where A is normal charges in upper part; B is charges of empty coke.

The assembly filling volume of normal charge and empty coke could be calculated with the following formula:

$$97.04\,A + 63.03\,B = 1728.571$$

According to the principle of burdening increasing gradually, the combination in upper part should be two charges of normal burden + 3 charges of empty coke + 5 charges of normal burden + 3 charges of empty coke + 7 charges of normal burden.

10.4 Furnace Blowing on with Full Coke

As early as in 1956, blowing on with full coke had been tried to practice at No. 9 BF of An Steel but was interrupted because of the lower temperature of hot metal and too stick slag, then changed over to timber. It was tried again in 1964 with very successful result to be evaluated by the famous iron-making expert Mr. Li Guoan as technology mature, economy, convenient and suitable to create the first example of large BF blowing on with full coke [1] (Table 10.6).

The advantages of blowing on with full coke:

Table 10.6 An example of blowing on with full coke

Plant	Furnace Number	Volume, m^3	Open date	Charging			Charging order				Normal burden		Literature
				Coke rate, t/t	CaO/SiO_2	Slag volume, t/t	Segment	Compressibility, %	Hearth	Bosh	Coke rate, t/t	Basicity CaO/SiO_2	
An Steel	10	1513	December, 1963	2.5	0.95		7		Coke		1		[1]
Capital Steel	4	1200	October 15, 1972	2.5	0.95	0.93	6	13	Coke	Semi-coke	1		[2]
Shexian Steel	3	569	July 2, 1975	3.5	0.92	1.51	7	Blowing & charging	Coke	Empty coke	1.2	1	
Ma Steel	12		March 10, 1972	2.63			5	12	Coke				
Ben Steel	5	2000	July 24, 1990	2.1	1.05	0.67		14	Coke	Coke blank	0.7	1.1	
Anyang Steel	8	2200	June 28, 2007						Coke				[3]
Xin Steel	6	1260	June 12, 2009	2.5	1/0.8		7	13	Coke		0.7		[4]

(1) A huge amount of timber could be saved for blowing on with full coke, just as Prof. Tang Qinghua said that 2500–3000 railway crossties would be used up for once BF blowing on, where a BF stood, but a large field of forest destroyed! This is not proper conduct for BF man. Blowing on with full coke technology should be transmitted from predecessors [5]. In the recent years, more and more larger BFs filled hearth with timber (usually once 400–600 m^3) even there were rich experience for full coke blowing on, especially at 2200 m^3 BF in Anyang Steel and Xinyu Steel as well as many medium and small BFs experiences to show the direction for large BF operation.

(2) The existing facilities could be used to charge coke into hearth, but special facilities must be built for timber charging as well as removed away and tuyers re-fixing afterward.

(3) The temperature in furnace should be lowered to about 40 °C until personnel could enter into charge the timber. It would take a long time for furnace cooling.

10.5 Locations of Iron-Bearing Burden

The first hot metal taping temperature is influenced by the location of iron-bearing burden which would not be heated and reduced enough, if to be located at too low part of furnace to decrease the temperature of hot metal certainly. The relation between location of burden and the time reaching to tuyere level was studied in Russia (Fig. 10.7).

The high-temperature heat should be saved as possible to keep enough hearth temperature when blowing on. The indirect reduction of burden in blowing on should be as enough as possible to decrease direct reduction and heat consumption in hearth, so the burden ought to stay in the region of 600–1200 °C with enough residence. Figure 10.7 shows the diagram for isothermal surface at a BF (3200 m^3) in Russia

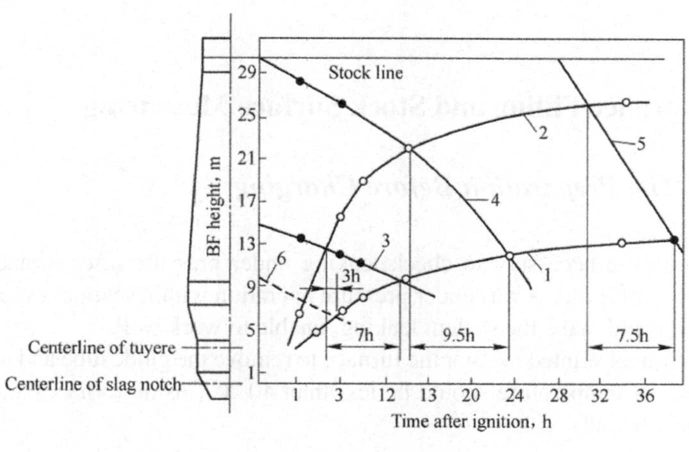

Fig. 10.7 Change of isothermal surface location when blowing on

Table 10.7 Original location of iron bear burden

BF volume, m^3	Charging		Iron-bearing burden position		
	Hearth	Lower stack, m	Above tuyere, m	Operational height, m	Rate, %
5705	Wood + coke	Empty coke, above belly 5.63	14.41	32.8	44
5705	Wood + coke	Empty coke, above belly 6.32	13.42	32.8	41
3952	Wood + coke	Empty coke, above belly 6.5	13.5	28.8	47
1780	Wood + coke	Empty coke, above belly 4	9.7	23.5	41

blowing on in 1985. The location of isothermal surface and the features of burden layer be obtained on the base of measuring data, blast index and burden analysis. The hypothesis of isothermal surface would be plane. The curves 1 and 2 correspond to isothermal surface of 1200 and 600 °C where the burden should stay 4–5 h for indirect reduction. The burden charged at 11 m above tuyere level stayed there 7 h with enough indirect reduction as curve 3 showing.

The curves 4 and 5 show last charge location change of first and second smelting cycle, respectively. It is obvious that last charge of burden (first smelting cycle) stayed in this region for 9.5 h (curve 4) and the last charge in second cycle stayed 7.5 h (curve 5) [6]. These important study results had been proved afterward in Soviet Union that iron-bearing burden should be located in shaft above barrel for enough reduction.

Table 10.7 shows the practical experience of blowing on at 4 BFs of Capital Steel. The iron-bearing burden was located in lower shaft with 41–47% of the work height (from tuyere centerline to the upper edge of throat). The temperature was quite enough with good fluidity.

10.6 Furnace Filling and Stock Surface Measuring

10.6.1 The Preparation Before Charging

(1) It would be necessary to check leaking under pressure after furnace drying because BF process was under pressure operation within seamless system. Any seam would make the system leaking, unable to work well.
(2) If personnel wanted to enter the furnace to remove the guide tube and make mud drum, the temperature should be less than 40 °C (inside furnace) and <70 °C (at hearth wall).

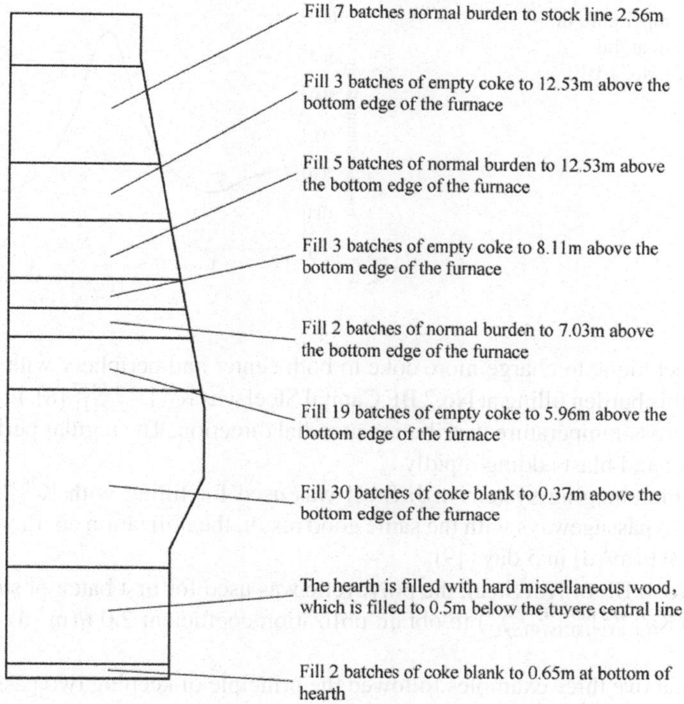

Fig. 10.8 Practical example charging of blowing on at No. 1 BF Jingtang Capital Steel

(3) To cover the off-take end with a network to prevent some fallings hurting the workers in the furnace.

10.6.2 Charging According to the Arranged Sequence

The experience of successful blowing on at No. 1 BF Jingtang Capital Steel in the past decades was created by the new generation of Capital Steel BF men [7]. Figure 10.8 shows the results calculated according to the data in Sect. 10.3.

10.6.3 The Proper Burden Distribution Should Begin from Furnace Filling

The blast volume increasing should speed rapidly after blowing on to guarantee enough blast for first tape of hot metal. Since the iron-bearing burden began to charge, burden distribution was arranged for two passageways of gas stream (peripheral, central), which could pledge regular performance and blast adding by the use of

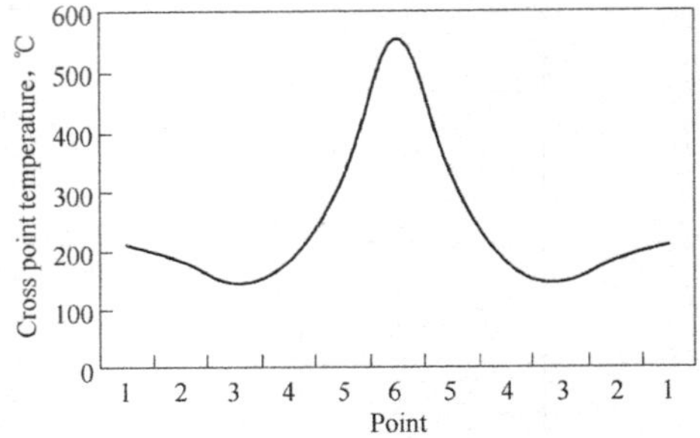

Fig. 10.9 Temperature in radial direction at the blowing on of No. 1 BF Capital Steel

bedding technique to charge more coke to both center and periphery with iron ores between, this burden filling at No.2 BF Capital Steel was $K_{333}^{654}J_{3222111}^{8765432}$ [8]. Figure 10.9 shows its cross-temperature distribution at radial direction. The regular performance was perfect and blast adding rapidly.

The same charging system in Ji Steel was used for filling with $K_{2332}^{7654}J_{422222}^{876541}$ to develop two passageways with the same good result, the utilization coefficient of BF reached 2.0 t/(m^3 d) in 5 days [9].

Also No. 6 BF of Xin Steel, the polycyclic was used for first batch of subsequent charging ($K_{020402}^{373432}J_{00200302002}^{36534532165}$) to obtain utilization coefficient 2.0 t/(m^3 d) in 5 days [4].

All the above three examples followed the principle of keeping two passageways of gas stream to gain blast adding smoothly and enough blast for first tapping of hot metal.

10.6.4 Stock Surface Measuring

Usually, the stock surface form and distribution measuring are arranged at the last 5 charges that would be very important for the production followed up. There are special companies in charge of such activities, including

(1) To measure the fall points, etc., the distributing feature for reference of followed production and proof the coefficients of falling points of burden equation to calculate the burden distribution afterward.
(2) To measure the shape of the stock surface especially the distribution of ores and coke as well as the deformation of stock layer.
(3) The burden width forming and hopper's depth.
(4) Particle distribution and thickness of stock layer.

Attention should be paid to the accuracy for measuring, any fault would hurt production afterward. Once a BF made mistakes of stock surface measurement, the adverse gas stream distribution happened at the time of subsequent charging for blowing on, then blowing out to treat (there were different opinions from participants) without success all the day, then re-measured to certify the formal measuring mistake of burden falling point. The originally regular performance with enough hearth temperature suffered setbacks followed up for such measurement fault.

10.7 Igniting and Blasting

Before igniting, it must be provided with the qualification of procedures from beginning to end, especially hot metal accepting, burden production, applying and transportation, then prepare to ignite and blast.

10.7.1 Burden Heating and Igniting Operation

Burden heating before igniting was the common view of BF men. F. C. Rorick(Bethlehem Steel Corp)presented in his book "Challenging Blast Furnace Operation" that "not all the operators recognize burden heating must be necessary for blowing on. It would be safe to heat with blast at less than 600 °F (about 300 °C) with 1/3 normal blast volume and the coke would be ignited if temperature rose continuously to 900–1000 °F or 500–600 °C. It would have 8–10 h for heating process." This was introduced by 9 experts of iron-making including J. J. Poverromo (Bethlehem Steel), J. H. Duncan (Armco Steel), J. E. Holditch (Dofasco Steel) et al.

There are a lot of methods and experiences about blowing on burden heating in China. Table 10.8 shows the practical cases of blowing on with different methods and roles.

1. In early An Steel, the hot sinter (cooled to 300–400 °C in the atmosphere) was used for blowing on with full coke to be charged in the middle shaft to utilize its latent heat. But the hot sinter was harmful for charging facilities and environment.
2. In Ma Steel and She Steel, the blowing on burden was charged with blasting to make the stock porous and blow the powders out of furnace that would benefit BF to receive blast volume adding. In addition, the lower temperature blast could heat and dry the burden to save heat as well as benefit blowing on operation.
3. For the BF of Capital Steel blowing on with timber the blast lasted 48 h after furnace filling to control the temperature < 50–180 °C to dry and heat the burden very well.
4. The cold blast (<100 °C) could be used for small BF burden heating because of the short stock column, but not for large one because the water might be frozen in the upper stock column. The temperature of cold blast and top for small BF just 80–85 °C and 29 °C, respectively, that could not ever heat the burden. It should be more than 130 °C for large BF to heat the blowing on burden

Table 10.8 Practical cases for burden heating of blowing on

Plant	Furnace Number	Volume, m^3	Open time	Burden heating		
				Time	Temperature, °C	Blast volume, m^3/min
An Steel	10	1513	December, 1963		Hot sinter 300-400	
Ma Steel	11	300	May 15, 1973	Charging process	250	310
She Steel	3	568.5	July 2, 1975	Charging process	<400	300–400
Jingtang Steel	1	5705	May 18, 2009	48 h	≤180	3000–5000
Jingtang Steel	2	5705	June 26, 2010	48 h	≤150	3000–5000
Shouqian Steel	3	3952	January 8, 2010	23 h 14 min	80–85	2700
Ji Steel		3200	August 2, 2010		Cold blast ≤ 180	
Xuan Steel	2	2500	September 18, 2010		80	800
Han Steel		1080	December 22, 2011	71 min	Cold blast	

5. Charging with hot blast had been practiced at No. 2 BF Jiu Steel in 2000. The total charges for filling full furnace was 80, when 53 charges filled, first charge iron ores bearing burden at stock line 13 m, igniting started with charging continuously, while blast temperature 730 °C, blast pressure a litter higher (0.0148 MPa). The blast pressure fell down in one hour and blast volume to 1500 m^3/h and charging continuously. This might be the fantastic blowing on in the history of Jiu Steel all-time.

10.7.2 To Choose Tuyere Parameters

It would be necessary to choose tuyere parameters properly before igniting to pledge the blast speed at the beginning of blowing on(less blast volume) with tuyere blending (plugging) or jacketing inside (minority). It was popular for igniting in blowing on without tuyere blending that keeps blast around hearth well-distributed and all combusted and bright in more than ten or even dozens of minutes to create conditions of steady and equality for BF operation (Table 10.9).

Table 10.9 Allocation of tuyeres during blowing on period

Plant	Volume, m^3	Open time	Tuyere			Works cited
			Open	Block	Total	
An Steel	599	September 7, 1949	8 × 180 reinforce 8 × 150		8	Zhuang Zhen'e, 1st episode of iron-making material, 1951
Capital Steel	1200	October 15, 1972	10	8	18	
Ma Steel	300	May 15, 1973	10	0	Add thimble 60 mm, occupy 32%	
She Steel	568	July 2, 1975	12	0	12 × 120 = 0.1356	Foreign Iron and Steel Technology of Kunming Iron and Steel Co., 1988(1):81–90
Capital Steel	1327	December 15, 1979	13	9	22	
Chelebowitz	5500	April 13, 1986	20	20	40 × 150	Сталь, 1988, 1: 12–18
Ben Steel	2000	July 24, 1990	11	11	22	
Capital Steel	2536	June, 1994	15	15	30	
Mei Steel	1250	December 16, 1995	7	7	14	
Bao Steel	4063	May 25, 1997	36	0	18 × 120 + 18 × 130 = 0.4432	Bao Steel Technology, 1997(6): 5–7,36
Han Steel	2000	June 28, 2000	18, 0.222 m^2	10	120 × 9+110 × 19 = 0.2823 m^2	Hebei Metallurgy, 2009(2): 17–19, 43
Jiu Steel	1000	October 25, 2000	12	6	18	
Ma Steel B	4000	May 24, 2007	24 × 120 = 0.2716	12	30	Journal of Anhui Metallurgical Technology Vocational and Technical College, 2008(2):8–10, 18

(continued)

Table 10.9 (continued)

Plant	Volume, m^3	Open time	Tuyere			Works cited
			Open	Block	Total	
Capital Steel	1780	September 9, 2008	12	12	24	
Ben Steel	4747	October 9, 2008	38	0	$15 \times 120 + 23 \times 125 = 0.4519$	
Jingtang Steel	5705	May 18, 2009	$13 \times 110 + 29 \times 120 = 0.4515$	0	42	Iron-making, 2010(2): 7-10
Xin Steel	1260	June 12, 2009	$20 \times 100 = 0.1392$	0	$20 \times 130 = 0.1468\ m^2$	Iron-making, 2010(2): 28-30
Wu Steel		August 1, 2009	$28 \times 130 + 14 \times 80 = 0.2964$	0	0.4778	Wuhan Steel Technology, 2011(6):14–17
Ji Steel	3200	August 2, 2010	$24 \times 130 \times 580 = 0.3184\ m^2$	8	$32 \times 130 \times 580 = 0.4245\ m^2$	Iron-making, 2011(1):16–19
Xuan Steel	2500	September 18, 2010	20	10	$28 \times 120 + 2 \times 130, 585 = 0.3732\ m^2$	Iron-making, 2011(2): 44–47; (6):25–28
Jiu Steel	2500	March 6, 2011	18	10	$23 \times 130 + 5 \times 120, 500 = 0.3616\ m^2$	Iron-making, 2011(5): 22–25
Han Steel	1080	December 22, 2011	$8 \times 115 + 12 \times 75 = 0.136$	0	20×115	Iron-making, 2012(6): 41–43
Mei Steel	4070	June 2, 2012	$24 \times 130 + 12 \times 90 = 0.3949$	0	0.4776	Compilation of AC Materials, 2013,10: 63–67
Anyang Steel	4836	March 19, 2013		0		Compilation of AC Materials, 2013, 10: 78–95

10.7.3 Igniting

The hot blast of 700 °C is quite enough to be used for igniting with coke or timber blowing on, may be higher temperature could be available with new hot stoves.

10.7.3.1 Operation for Igniting

See Table 10.10.

10.7.3.2 Igniting Blast Volume and Its Ratio

Igniting blast volume influenced not only igniting speed but also stock moving, charging time and the first taping. The "blast volume ratio" in Table 10.10 means the ratio of "taping blast volume" to BF volume. When this blast volume ratio is bigger than 1, the first taping would be OK with regular performance and speeding blast. The "blast volume ratio" in Fig. 10.10 involves 30 BFs, which show that the ratios of most small BFs were higher (some >1.0) than large BFs (most 0.4–0.6), so the former indicated the direction for later, it should be bigger than 0.5.

10.7.3.3 Gas Guide Tube for Iron Notch

The gas would eject from iron notch after igniting, so a flame should be put in front of notch to ignite the gas in advance and be observed to prevent gas poisoning and fire. The guide tube should be taken out when hot metal and slag eject a while, then tap hole plugged with mud.

Before years, the guide tube used to be a whole one which was very hard to take out that would be possibly plugged with slag; sometimes, oxygen must be used for tap hole opening. It wastes oxygen, personnel and time. Now the guide tube changed to two sections that would be easy to pull out.

The gas ejecting period should last as long as possible, which is the best way to heat the bottom, hearth and mud drum. It should be concerned during design and install the guide tube that the gas ejecting time should be enough until hot metal surface rising to tap hole, generally in 6 h, while any slag must be braised out of guide tube. When ejection last over 6 h and slag appeared, then the external section of guide tube could be pulled out, otherwise would be very hard to remove and hurt taping afterward.

10.7.3.4 Observation After Igniting

The brightening time and number of tuyeres:

The first tuyere brightened in 10–20 min when blowing on with timber. It might be faster with all tuyeres opening and igniting than 10–20 min usually. At some BF igniting with tuyeres plugging, it might take 1 h for all tuyere brightening, and it might be also for blast and charge speeding without matching each other.

Table 10.10 Operation for igniting

Plant	Volume, m^3	Open time	Igniting time	Igniting system					Tuyere after igniting		Pull out nozzle and close up breach	
				Blast volume	Blast pressure	Blast temperature	Blast rate	Blast speed	Start shining	Whole tuyere shining	Time	Utilizing mud
				m^3/min	MPa	°C	m^3/m^3	M/s	min	min		kg
An Steel	599	September 7, 1949	10:00		0.066	600		76.4	5 min	43 min	90 min	
Capital Steel	576	August 8, 1962	18:20	550	0.5	700	0.95				4 h 55 min	
An Steel	1513	December, 1963			0.1	700				12 min		
Capital Steel	515	July 27, 1964	12:35		0.40	700						
Capital Steel	576	May 30, 1965	4:40	650	0.50	700	0.46	100			2 h 50 min	
Capital Steel	1036	April 25, 1970		1200	0.70	650	0.7				Igniting without pulling out nozzle	
Capital Steel	1200	October 15, 1972	5:00	1100	0.75	800	0.62	120		1 h	8 h 20 min	40
Ma Steel	300	May 15, 1973	10:00	330	0.008	580	1.1					
She Steel	568	July 2, 1975	9:48	800	0.37	600	1.41				8 h 40 min	Iron flowed out and the nozzle flattened

(continued)

Table 10.10 (continued)

Plant	Volume, m^3	Open time	Igniting time	Igniting system					Tuyere after igniting		Pull out nozzle and close up breach	
				Blast volume	Blast pressure	Blast temperature	Blast rate	Blast speed	Start shining	Whole tuyere shining	Time	Utilizing mud
				m^3/min	MPa	°C	m^3/m^3	M/s	min	min		kg
Capital Steel	1327	December 15, 1979	11:15		0.30	700				1 h 25 min	6 h 11 min	
Dofasco Steel	3795	February 18, 1981	10:35	1800		700	0.47		12 min			
Bao Steel	4063	September 15, 1985	12:00	1776		700	0.44				18 h 25 min	
Chelebowitz	5500	April 13, 1986				700					26 h	
Ben Steel	2000	July 24, 1990	23:05	2500	0.097	870	0.44			40 min	6 h 55 min	150
Capital Steel	2536	June 2, 1993	22:50							1 h 27 min	30 min	300
Mei Steel	1250	December 16, 1995	10:45	1000		800	0.8					
Ling Steel	380	March 10, 1995	6:24	480		650	1.26				6 h 18 min	

(continued)

Table 10.10 (continued)

Plant	Volume, m^3	Open time	Igniting time	Igniting system					Tuyere after igniting		Pull out nozzle and close up breach	
				Blast volume	Blast pressure	Blast temperature	Blast rate	Blast speed	Start shining	Whole tuyere shining	Time	Utilizing mud
				m^3/min	MPa	°C	m^3/m^3	M/s	min	min		kg
An Steel	2580	February 12, 1995			0.1	700	0.27					
Bao Steel	4063	May 25, 1997	11:30	2000		750	0.49	250				
Pan Steel	1200	June 14, 1997	5:25	1750	0.068	710	1.46				40 h 5 min	
	2000	June 28, 2000	15:15	1976		800	0.99					
Jiu Steel	1000	January, 2000	7:48	730	0.148	678	0.73	200			10 h	
Bao Steel	4747	April 27, 2005		2400			0.51					
Ma Steel B	4000	May 24, 2007	16:28			700			12 min		10 h52 min–1 h 6 min	
Capital Steel	1780	September 9, 2008	17:28							17 min	5 h 32 min	

(continued)

Table 10.10 (continued)

Plant	Volume, m^3	Open time	Igniting time	Igniting system					Tuyere after igniting		Pull out nozzle and close up breach	
				Blast volume	Blast pressure	Blast temperature	Blast rate	Blast speed	Start shining	Whole tuyere shining	Time	Utilizing mud
				m^3/min	MPa	°C	m^3/m^3	M/s	min	min		kg
Ben Steel	4747	October 9, 2008	8:58	2500		800	0.53	92	3 min	15 min	10 h	
Xin Steel	1260	June 12, 2009	16:36	600	0.11	630	0.48			39 min	2 h9–15 min	
Ba Steel	2500	February 27, 2009		1250		700	0.5	78				
Jingtang Steel	5705	May 18, 2009	13:18	3370	0.055	711	0.59	271		7 min	22 h–22 h 40 min	
Shouqian Steel	3952	January 8, 2010	12:16	2700	0.5	750	0.66		12 min	36 min	20 h 24 min	
Ji Steel	3200	August 2, 2010	16:16	1600	0.042	800	0.5	250		9 min		
Xuan Steel	2500	September 18, 2010	9:58	1500		750	0.72		10 min	38 min	9 h	
Long Steel	1800	November 15, 2010	8:00	1300		700	0.47				6.5 h	

(continued)

Table 10.10 (continued)

Plant	Volume, m^3	Open time	Igniting time	Igniting system					Tuyere after igniting		Pull out nozzle and close up breach	
				Blast volume	Blast pressure	Blast temperature	Blast rate	Blast speed	Start shining	Whole tuyere shining	Time	Utilizing mud
				m^3/min	MPa	°C	m^3/m^3	M/s	min	min		kg
Jiu Steel	2500	March 6, 2011	10:15						20 min	1 h 42 min	16–18 h	
Han Steel	1080	December 22, 2011	19:58	800	0.08	800	0.74		8 min			
Mei Steel	4070	June 2, 2012	10:58	2100		700	0.5					

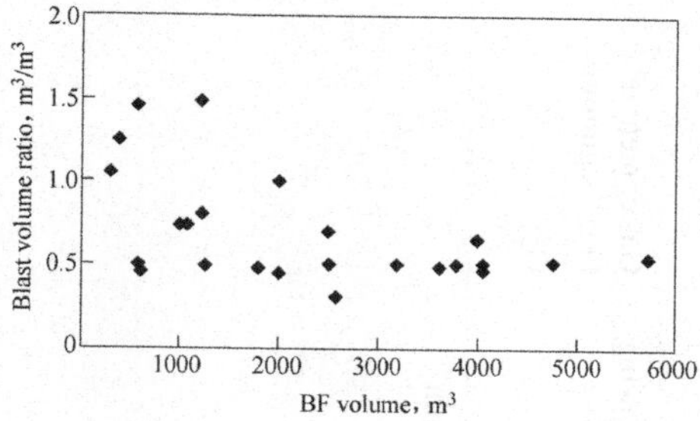

Fig. 10.10 Igniting blast volume ratio of blowing in

10.7.4 Burden Charging

The suitable relations between blast pressure and volume as well as lower pressure difference indicate the good permeability of stock and began to charge after stock line moving. For some BF, blasting and igniting began just when the furnace got half-filling, although the pressure difference could be lower temporary, but the subsequent charge could not be heated and reduced enough leading to higher total coke rate of whole furnace and some risk. It did not advocate generalization, even though some success at few BFs. Table 10.11 shows the charging after stoke line moving.

The data from Tables 10.9 and 10.11 indicate that stock line moved faster after igniting when all tuyeres opening, but there were some exceptions sometimes.

Blowing on with full coke and all tuyeres opening at No. 6 BF Xin Steel, it took 39 min after igniting for all brightening in front of tuyeres and 2 h for gassing. It was propitious for blowing on with full coke, but stock line did not move until checking in 2 h to improve the relation of blast pressure and volume and began to charge subsequent burden. The first tape of hot metal was at 18 h.

The blowing on with full coke was successful for No. 6 BF (some blowing on with timber had the same result), but the proper relation of blast pressure and volume did not realize until checking or slipping. The reason, generally to say, is that the coke in stock started to combust without cohesive zone and gas pass ("gas window"). If the relation of blast pressure and volume did not well control, it would have irregular situation. The experienced BF person cut down blast temperature immediately after finding out some abnormal relation of blast pressure and volume that would be improved. But some BF men concerned for fear that the hearth temperature might be not enough and might not bridge with coke stock, so rising blast temperature instead of cutting down, then adverse furnace condition worsening. When blast pressure was getting higher with volume decreasing automatically, the blast temperature must be cut down and it should not scary about slipping (if necessary), which should be as early as possible. There was a BF to blow on without proper relation of blast pressure and volume for a few hours (without cutting the blast temperature and volume down), then forced to checking but lost the opportunity, leading to blooding of tuyeres, while more hot metal and slag accumulating in hearth, so it would be dangerous if checking in 3 h after igniting.

Table 10.11 Stock line moving and subsequent charging

Plant	Volume, m^3	Open time	Follow-up charging system					Gas-blowing time	Gas content of blast beginning	
			W_K, t	W_J, t	Stock line, m	Distribution of burden	Charging time		O_2	H_2
Mei Steel	1250	December 16, 1995	16		L_K1.4 L_J1.0	K^{76543}_{22333} J^{76543}_{22222}		6 h 7 min		
An Steel	2580	February 12, 1995	43		2.1					
Ma Steel		April 25, 1994				K^{76543}_{22220} J^{765432}_{222222}				
She Steel	568.5	July 2, 1975		4.6	1	2A + 3B A = KKLL, B = JKKJ				
Han Steel	2000	June 28, 2000	24.5		1.5	$K^{40}J^{29}$		7 h		
Dofasco Steel	3795	February 18, 1981					70 min	8 h 14 min	<0.3	
Ji Steel	3200	August 2, 2010	91	21	1.3	K^{353330}_{333} $J^{393735333014.5}_{422223}$		7 h 14 min	0.04	3.4
Xin Steel	1260	June 12, 2009				K^{373433}_{243} $J^{36.534.53316.514.5}_{2322}$		2 h		
Ba Steel	2500	February 27, 2008				K^{76543}_{32222} J^{76543}_{22222}	123 min	5 h 35 min		

(continued)

Table 10.11 (continued)

Plant	Volume, m^3	Open time	Follow-up charging system					Gas-blowing time	Gas content of blast beginning	
			W_K, t	W_J, t	Stock line, m	Distribution of burden	Charging time		O_2	H_2
Mei Steel	4070	June 2 2012				$J_{333221}^{1098765}$ K_{23322}^{109876}	11 h	2 h		
Capital Steel	576	August 8, 1962				3A + 2B, A = PPKK, B = KPPK				
Capital Steel	515	July 27, 1964			L1.0 m	A = KKJJ, B = JKKJ; adjust to JKKJ				
Capital Steel	576	May 30, 1965			1.5 m	KKJJ	15 h 40 min			
Capital Steel	1036	April 25, 1970			1.5	2KKJJ + 3JKKJ	7 h 10 min			
Capital Steel	1200	October 15, 1972		7.32	1.5	2KKJJ + 3JJKK	9 h 20 min			
Capital Steel	1327	December 15, 1979				$K^{37}J^{37}$	18 h 10 min			
Capital Steel	2536	June, 1994				K_{332}^{323936} $J_{22323}^{3432292622}$				
Capital Steel	2536					K_{44}^{2822} $J_{22323}^{3432292622}$				

(continued)

Table 10.11 (continued)

Plant	Volume, m^3	Open time	Follow-up charging system					Gas-blowing time	Gas content of blast beginning	
			W_K, t	W_J, t	Stock line, m	Distribution of burden	Charging time		O_2	H_2
Capital Steel	1780	September 9, 2008				K_{333}^{654} $J_{3222111}^{8765432}$	2 h 27 min			

10.7.5 Gas Applying

The top gas admitted to atmosphere through bleeders at the beginning of blowing on, that polluted environment and hurt human health, also to be fined by government agency. On the other hand, the gas should be applied for heating hot stoves as soon as possible after blowing on, also for shifting to high top pressure operation.

BF gas sampling analysis requires $H_2 \leq 4\%$, $O_2 \leq 1\%$ for safe.

For dry gas cleaning, the temperature should be higher than 70 or 100 °C (different opinions) to prevent water freezing. Actually, 70 °C would be enough for bag collector. It was indicated that the freezing point of BF gas was different from pure water because of the H_2 and CH_4 content.

For large BF blowing on, the top gas temperature was always lower (just 40 °C) and the countermeasure was to increase blast volume rapidly, so blast volume could not be too low for igniting that was why top temperature of small BF rising faster than large BF, the main reason might be blast volume rate >0.6 (Table 10.10).

Pay attention to sampling and analysis for top gas, to record once an hour.

10.8 Hot Metal Taping

The first tapping of hot metal after blowing on should be determined according to the calculated height of accumulated hot metal and slag surface in hearth and delayed as long as possible without suppressed wind and tuyere damaging that not only benefit hearth heating but also reduce labor intensity because more hot metal could flow through the sow easily.

10.8.1 To Determine the Time for First Tape

A certain amount of hot metal must be accumulated for first tapping that related to the ratio of hot metal quantity/BF volume, which should be >0.03 m^3/m^3. Modernize large BF has deeper sow, less hot metal could not flow through the skimmer (slag separator), as Table 10.12 shows that this ratio of some BFs just 0.01–0.02, so the first tape is easy out of gear.

10.8.2 To Determine When the Tap Hole Opened for the First Tapping

It would be considered how long for tap hole opening before first taping, first of all, when ejecting tube would be pulled out (Table 10.10). Generally, tap hole would be

Table 10.12 First tapping status of some BFs

Plant	First taping										Days of Si < 1%
	Time after igniting	Blast volume while tapping	Blast rete	Tapping time	Iron amount	Iron rate	Si	Mn	S	Hot metal temperature	
		m^3/min	m^3/m^3	min	t	t/m^3	%	%	%	°C	
An Steel	32 h 15 min	750	1.25		8	0.01	12.007	1.01	0.28		Casting iron
Capital Steel	19 h 18 min						3.78		0.049		
Capital Steel	17 h 6 min				75.3		1.18	0.92	0.04		
Capital Steel	13 h				40		0.43		0.07		
Capital Steel	13 h 56 min				154		1.6	0.65	0.063		
She Steel	15 h 27 min	940	1.65		46	0.08	2.13	0.65	0.051		4
Capital Steel	32 h 15 min				135.6		5.36	0.85	0.004		
Dofasco Steel	22 h	3600	0.95		490	0.13	4.03		0.01	1475	6
Bao Steel	22 h 15 min	5000	1.23		178	0.04	5.56	0.69	0.009	1430	
Chelebowitz	38 h				10	0.002	6.25		0.007		
Muroran Steel	21 h				73	0.03	6.08	0.55	0.006	1382	
Fukuyama Steel	24 h 35 min				700	0.16	3.6		0.029	1451	

(continued)

Table 10.12 (continued)

Plant	First taping										Days of Si < 1%
	Time after igniting	Blast volume while tapping	Blast rete	Tapping time	Iron amount	Iron rate	Si	Mn	S	Hot metal temperature	
		m^3/min	m^3/m^3	min	t	t/m^3	%	%	%	°C	
Kashima Steel	22 h 38 min			266	558.9		3.25		0.025	1410	
Ben Steel	17 h 5 min				344.8	0.17	3.29	0.44	0.03		
Mei Steel	20 h 15 min						3.32	0.61	0.042		
An Steel	24.5 h						5.26		0.032		
Bao Steel	24 h 30 min	5200	1.28		573	0.14	3.76	0.56	0.031	1448	6 days 0.64%
Jiu Steel	16 h 14 min				70.86	0.07	3.78	1	0.052	Poor fluidity of iron slag	6
Ben Steel	16 h 20 min			152	340	0.07	1.67	0.29	0.1184	1380	
Xin Steel	16 h 84 min				60	0.048	5.19				
Jingtang Steel	31 h 41 min				198	0.034	4.04		0.01	1426	3

(continued)

Table 10.12 (continued)

Plant	First taping										Days of Si < 1%
	Time after igniting	Blast volume while tapping	Blast rete	Tapping time	Iron amount	Iron rate	Si	Mn	S	Hot metal temperature	
		m^3/min	m^3/m^3	min	t	t/m^3	%	%	%	°C	
Shouqian Steel	29 h				380	0.096	5.46	0.6	0.024	1496	3 days 0.95%
Ji Steel	27 h	4500	1.41	304	400	0.13	4.02	0.3	0.028	1300 ~ 1475	3
Xuan Steel	23 h 220 min	3200	1.28		294	0.12					
Long Steel	19 h 28 min	2400	1.33		226	0.13					
Jiu Steel	25 h 57 min						4.1	2.2	0.047	1350	9
Mei Steel	About 29 h	About 4600	1.13	300	634	0.16	4.65	0.996	0.01	1467	3

easy to open in 6 h after igniting that depended on the quality of mud plugging even just 2 h for ejecting. In some cases, the ejecting tube did not be pulled out without mud plugging, then oxygen had to be used to burn through the tap hole again and again wasting time of several hours. If estimated the tap hole would be hard to open, the preparation should be in advance.

It should be considered to shift first tapping at earlier time, if check would be possible.

10.8.3 Some Innovations for Blowing on Taping

(1) In An Steel blowing on, a blowpipe was put through tap hole for oxygen blowing into hearth (its pressure is higher than inside) till the hot metal came to melt the pipe and taping continuously and automatically while unnecessary for any tap hole opening.
(2) In Han Steel blowing on, a bypass pipe was put aside the ejecting tube to blow compress air (0.8 MPa) till tapping, then cut down the air let hot metal taping along the pass automatically without tap hole opening.

10.9 The Subsequent Operation

The subsequent charge of blowing on should be on the basis of regular charge after stock line moving. If the stock line did not move in 3 h, check would be necessary without any delay.

10.9.1 Change the Charging

(1) It should be careful to change the charge during blowing on, when burden (ores/coke) and blast volume increased to a certain extent.
(2) The charging might be changed to release peripheral load for regular performance. Some BF men intended to increase peripheral load for wall protecting and gas utilizing better, but resulting irregular performance, channeling and scaffolding at last.

The furnace wall might be very cool at the beginning of blowing on, so it would be necessary to develop peripheral gas stream to heat the wall.

Regular performance would be the first consideration at the beginning of blowing on, that should be guaranteed by all adjust and regulation.

Table 10.13 Early four tapes of hot metal during blowing on

Project	Tap position	Hot metal volume, t	Physical heat of hot metal, °C	[Si], %	[S], %
1st	2	198	1426	4.04	0.010
2nd	2	1351	1496	3.23	0.012
3rd	4	915	1488	2.42	0.015
4th	2	574	1472	2.13	0.016

10.9.2 Increasing Burden and Its Conditions

It would be possible to increase the burden (ores/coke) when the first tapping of hot metal [Si] > 2.0% with regular performance, enough heat, enough blast temperature and coal injection. If [Si] content is more than 2%, the burden could increase continuously.

10.9.3 The Speed of [Si] Decreasing in Hot Metal

[Si] should meet the requirement of steelmaking, so higher [Si] (>2%) should be decreased to normal level as soon as possible. Table 10.13 shows [Si] decreasing experience of a large BF. It was unable to speed [Si] decreasing because of steelmaking restrained.

10.9.4 The Speed of Blast Volume Increasing

The speed of blast volume increasing would be depending on the situation of BF regular performance and steelmaking requirement. Volume increasing indicated the process of blast velocity adjusting to proper level, which would be a necessary operation for volume increasing continuously. The tuyere area should enlarge in the process of volume increasing to pledge proper blast velocity whether all tuyeres open or a part of the blending.

10.9.5 The Blast Volume Increasing Process Should Be Geared to Charging System

The ores charge weight should increase to coordinate with blast volume increasing; otherwise, it would result in gas stream instability, channeling unceasingly, regular performance destroying and chilled hearth.

References

1. G. Li, *The Formation and Development of Iron-making Technology in An Steel* (Metallurgical Industry Press, Beijing, 1998), pp. 387–391
2. S. Ma, Capital Steel Sci. Technol. Inf. **6**, 24–36 (1973)
3. X. Guo, H. Yu, S. Gu et al., Innovative practice of full-focus boiler opening in Anyang Steel 2000 m^3 grade blast furnace. Iron-Making **3**, 17–20 (2011)
4. C. Guan, G. Liu, Y. Mo, et al., Innovative practice of full-focus boiler opening in Xin Steel No. 6 blast furnace. Iron-Making **2**, 28–30 (2010)
5. Q. Tang, Problems worthy of attention in the operation of newly built blast furnace, in *Compilation of Materials for the Expert Committee of Blast Furnace Production Technology* (2013), pp. 40–45
6. B.B. Kaserolin, Some problems in the blast furnace opening process. Foreign Iron-making Steel-Making **3**, 11–14 (1990)
7. T. Wang, W. Zhang, L. Ren, et al., Practice on startup of No. 1 blast furnace of Capital Steel Jingtang Joint Co., Ltd. Iron-Making **2**, 7–10 (2012)
8. H. Zhang, S. Sheng, H. Ma, Practice on startup of No. 2 blast furnace of Capital Steel. Iron-Making, **2**, 1 (2009)
9. Z. Zhang, D. Zhang, L. Dong et al., Practices of achieving target output in 3200 m^3 blast furnace of Ji Steel. Iron-Making **1**, 16–19 (2011)

Chapter 11
The Operation of Blowing Out, Blanking and Furnace Blowing Off

Abstract The safety of blowing out and blowing on must be guaranteed to strive for shortening the blowing out period to create a condition of production. There are different management modes for blowing out in operation instructions of every plant, in general: (1) The plan of blowing out of BF should be decided by the head of the plant and approved by the general dispatching center of company. Temporary blowing out for renewing some cooling facilities should be decided by the controlling center of plant. If some serious accidents happen, such as tuyere and blowpipe burn out, the foreman could stop blast independently, while report to relative departments without delay. If the period is longer than 4 h, the plant control center should report to the head of plant after blowing out. (2) The blowing out must be after hot metal and slag tapping cleanly and stop slowly in some special situations to prevent blooding of tuyere. The shift foreman reports for the plant control center to make a decision. (3) It is necessary to contact individual units including plant control center, gas washing, blower, hot stove, coal injection, bunker (bin), TRT, dry dust collection, etc. (4) The longtime blowing out with descending charge level, top igniting and driving out the gas for relining or maintaining should be planned in advance to assign technical coordinator to direct. (5) To keep furnace regular performance and enough furnace temperature before a longtime blowing out. (6) If the blowing out is for repairing steam system, gas system, top system or someone working on the top, the top igniting should be necessary to prevent poisoning and explosion in charge of foreman or shift head. (7) When longtime blowing out or blower fault, the cold blast valve and flue valve should be opened to prevent gas back draught explosion. (8) In general, the minimum hot blast pressure sets as 0.05 MPa (0.5 kg/cm^2) to permit charging and mixer selector valve opening.

Terms of Iron-making

Blowing out (or down): to stop the blast into the furnace.

Blowing out in a short time: usually less than 4 h.

Blanking: to halt production under normal conditions to be recovered on schedule without salamander tapping and burden braising after blanking.

Blowing off: to halt production for relining (with salamander tapping and burden braising) or intermediate reline (without salamander tapping).

Y. Liu, *The Operation of Contemporary Blast Furnaces*,
https://doi.org/10.1007/978-981-15-7074-2_11

Salamander: the hot metal reserved in the hearth below the tap hole, so it could not be tapped out.

11.1 Short Blowing Out

11.1.1 The Feature of Short-Time Blowing Out

It is often to stop blast for a short time to renew those burned tuyeres or slag notch (monkey), etc. Before blowing out, the regular performance and enough hearth temperature are necessary and the bridging, if any, should be sorted in advance; also, the hot metal and slag must tap cleanly to prevent tuyere blooding.

11.1.2 The Operation Procedure of Short Blowing Out

(1) The top watering must stop immediately. The high top pressure should be changed to normal pressure. The blast volume ought to be reduced.
(2) Steam or nitrogen should be introduced to the top, gas down-take and gas shutoff valve.
(3) To cut up coal and oxygen injection.
(4) To cut off gas, open the top relief valve and close the shutoff valve when the blast pressure reduces to 0.05 MPa (0.5 kg/cm^2).
(5) To close the mixer selector valve and change automatic adjusting mixer selector valve to manual operation as well as stop charging, when blast pressure snorting to 0.05 MPa (0.5 kg/m^2). If blast could be stopped, the cooling water might be reduced more, and if not, the cooling water for broken tuyeres and plates should be recovered in time.
(6) Snorting to 0.01 MPa (0.1 kg/cm^2) of blast pressure, to keep positive pressure and observe tuyeres without the risk of tuyere blooding, then send the signal of blowing out and rise up the stock rod.
(7) When back draught is necessary, more than one-third peepholes ought to open evenly at regular intervals and inform back draught. If the hot metal and slag tapping are not clean and require blowing out, the time from blowing out to back draught signals sending out must have 3–5 min difference.

11.1.3 The Operation Procedure of Blast After Short Blowing Out

(1) To close up the peephole and stop back draught;

(2) To send out the signal for blast on, prepare well at hot stove and then close the snort valve after getting in touch;
(3) To check the tuyeres and slag notch to make sure at low blast pressure and then permit to increase blast;
(4) When blast reaches 1/3 of full volume, open the gas shutoff valve and then close the top bleeding valve, so that the gas would be carried in;
(5) To stop the steam or nitrogen supplying for down-take and gas shutoff valve;
(6) Turn on high top pressure operation and recover the normal blast volume, open the mixer selector valve, and recover the automatic control for blast temperature, then oxygen and coal injecting according to the requirement of furnace.

11.1.4 Attention for Blowing Out and In

(1) Blowing off should be at the situation of the regular performance of BF. Do not hurry to stop blast for renewing slag notch and tuyere within a few minutes.
(2) It is necessary to observe tuyeres in checking operation, blast back against the slag blooding of the tuyere and then continue to check slowly to the end with the slag level descending. If there is individual tuyere still blooding, the blooded slag could be braised out from the tuyere holder after blowing in.
(3) If the fire breaking out from tuyere after back draught, the following items should be checked up:

1. If the relief valve opens fully?
2. If there is any water leaking?
3. If the back draft valve opens fully?
4. The procedures of blowing in and out must be step by step. The mixer selected valve and its adjusting valve could not open (strictly forbid) before blowing again.

Figure 11.1 shows the blast furnace–hot blast furnace valve position.

11.2 Blowing Out Suddenly

At 22:28 on August 31, 1982, No.2 BF of Capital Steel accidendly blew out in a sudden and sounded like thunder. The foreman found No. 22 tuyere breaking out seriously, and a lot of fire with hot coke ejected out. Fortunately, the hearth worked actively with enough temperature so the operators with confidence do their best step by step, change high top pressure to normal pressure, reduce blast and blow out gradually, tape the hot metal and slag cleanly, renew the broken facilities, etc. Finally, reducing blast 1 h 53 min, blowing out 4 h 15 min, one blowpipe and one penstock burned out.

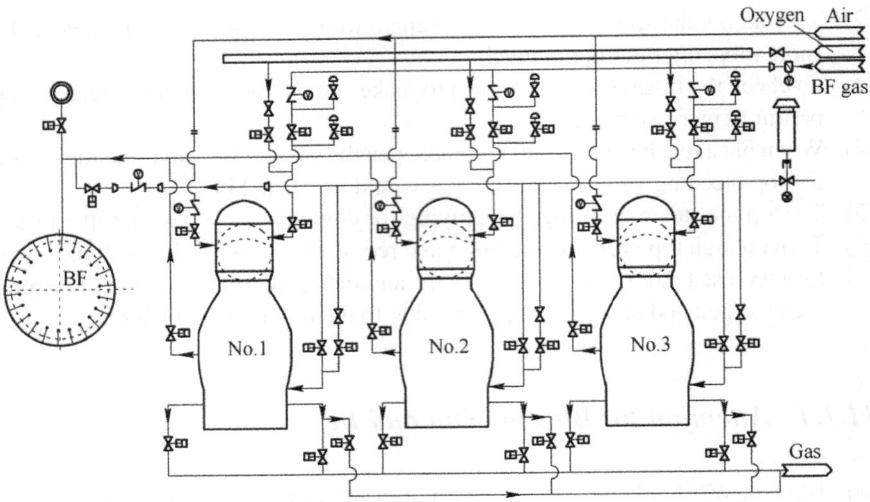

Fig. 11.1 Blast furnace–hot blast furnace valve position

It was found afterward that the slag shell fell down to knock out the tuyere, which was first named "tipping tuyere" in Japan. The original reason of "buckling" might be that there is too little flow around the peripheral gas, which make the slag shell fall off.

The real intention of slower reducing blast was to renew some broken facilities before tapping, but moreover it also prevents more hot coke ejecting out and tuyeres blooding and refraining from more accidents.

The similar "buckling" also happened after blow on at No. 1 BF of Bao Steel in 1985.

11.2.1 Blowing Out for Sudden Accident

(1) The blasting must be stopped urgently when the accident happens, while the shift foreman should operate at once and then report.
(2) If the snort valve does not work, the foreman could inform the operator of blower to blow out or let the hot stove open the blow-off valve.

11.2.2 Water Interrupting Emergent

It should prepare well when alarming and the blast pressure ought to reduce less than water pressure at least 0.1 kg/cm^2.

The emergency steps should be following after water pressure reducing:

(1) To reduce watering for shaft cooling, if possible, to guarantee the cooling water supplying for tuyere and slag notch;
(2) To stop the injection of coal and oxygen and shift from high top pressure to normal and blowing out to the minimum blast pressure without tuyere blooding;
(3) To organize tapping actively;
(4) To stop gas supplying;
(5) If the water pressure is too low and could not recover soon or interrupted, blowing out must be immediate.

The operation for water pressure recovering to normal level should be accordant to the following steps:

(1) To turn down the main water valve;
(2) To turn off the water cutting down valve of the tuyere, if inside water dries up;
(3) To re-watering for tuyeres should be one by one and as slow as possible to prevent steam explosion inside the tuyere;
(4) To re-watering the cooling plates and staves should be slow and differentiate one from another part or zone of the furnace location;
(5) To recover the water pressure to normal level gradually, while all the outflow of water is normal after checking;
(6) All of the cooling facilities, especially tuyeres and slag notches, should be OK after checking;
(7) To renew the broken tuyeres and slag notches;
(8) To fix the broken staves.

Only after making sure that the water break factor is eliminated and the water pressure is restored to normal can the Blast furnace reblow at last.

11.2.3 The Blower Does not Work Suddenly

The main dangers are:

(1) The gas flows backward to the blast system, where explosion might happen.
(2) The pressure in the gas pipe might be negative, and air could be sucked in, leading to an explosion.
(3) All tuyeres, blowpipes and penstocks might be blooding.

The operation for blower fault should be accordant to the following steps:

(1) When there is no blast through tuyeres, the snort valve could be open fully. The foreman could inform the hot stove to stop blowing, and open the cold blast valve and the flue valve of a hot air stove to drive out the gas in blast pipe;
(2) Close the mixer selector valve and mixer adjusting valve, and stop coal and oxygen injection;
(3) To stop charging and the automatic controlling valve working;

(4) To steam in top, dust catcher, gas down-take and gas shutoff valve;
(5) To shift high top pressure operation to normal and close relative valves, according to relative regulation;
(6) To clean up the blooded tuyeres, if any, all of the above should be operated by the foreman and reported to the plant control center.

11.2.4 Power Cutting Suddenly

When power cutting happens suddenly, the reason and extent should be considered to deal with different measurements. The final basis of judgment would be the volume of watering and blasting out from tuyere.

If the blowing out and water cutting happen separately, we can treat them respectively; if both occur at the same time, we should deal with the blowing out first and then the water interrupting. If power cutting at the charging system, shift to emergency power, and if both are not available turn off the power source in all systems and prepare to turn on one by one when power recovery.

11.3 Long-Term Blowing Out

The difference between long-term and short-time blowing out is not only the time difference but also the more important situation changes for BF. For long-term blowing out, it must ensure the BF passes the period safely including the maintenance personnel and also consider how to recover the furnace operation afterward well.

The gas and blast pipe system are connected through BF and hot stove, but must be separated from each other; otherwise, explosion would happen, during long-term blowing out. In addition, it is also necessary that to ignite at the top to burn up the gas yield after blowing out, the tuyeres and slag notches must be sealed to reduce gas yielding and some parts of gas system should be cleaned up or replaced by air, for convenience of maintenance.

11.3.1 Preparation of Blowing Out

(1) Check thoroughly all of the cooling facilities without water leaking.
(2) Prepare the materials for sealing.
(3) Clean up the dust in catcher and furnace top.
(4) Prepare oil cotton silk or coke oven gas pipe and burner, to check if the steam taps are available.

11.3.2 The Procedure of Blowing Out (the First Step)

(1) Stop watering at the top in advance, shift top high pressure to normal operation, and reduce blast volume;
(2) To steam (or nitrogen) at the top, gas down-take, gas shut down valve and dust catcher;
(3) To stop coal and oxygen injection;
(4) To stop gas sending, reduce blast pressure to 0.05 MPa (0.5 kg/cm^2), then open the top bleeder valve and close gas shutoff valve;
(5) After openning the tap hole according to the plan, the foreman informs the loading workers to fill charge batch by batch without accumulation in the charging line. Any special situation should be explained in the plan;
(6) No more charge should be added after the particular charge for blowing out was filled completely, and at the same time, opening the lower seal valve is forbidden. Then, continue to snort with reducing the cooling water of broken tuyere and staves to the end to prevent the water leaking from seal box into the furnace during the period of blowing out;
(7) Snort to 0.01 MPa (0.1 kg/cm^2) to keep positive pressure, send out the signal of blowing out, rise the stock rod, cut off the full gas, and prepare to ignite at the top.

11.3.3 To Ignite at the Top of the Furnace

After the application of bell-less top in Capital Steel, the operation of top igniting is improved and simplified as follows [1]:

(1) To fully open the top relief valve, open the relief valve of dust catcher, close the shutoff valve of dust catcher, steam for top and dust catcher, close the lower seal valves of left and right hoppers, open the upper seal valve and relief valve of hoppers, close the first (for gas) and second (for nitrogen) equalizer valves of hopper;
(2) Shift the top nitrogen to compressed air;
(3) To open manholes;
(4) To stop top steaming and nitrogen for sealing chamber;
(5) If the top gas could not burn automatically, the oil cotton silk or burner could be used for igniting;
(6) After ignition, the foreman designated one person to the top of the furnace to watch the fire. See the fire should ignite the coke oven gas pipe, on the material surface. Watch the fireman keep the flame on the surface;
(7) No permission for blowing back.

Clean out the residual gas from the gas pipe of BF, which is the same as the Bell blast furnace. Close the lower seal valve to separate the feed tank from the blast furnace, open the upper seal valve and relief valve, so that the feed tank and the gas can establish communication. Then, close the first and second equalizer valves to pledge gas and nitrogen could not enter into the hopper.

The second stipulation is that the compressed air is introduced to the hopper to replace nitrogen for the maintenance person entering.

The third stipulation is to open the manholes, from which the air could enter into top space for combustion.

After top steam and seal chamber nitrogen closing, the gas above the stock surface ignites automatically with short flame, sometimes the cotton silk or burner to be used if no auto-igniting.

Sometimes in the final campaign of the furnace, the coolers leak too much water into the furnace to yield a lot of steam, then it is hard to ignite, and therefore close the taps of income water after blowing out to reduce the inside steam. After igniting, the operation is still according to the regulation. The igniting process with 6 steps could accomplish in 5–15 min with convenience.

11.3.4 The Procedure of Blowing Out (the Second Step)

(1) Open the backpressure valve to make sure no hot gas ejecting, then demount the blowpipes, renew the broken tuyeres and seal tuyeres.
(2) Open cold blast valve and flue (smoke) valve, and clean up the residual gas.
(3) After sealing, inform hot stove to stop backpressure and back draught, and then inform the blower to stop.
(4) After blowing out, the gas system must clean up according to the following steps under the permission of gas clean agency:

    1. Open the relief valve of the dust catcher.
    2. A lot of steam appears from the relief valve of the dust catcher, and open the upper and lower manholes as requiring.
    3. Steaming could not stop until 2 h after above operation, and then steam again at 1 h before blowing in.
    4. To reduce water pressure to 0.15 MPa (1.5 kg/cm^2).

11.3.5 Attention for Long-Term Blowing Out

(1) Before top igniting, nobody can stay or do some other works over there.
(2) The backpressure might be available for tuyere renewing or sealing.
(3) Set up a special person to take care of top igniting.
(4) If the top fire is cut off, it should ignite immediately, but if there is a delay for a long time, it should be reported to higher authorities to re-ignite or not. When re-ignited, steam was conveyed into BF to clean out the residual gas. During

this time, the workers must leave the top and nobody is allowed to stay around the tuyere area to avoid hurting by the flame and knock.

(5) When press burden is necessary, the top igniting could not be cut off.

11.4 Descending Charge Level (Stock Line Descending)

Sometimes, it is necessary for maintenance or blowing off by descending charge level with the method like line spraying, skull blowing up or reline blowing out.

11.4.1 The Disadvantages of Descending Charge Level with Gas Closing

The gas closing used to be operated with descending charge level in Capital Steel. In order to reduce noise, the blast is stopped once a time generally before the beginning of decending charge level, and a relief valve with 800 mm diameter is cut off to increase the relief area of bleeder stack from 0.256 to 0.52 m^2, so as to depress the speed of the gas stream.

For top gas temperature controlling, water was sprayed into the furnace during descending charge-level process or combining watering with charging in Capital Steel. The blast volume decreased gradually along with descending charge level to prevent top temperature too high and watering too much.

Such operation used to be popular in many plants, but there are so many disadvantages.

(1) The gas relief speed is much over 100 m/s, and even blast is reduced, to make noise pollution all around the area with 5 km radius.

(2) To make air pollution because of a lot of harmful gas such as CO and CH_4. There is 150,000 m^3/h of harmful gas release to atmosphere from a BF with 1000 m^3 volume.

(3) Because of gas closing operations, the gas utilization rate decreases and its pollution delays. Productivity decreased over time.

(4) To waste fuel because of a lot of gas relief. The loss of heat reaches 5000 GJ, equal to coal 18t/h.

11.4.2 Practice of Descending Charge Level with Full Blast

The new No. 1 BF in Capital Steel is the first BF with bell-less (1327 m^3), which went into operation on December 15, 1979, and prepared to maintain for 5 days to descend stock to 12 m. It began at 0:00 February 12 to start descending charge level with full blast to prepare 7 charges of burden (14 buckets) for top gas temperature

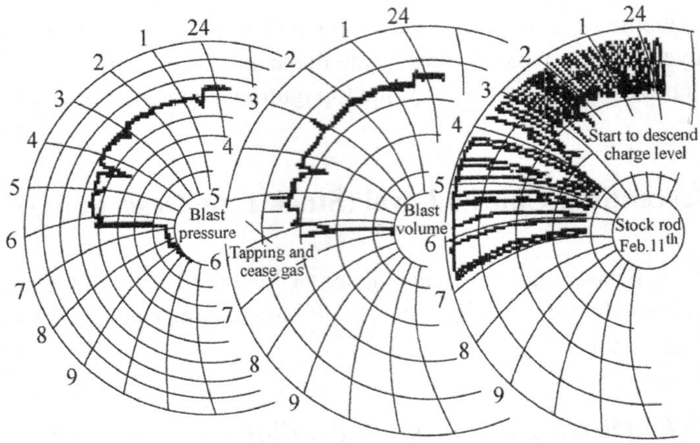

Fig. 11.2 Automatic records for descending charge level in the night shift on February 11, 1980

lowering. The descending charge level started from 3:00, and the depth of stock line at that time was 1.4 m. The top of the BF began to spray water at 3:50, while the depth of stock line was 4 m. Until 6:00, the stock line was reduced to 12 m (calculated).

The last tapping of hot metal was at 5:35 with checking to reduce blast pressure to clean up the hot metal, then to cease gas at 6:17, to plug iron notch at 6:26, blowing out at 6:31, to finish top igniting at 6:50 and to shift the BF to maintenance agency.

The actual stock line was 13.5 m after calculating with burden volume. All the operation for descending charge level (include igniting) lasted 3h31min. The average descending velocity of the stock line was 3.86 m/h with 7 charges (14 buckets, every 10–20 min for 1 bucket), which is dependent on top temperature level (after every one bucket burden, the top temperature decreases to < 100 °C from 400 °C) and then rises slowly again.

The blast volume and top gas pressure could keep the original level in the process of descending charge level. It worked very smoothly. Figure 11.2 shows the automatic records of the process of descending charge level. The blast volume, blast pressure and stock rod were very steady [2].

No. 2 BF was maintained again on October 13, 1980. It was still operated with full blast and gas ceasing with the blast volume of 266–2750 m^3/min, the top gas pressure of 1.3 kg/cm^2, and the charge level reduced to 8 m. The charge level began to fall at 18:13 and blow out at 20:50, lasting 2 h 37 min. During the descending charge-level process, the 4 buckets of burden were charged and combined with watering to prevent the top temperature rising effectively. Figure 11.3 shows the process was very steady [2].

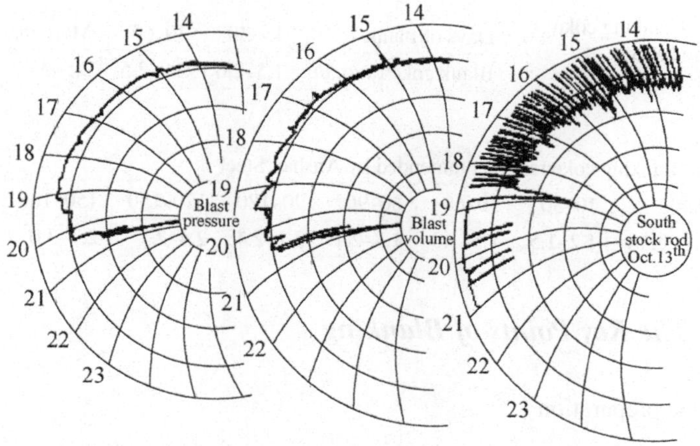

Fig. 11.3 Automatic records for descending charging level with full blast

11.5 Blanking

According to the requirement of market or the balance of iron and steel yielding, the BF has to blow out for a period of time to keep everything in order and is easy to recover yielding. This is so-called blanking, which happened 5 times in Capital Steel in the 1980s and 1990s (Table 11.1) [3] with the periods from 18 days to 176 days.

Table 11.1 Information of blanking in Capital Steel

Furnace number	Furnace volume, m^3	Date	Days of planned blank	Days of actual blank	Blank coke ratio, t/t	Silicon content in pig iron before blank, %	Silicon content of iron in the first blast furnace, %
Old 2	413	July 29, 1961	60	19	3.0	1.9	3.08
Old 2	413	February 16, 1977	60	27.3	4.0	1.30	6.56
1	576	November 29, 1967	61–90	167	1.9	2.35	0.64
4	1200	April 1, 1981	60	75	1.8	0.95	0.34
New 2	1327	July 1, 1981	60	18.75	1.8	0.53	0.08

Table 11.2 Blanking coke rate of Capital Steel

Days of blank	15–30	30–60	More than 60 days
Blank coke ratio, t/t	1.5–2.0	2.0–2.5	2.5–3.0

Table 11.3 Blanking coke rate recommended by Anshan Steel

Days of blank	10–30	30–60	60–90	90–120	120–150	150–180	>180
Total coke rate, t/t	1.2–1.5	1.5–1.8	1.8–2.1	2.1–2.4	2.4–2.7	2.7–3.0	3.0–3.5

11.5.1 The Key Points of Blanking

(1) Burden preparation

The requirement for blanking is similar to blowing on to use qualified burden with better strength and larger size especially coke, of which CSR > 62%, CRI ≤ 25% (for > 2000 m^3 BF) and CSR > 60%, CRI ≤ 28% (for smaller BF). The reaction destructive effect for coke inside the furnace for a long time is unavoidable.

(2) To prevent and check water leaking and close every water valve
(3) The tuyeres and slag notches should be sealed strictly. In Capital Steel, the tuyeres and slag notches with coolers were removed and then sealed with multilayer materials
(4) Coke rate for banding

The situation of hearth temperature after blasting is determined by the coke rate of blanking, which is always the same as blowing on. The experience in Capital Steel was less than AN Steel with more BFs and blanking numbers, so they always learn from AN Steel. Table 11.2 shows the blanking coke rate of Capital Steel and the blanking coke rate recommended by AN Steel in Table 11.3 [4].

(5) The location of iron containing burden is similar to blowing on reaching the lower part of the shaft.

11.5.2 The Case of Blanking [5]

No. 4 BF (1200 m^3) of Capital Steel was blowing on in July 12, 1977. It used to have blown out for two months for a lack of raw materials to balance the production in 1981. To stop blast for blanking since April 1, 1981, with the full coke rate of blanking 1.8 t/t comparing with normal 0.7 t/t, coke blank (straight coke burden) in half lower bosh and coke blank with flux over the barrel, the first charge with sinter charged to the lower part of the shaft. Altogether, 6 batches of coke blank charging and 14 batches of coke blank with flux charging were counted. The coke charge weighted 8.7838 t, wet coke weighted 9.2 t, and ore charge weighted 20 t. It was

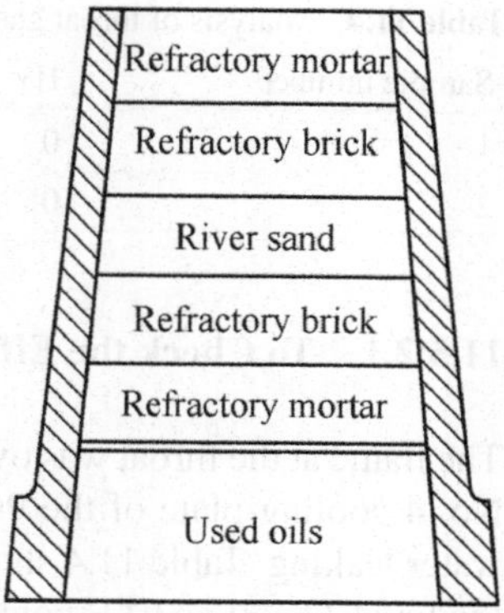

Fig. 11.4 Tuyeres and its coolers sealing with layers

planning to stop blast when the blanking stock line down to the tuyere level. Sealing of tuyeres and slag notch are shown in Figs. 11.4. Figure 11.5 shows the planning location of the burden after blowing out.

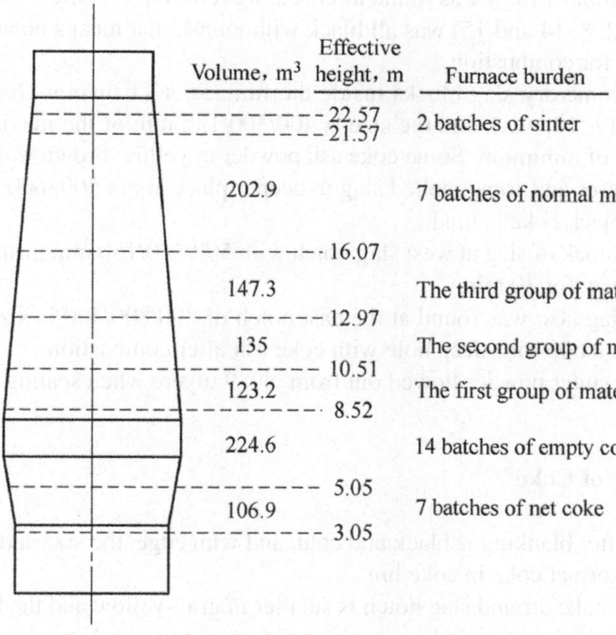

Fig. 11.5 Planning location of burden after blowing out

Table 11.4 Analysis of throat gas samples on April 20

Sample number	H_2	CO_2	CO	O_2
1	0	0	0.6	21.4
2	0	0.2	0.4	21.4

11.5.2.1 To Check the Effect of Sealing

The flame at the throat was over in 4 days, just a litter left at the east side of the stock. No. 4 cooling plate of the 9th layer was plugged and turned off steaming without water leaking. Table 11.4 shows the analyses of throat gas samples on April 20, and the low CO and no CO_2 contents mean no water leaking and effective sealing.

11.5.2.2 State Before Tuyere

It was planning to blank within 60 days but actually 75 days, and then the stock line descended about 1 m. It was found after sealing removed that the coke before 5 tuyeres (No. 1, 2, 8, 14 and 15) was all black without ash that means no air entered into the furnace for combustion.

There were some dry slag blocks inside the furnace at 11 tuyeres (No. 3, 6, 7, 11, 12, 13, 16, 17, 18, etc.), and the size is 400/300/150 mm of the maximum and 100/100/50 mm of minimum. Some coke ash powder in yellow-brown was found at No. 4 and 5 tuyeres and some melted slag in deeper place about 500–600 mm from tuyere with all black coke behind.

There was a block of slag at west slag notch with 500/400/150 mm in size, which was removed away for 30 min.

A block of slag also was found at the east notch of 800/800/400 in size, behind which there was a 600-mm-deep hole with coke ash after combustion.

A lot of cold sinter powder flowed out from No. 9 tuyere when sealing removed.

11.5.2.3 State of Coke

(1) The coke after blanking is black and cold, and with edge, the size and color are similar to normal coke in coke bin.

(2) The size of coke around slag notch is smaller in gray-yellow and the lower, the breaker. The coke layer is looser, so easy to penetrate in 2 m by a steel bar, which just heats to 30–40 °C after pulling out, but is hard to penetrate in with hammering over 2 m and the pulling out temperature also 30–40 °C.

11.5.2.4 The Situation for Tap Hole Drilling

The tap hole drilling began at 16:00 on 14th. It was easier before 1.2 m (deep), but very difficult over 1.3 m to 2.1 m in 80 min in total. The coke in the tap hole is dense and the temperature is very low (about 3 °C).

11.5.2.5 Operation of Blowing

The sealing was removed since 8:00, June 14, and finished at 10:30, 2.5 h in total to keep 7 tuyeres (No. 1, 2, 4, 14, 15, 17 and 18) blowing with tuyere area 0.1337 m^2.

The blast volume was 1450 m^3/min, the blast temperature was 650 °C, and the blast pressure was 0.7 kg/cm^2 when the operation of blowing was taken at 1:20 on 15th. The No.17 tuyere was bright and fire was ejected from iron notch at 1:27. When the blast temperature rised to 680 °C, No.4, 14 and 15 tuyeres were bright at 1:31. Then, No.1, 2 and 18 tuyeres were bright at 1:35 with the temperature rising to 700 °C. Two stock rods were working at same time at 2:09 and gas was sent at 4:30. Finally, hot metal appeared at iron notch without slag, then iron notch plugging with 60 kg clay at 6:30. The analysis of the hot metal samples showed [Si] = 0.34% and [S] = 0.20%.

11.5.2.6 Explosion of Slag Notch

There was a bit of "nest" slag at No. 4 tuyere at 4:00, also at No. 7 and No. 8 when 6:00, No. 1 tuyere broken at 9:42 and then opened the tap hole to prepare to renew the broken tuyeres. The western slag notch burned out and then exploded running out 2.5 ladles of slag (40 t estimated) and hot metal 30 t.

The blast was stopped to renew No. 1 tuyere and western slag notch's second cooler with carbon brick to be the temporary taphole, put a new runner on, plug 4 tuyeres (No. 1, 4, 14 and 15), keep 4 tuyeres (No. 2, 3, 17 and 18) blowing out for 9 h 37 min and then blast at 20:04.

11.5.2.7 Hot Metal Tapping Through Slag Notch

It was expected that the iron would come out from tap hole. After blasting, the tap hole was drilled to a depth of 2.2 m at 23:20 on 15th and 0.6kg of explosive charge was inserted for blasting, but it did not work successfully. On the 16th, 1.2 kg of explosive charge was released at 0:05, 1.95 kg of explosive charge was released at 0:50, and two cannons were fired successively. None of them worked, so iron was forced to be discharged from slag notch instead. Then, 50 t of hot metal and about 30 t of slag tapped through slag notch from 1:20 to 2:50.; 15 t of hot metal and about 15 t of slag from 3:57 to 5:00; 15 t of hot metal and about 30 t of slag from 9:50 to 11:03; and 20 t of hot metal and about 30 t of slag from 14:16 to 15:08.

From then on, the tap began to come out of the tap hole. At 16:01 on 16th, the depth of the tap hole was 2.5 m with explosive charge of 1.95 kg to open tap hole. The hot metal contained [Si] 0.66% and [S] 0.16%, resulting in normal level gradually.

11.5.3 *Experience and Lesson*

(1) The proper burden arrangement and coke rate determination;
(2) The successful sealing;
(3) To drill tap hole before blasting and mediate the pass between the tap hole and around tuyere area successfully so there is fire ejection from iron notch in 7 min only after blasting and tap hole blowing in 6 h after plugging;
(4) Although it was known that the temperature under the tuyeres was very low, it is certain that the hot metal and liquid slag below the tuyere have been frozen, so it is necessary to decide to discharge the slag. As a result of this negligence, the blast supply generated by the iron slag can not go, so tuyeres damaged and the slag cooler exploded as well. If the slag outlet iron is decided at the beginning, and the modification of slag outlet iron is completed before air supply, the result may be completely different.

11.5.4 *The Brief Summary for Blanking*

(1) Blanking was originally for smaller BF, but now it is the time of large BF, so blanking would not be available extensively. For large BF, the distance from tuyere to iron notch is more than 4 m with low temperature over there so it would be hard to be mediated between the freeze hot metal and slag that is very important for a blast after blanking.
(2) There is no slag notch of large BF, so no tapping through it.
(3) Large BF, more than 2000 m^3, also has some advantages for blanking, and the condition of well lining with good sealing to prevent water leaking results in the possibility to keep the temperature under tuyere level without hot metal and slag freezing in one week after blanking, though sticky unavoidable.

No.1 BF of Baoshan Steel used to stop blast without a plan for 6 days because of blower fault. Then, it blew in and recover production successfully.

11.6 Blow-Off

"Blow-off" is the preparation for the relining and reconstruction of BF. The difference between blowing off and blanking is with or without descending charge level, salamander cleaning up and burden raking out. The key points for blowing off are descending charge level safely and salamander cleaning, which would be discussed in detail here.

11.6.1 Descending Charge Level Safely and Rapidly

The main danger is knock, which will be discussed in Chap. 12.

In the past, it was usually to stop blast before blowing out and cut off 1 or 2 top relief valves (800-mm diameter) to release top gas into the atmosphere. The relief area increased by 1.4–2 times. Now, it is improved to control the H_2 and O_2 in gas for safe and rapid descending charge level at full blast. The multiple purposes are most top gas could be recovered, noisy and air pollution-reducing.

11.6.2 To Judge the Stock Line with Gas Analysis Changing

The relationship between gas analysis changing and stock line descending depth at No. 4 BF of Capital Steel in the blow-off operation in 1983 is shown in Fig. 11.6, in which the cross of CO and CO_2 corresponds with the stock line down to bosh. This is an important feature, which was found first by AN Steel and thereafter accepted by BFs all over the country, and then the descending charge level operates under the guidance of this feature without special stock line rod, which used to be set with blowing out.

11.6.3 To Control the Content of H_2 and O_2 to Completely Eradicate the Knock

The descending charge level with blowing on at No. 2 BF began at 10:40, October 16 [7].

11.6.3.1 Blast Volume Controlling

In the earlier stage of descending charge level, the blast volume would increase rapidly to 90% of the normal level, top pressure to 95% of normal, as shown in Fig. 11.7, and maximum saving time for descending charge level and reduce the blast volume marginally along with the stock becoming thinner and thinner to lower down top pressure steadying the gas stream. So, it was unnecessary to decrease blast much for knock preventing.

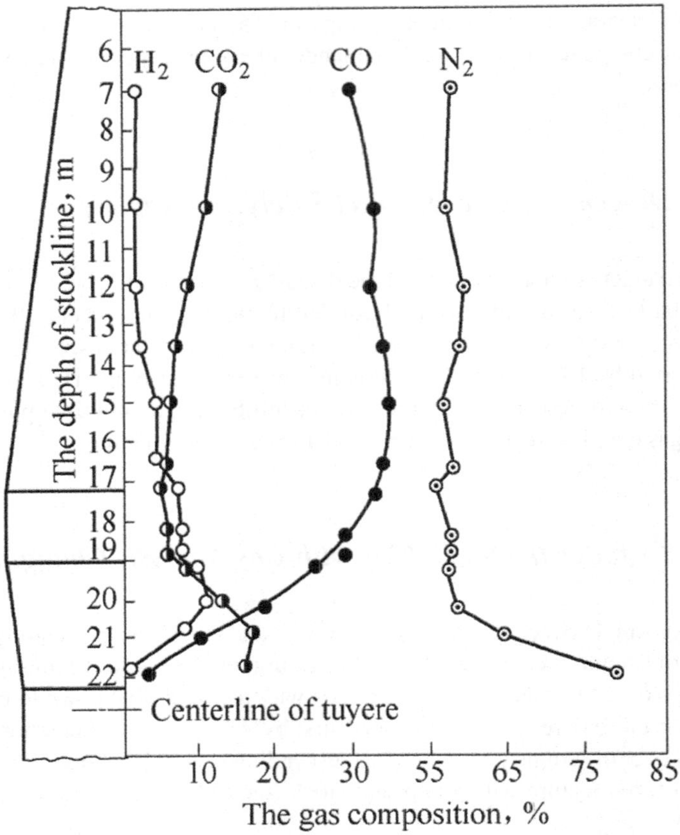

Fig. 11.6 Composition of gas versus stock line changing [6]

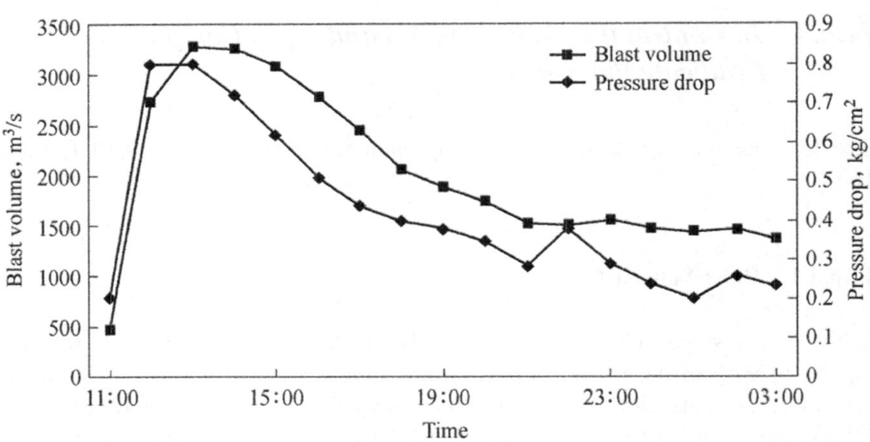

Fig. 11.7 Blast volume of descending charge level versus pressure drop controlling

The gas analysis at 14:30 showed that the content of $H_2 > CO_2$ to indicate the stock line reaching barrel, so blast volume reduces greatly to lower top pressure to control gas stream.

The temperature of copper cooling stave decreased, and CO_2 % increased obviously at 21:00 to indicate the stock line reaching the bosh, so open the snort valve, reduce blast volume to 1000 m^3/s and blast pressure to 0.55 kg/cm^2 and then cut off the gas, while the relief valve on fire then reduced blast pressure to 0.35 kg/cm^2, then flame out and keep the blast volume at 1500 m^3 until blow-off.

11.6.3.2 Top Temperature and Water Controlling

The average top temperature was 480 °C in this descending charge level. The temperature of cooling staves at the 9th layer began to rise at 17:50 and lower down at 20:10 with the maximum of 112 °C. The temperature of staves at 8th layer also began to rise at 16:30 and lower down at 20:10 with the maximum of 148 °C. The temperature of staves at 9th layer rose a little with the maximum of 81 °C and without big change as an entirety. Every parameter was in the reasonable range without damaged staves. Watering for descending charge level was 1198 t in total, which might be a little bit more because of the bigger blast volume and higher top temperature in the early stage.

Figure 11.8 shows that during descending charge-level process, the maximum of H_2 was 9.3%, O_2 was 0.6%, CO_2 returned to 2.4% at 20:30, while the stock line reached to bosh, then gas cuts off and the N_2 reached to 69% close to the atmosphere while empty in front of tuyeres.

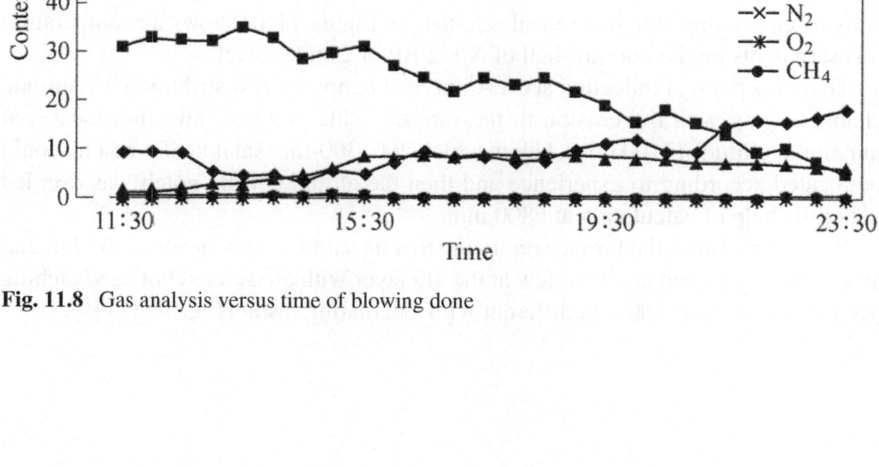

Fig. 11.8 Gas analysis versus time of blowing done

11.6.3.3 Consumption of Blast Volume in the Process of Descending Charge Level

The consumption of blast volume in descending charge level was used for combustion of the coke on the base of compression 13%. The quantity of coke of normal stock line is 540 t (including 54 t for covering), which needs 2800 m^3/t of air (excess air rate 1.35). The calculated total air consumption would be 2,041,000 m^3 and real consumption was 2,043,000 m^3 up to blowing out. At 21:00 on the 16th, the blast consumption was 1,514,000 m^3; at 23:25 on the 16th, the blast consumption was 1,740,000 m^3 with early empty in front of tuyeres; at 0:15 on the 17th, the blast consumption increased to 1,900,000 m^3 with all empty in front of tuyeres.

11.6.4 Clean up All the Salamander

11.6.4.1 To Determine the Location of the Tap Hole for Salamander Tapping

First of all, the erosion of the furnace bottom could be calculated with many methods mainly finite element programs to solve the heat transfer equation. A method created by Mr. Ma Songling has been used up to date that combines the calculation and temperature measuring at some points on furnace bottom shell, where the temperature is influenced by the salamander, and the higher at shell, the closer to the salamander and vice versa [8]. Figure 11.9 shows the measurement results of No.1 BF on December 27, 1963, the third day after blowing out and closing the cooling staves of 1 and 3 Sect. 3 h later. At the circumference of the furnace bottom, six points are measured vertically in each of the ten directions. A total of 60 points are measured at a distance of 500 mm from each point.

In accordance with the historical experience of salamander tapping in Capital Steel, the point of inflection (peak value) in the temperature distribution of bottom shell could be found from measurement to estimate the maximum erosion location, and the location of the notch for salamander tapping was selected appropriately by combining with theoretical calculation. Figure 11.10 shows the temperature measurements on the bottom shell of No. 2 BF of Capital Steel.

There is a point of inflection at every curve of temperature distribution. Its altitude changes along with the erosion in the furnace. This point of inflection locates at maximum altitude (7100 mm), below which 200–300-mm salamander notch should be located according to experience and then the altitude of the notch was decided under the help of calculating at 6800 mm.

It was found after the furnace removing that the carbon brick heads of the 1st, 2nd and 3rd layers etched at all but few at the 4th layer with an isolated hole and etching deep 1600 mm, just 200 mm different with calculating result (Fig. 11.11) [9].

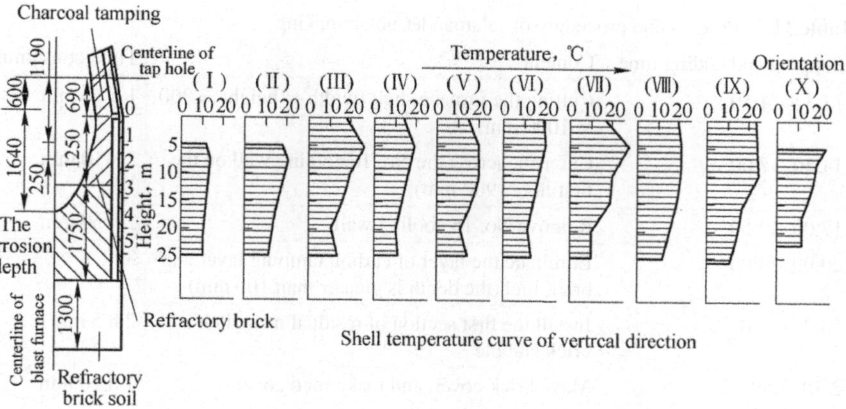

Fig. 11.9 Temperature measurement on the shell of bottom

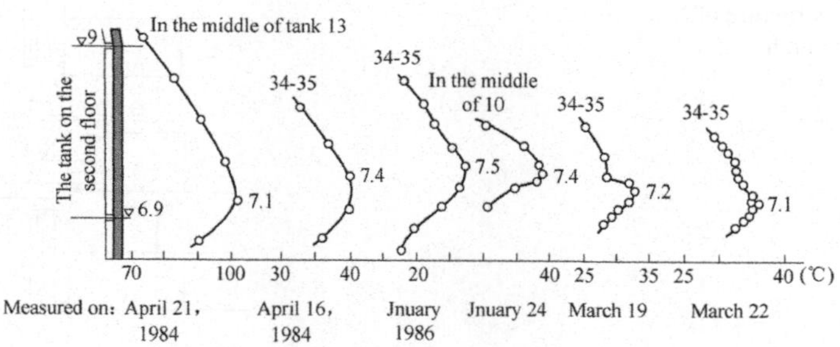

Fig. 11.10 Temperature measurements on the bottom shell of No. 2 BF of Capital Steel

Fig. 11.11 Scene of erosion of the bottom brick and salamander tapping

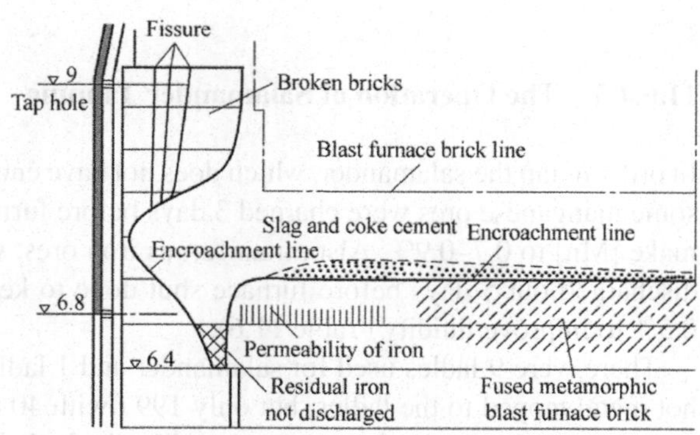

11.6.4.2 The Salamander Notch Making and Punch-Out Runner

Since the beginning of descending charge level, the salamander notch started to determine the location and cut off the steel shall in the bottom part. The procedure and time are quite different for the different furnaces. Table 11.5 shows the schedule

Table 11.5 Process and procedure of salamander notch making

Begging and ending time	Content	Time-consuming
13:34–14:50	Cutting the furnace shell (width × height = 900 × 1000 mm)	1 h 16 min
14:50–17:00	Cut a slit across the No. 16 cooling wall on the first floor (900 mm)	2 h 10 min
17:00–19:50	Remove No. 16 cooling wall	2 h 50 min
20:00–23:00	Eliminate the layer of carbon tamping layer and brick fuel (the depth is greater than 100 mm)	3 h
23:25–2:30	Install the first section of residual iron ditch and brick stubble	2 h 5 min
2:30–3:50	Make brick cover, and make mud cover	1 h 20 min
3:50–6:40	Bedding groove and baking	2 h 50 min

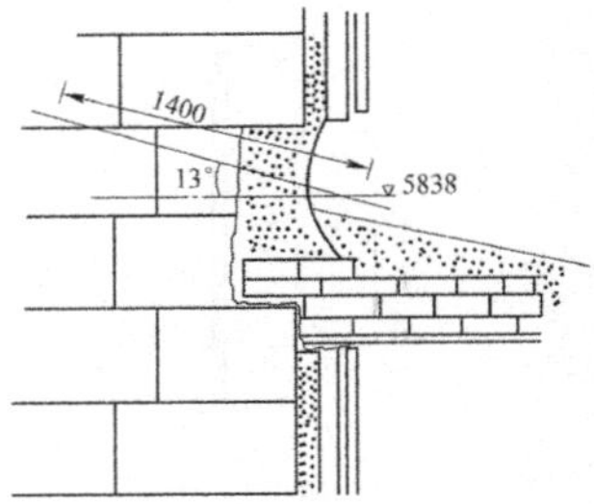

Fig. 11.12 Structure of salamander notch

for notch making at No. 4 BF of Capital Steel for reference. Figure 11.12 shows the structure of salamander notch [6].

11.6.4.3 The Operation of Salamander Tapping

In order to tap the salamander, which does not have enough temperature and fluidity, some manganese ores were charged 3 days before furnace shut done of No. 4 BF to make [Mn] to 0.7–0.9%. Also, titaniferous iron ores, which were used for relining, discharged out 5 days before furnace shut done to keep [Ti] < 0.45% and [Mn] > 0.5% to improve fluidity (Table 11.6).

There were 9 ladles used for salamander and 1 ladle for slag. About 300 t crude hot metal tapped to the ladles, but only 199 t with 40 t slag poured out. The reason was some of them were freezing and sticking in the ladle. The volume untapped was estimated 57.7 m^3, 60% of the volume above erosion line and under tap hole level.

It took 35 min from the beginning of iron tapping to slag appeared, during which 6 ladles were filled (about 250 t hot metal) with a current velocity of 7.1 t/min. In another 15 min, 50 t crude hot metal and 40 t slag were tapped out with the velocity

Table 11.6 Analysis of salamander

Element	1st tank	2nd tank	3rd tank	4th tank	5th tank	6th tank	7th tank	Average
[Si]	0.349	0.365	0.373	0.411	0.458	0.528	0.582	0.438
[Mn]	0.58	0.62	0.63	0.68	0.80	0.89	0.79	0.71
[Ti]	0.029	0.032	0.034	0.036	0.038	0.043	0.042	0.036
[P]	0.037	0.037	0.037	0.040	0.042	0.044	0.042	0.040
[S]	0.087	0.092	0.087	0.073	0.052	0.041	0.053	0.069

of 6 t/min. So far, all salamander were almost clean up and only small amount of it was left in the pitting on the erosive bottom (Fig. 11.11).

11.6.4.4 Lesson for Salamander Tapping

(1) No. 3 BF of Capital Steel blowing out for relining and everything was all right but interrupted because of freezing of tap hole at half of the salamander tapping. Actually, it should reopen the hole by oxygen burner to continue tapping, but watering blindly to freeze the salamander at furnace bottom with serious consequences. It could not help but used explosive charges 450 kg to break up more than hundred kg salamander in 8 days instead of a few minutes for tapping. The explosive bang came suddenly as an earth quick that shakes heaven and earth, and even shocked the police.
It is necessary to have an expert and sensible person in charge of salamander tapping and must clean up thoroughly. The instruments and oxygen should be prepared readily so as to be available to pledge the tap hole unblocked.

(2) No. 1 BF of Capital Steel blowing out for intermediate reline in June 1978. But no furnace flushing to clean up the accretion, which broke down from the wall into the coke layer in the hearth when the stock line descended to lower level of tuyere with watering. A lot of coke squeezed to the tuyere level and is hard to be taken off. And the accretion was hard and glossy, and very difficult to take out, to be bound to break up by manual work with steel hammer and drill.

(3) It should consider to choose the location of the salamander tap hole and the angle of punch-out runner in bottom design. Because of the lower temperature and less fluidity, the angle should be bigger than 15°. One of the faults of salamander tapping was too small angle ($< 10^{\circ}$) that would be easy to freeze. Once upon a time, the salamander tapped out with 9.5° punch-out runner haunted with fear by multiple measures.

References

1. Y. Liu, Bell-less top ignition operation. Capital Steel Technol. **6**, 12–13 (1982)

2. Y. Liu, Full air aeration and lowering surface operation. Capital Steel Technol. **1**, 1–3 (1983)
3. S. Wei, X. Chen, *The Production of Blast Furnace*, ed. by C. An (Capital Steel, 1983), pp. 287–290
4. C. Zhou, *Technical Manual of Blast Furnace Iron-making Production* (Metallurgical Industry Press, 2002), p. 455
5. Technical Section of Capital Steel, Summary of No. 4 blast furnace (1981) (internal information)
6. S. Shi, Y. Jin, The stop operation of No. 4 blast furnace. Capital Steel Technol. **4**, 1–5 (1984)
7. Technical Section of Capital Steel, Summary of the blast furnace shutdown in October 2008. 2008 (internal information)
8. S. Ma, Discussion on the disintegration of blast furnace bottom. Capital Steel Sci. Technol. Inf. **2**, 23–24 (1964)
9. Y. Jin, Determination of maximum erosion depth of water-cooled integrated furnace bottom and discharge of residual iron after filling vanadium-titanium ore. Capital Steel Technol. **3**, 1–3 (1987)

Chapter 12
The Explosion of Blast Furnace

Abstract In the theme park of Jiu Steel, there is a cenotaph (Fig. 12.1) to memorialize the martyrs who died in the explosion of BF and to report the story of this accident. It was on March 12, 1990, that the blast furnace (1513 m^3) disappeared and submerged in the strong smoke and fire suddenly. The supporting structures collapsed and the broken-up shell flew away (the farthest was more than 200 m). Even seriously, the disaster also spread to the hoisting machine house. Unfortunately, 19 workers died in an unprecedentedly serious accident.

12.1 Gas Exploding

The serious disaster was caused by gas exploding, which might be prone to happen inside BF, but the severity of this one must be unprecedentedly (Fig. 12.1).

12.1.1 The Prerequisites for Gas Exploding

(1) The concentration of gas ranges from 35 to 75%.
(2) There is air or oxygen mixed with the gas to form the explosive gas.
(3) To reach the ignition temperature of mixed gas or naked fire exists.

12.1.2 The Reasons for the Gas Exploding Inside no. 1 BF

(1) No.1 BF did not reline in time. The operation of the furnace had been disordered since May 1987, and a lot of cooling facilities was serious damaged. Up to the accident, 33 stave coolers were damaged (1 at tuyere belt, 32 at bosh) which was 66.7% of the total, and among the 590 cooling plates at shaft, 393 was full damaged, and 100 was half damaged. The total damage rate was 75.1%. In the interest of yielding, the high-pressure water was used to spray onto the shell for

Y. Liu, *The Operation of Contemporary Blast Furnaces*,
https://doi.org/10.1007/978-981-15-7074-2_12

Fig. 12.1 Cenotaph and inscription

cooling, but worsening the rupture of shell. Obviously, the BF is ought to be relined earlier.

(2) The furnace must have begun to crack and break of weld joint since June 1989 and got more and more serious. Up to the accident, 28.5 m of crack in total had been found and fixed, but it was not very essential to overcome the big temperature gradient. Partial stress concentrated and thermal fatigue emerged, so the shell and furnace body could not support the inside burden, which leads to brittle crack at many places when the explosion happened. Then some parts broke, supporting column pull down and finally the whole body fell down [1].

The furnace still worked online before the accident, although the condition was so bad. Because the pressure inside the furnace was positive, the air could not enter in, the explosion was impossible to happen without air or oxygen. But the inside pressure would change to negative after bridging and slipping, then the air could be sucked into form the explosive mixed gas and to explode.

(3) The burning out of tuyeres is very often. The total broken tuyeres reached 45 (from March 1st to 12th), which concentrated in two areas (No. 4-10, No. 14–17) to leak a lot of water into the furnace and more water leaked because of batch tuyeres renewing. As a result, too much water in the furnace built up some dull zones.

(4) The adverse furnace condition and thermal performance led to the temperature rise of hot metal. In the last tapping before the accident, the rate of [Si] rose from 0.62, 0.92% (last 2, 3 tapping) to 1.75% (last tapping). At the same time, the top temperature rose to 320 °C two times (4:20, 6:30), and two spray water at the top. The temperature of the shell (elevation of 17 m) rose from 37.5 °C to 56–70 °C after 7:20. But only two charges were added (36.4 t of sinter and 12.3 t of coke) which was not enough for combustion, thus to form an empty space (imaginary of 50 m^3), which might make possibility for slipping.

(5) According to the measuring record of stock rods on the top before the accident, the northern stock rod dropped from 2350 mm down to 2450 mm, then plummeted to 2860 mm and had no record anymore in final. The southern stock rod also dropped from 2250 mm down to 2400 mm, then plummeted to 3180 mm and lost at last. All of this showed that the slipping had happened inside the furnace [1].

To sum up, No. 1 BF was running before the accident, but there were 3 tuyeres (totally 20) leaked water into the furnace and another 5 tuyeres closed up. The tuyere area was not active, dull zones existed, and thermal and irregular performance happened. The rising of top temperature and spray water two times is not for cooling but to some extent pulverizing the burden to reduce the permeability and cause bridging and slipping. So many factors put together inside the furnace to make the water in furnace gasifying and expanding suddenly and finally got the bitter end—exploding.

12.1.3 The Explosion Occurs from the Explosive Mixed Gas

The Capital Steel has experienced the explosion after the formation of mixed explosive mixed gas [2]. Whether the gas mixes air or the air mixes gas at a certain concentration contact, as well as the open fire emerges, the explosion would happen immediately.

12.1.3.1 The Explosion at the Connection Between Dust Catcher and Gas Down-Take

An explosion occurred at the connection between dust catcher and gas down-take of No. 3 BF in November 1959, which breaks a hole with 10 m^2 on the pipe (Fig. 12.2). It was planned to maintain facilities for 24 h on November 24, 1959. Then, begin to lower the stock line at 6:50 and drop down to 3.0 m according to the regulation. Open the bleeder valve, close the gas isolating valve and then stop the gas supplying at 7:20. At the same time, keeping mild blast operation with 0.1–0.08 kg/cm^2 of blast pressure and charging 9 skips of granulated slag continuously. Next, close the steam valve at the top and gas down-take and ignite at the top at 8:12. Damping down (wind off) and opening the sampling hole at 8:22. But the top fire was found extinguished at 9:10, thus to ignite again by putting the oil-immersed cotton lighting into the sampling hole of furnace throat at 12:45 to lead the explosion right away to break a big hole (10 m^2) at the connect between dust catcher and gas down-take and lose more than 5000 t hot metal during the period of 101 h and 51 min of blast off.

The lessons of the accident are:

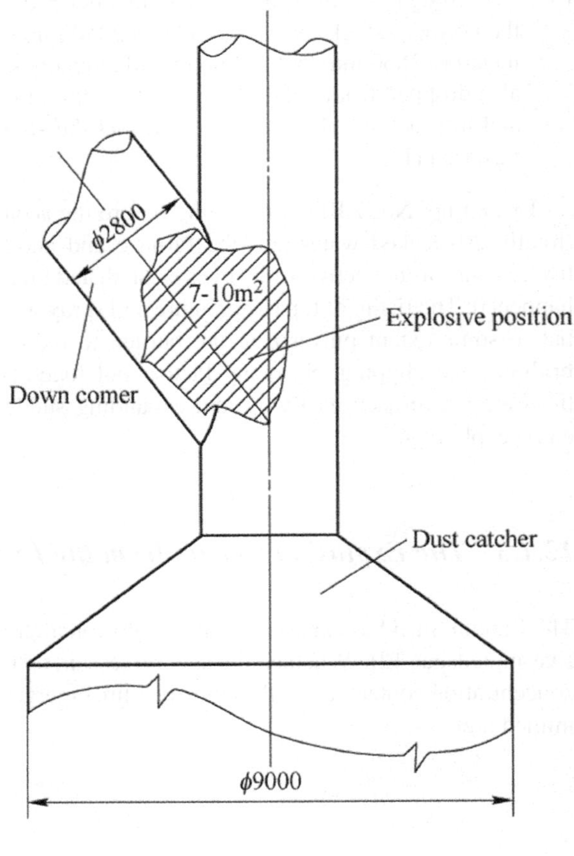

Fig. 12.2 Explosive position of No. 3 BF at the gas downcomer

(1) Nobody took care of the top fire after igniting and the combustion air was insufficient because only a small hole of the furnace throat sampling opened. It was wrong to open a manhole.
(2) After the flame went out, it was found that the second ignition was delayed for 3 h 35 min, which had accumulated a lot of mixed explosive gas.
(3) It was a wrong decision to close the manhole of gas uptake during the second ignition, reducing the discharge ability.
(4) The steam valve was broken up at dust catcher's isolating valve, whenever the existed explosive gas without steam would be bound to explode with igniting.

12.1.3.2 The Gas Explosion Between the Large and Small Bells

There was an explosion between the large and small bells of No. 1 BF of the Capital Steel on Oct. 22, 1964. The pressure over there was up to 1.06 kg/cm^2, which exceeded 0.45 kg/cm^2 (inside the furnace) that pressed the large bell open and the bar bent. The equalizer valve closed too late, and the negative pressure (267–533 Pa, i.e. 2–4 mmHg) appealed in the space between two bells during the time from small bell equalizer valve open to small bell open that the air was sucked up and mixed with gas entering when large bell was equalizing to make up explosive mixed gas. The temperature of the hot sinter in the large bell hopper was 600–700 °C, so the explosion occurs.

12.1.3.3 The Explosion at the Burner of Hot Blast Stove

In May 1968, when No. 3 BF of the Capital Steel was in the stove-changing operation, the hot blast blew into burner without shutting off the stove-burner shutoff valve to ignite the residuary gas, leading to the explosion and blowing up the burner 10 m away. Similar accidents also happened at No. 3 BF in June 1970 and No. 4 BF in Feb. 1979, which repeatedly blown up or deformed the fan rotor.

12.1.4 To Guard Against Gas Explosion

(1) To guard against gas explosion, the formation of mixed explosive gas should be controlled first, the limited concentration is shown in Table 12.1.
It is most important and effective to separate the gas and air without any mixture.
(2) It is not easy to avoid the contact between the explosive mixed gas and the fire resource (high temperature) in the BF system, but it must be done.
(3) To keep positive pressure in the gas pipe and container system to prevent air sucking into form explosive gas. If the pressure is less than 500 Pa, the steam must be introduced in.

Table 12.1 Gas firing point and explosion limitation

Type of gas	Ignition point, °C	Upper limit of mixing ratio, %	Lower limit of mixing ratio, %
BF gas	700	68	46
Coke gas	600–650	30.4	5.6
Gas fired regenerative furnace	650–700	65	44.5

(4) In the last phase of a campaign, all of the facilities should be well-maintained. The shell cracking (if some) should be welded in time. If the line and cooling system were seriously damaged, it is necessary to repair according to plan. If certainly needed, the reline should be in advance. Those fatal explosion accidents happened whenever the furnace did not reline on time.
(5) After the BF blowing out, the chemical reaction in the furnace still produced a small quantity of gas, which should be burned as getting out the stock surface. This is so-called igniting operation or top igniting—an important step in the long-term blowing out.
(6) It is necessary to fill the spaces with steam or other inert gases, such as in the large bell hopper (the space between small and large bells), to stop forming explosive gas.
(7) The facilities of gas system, including pipe, container valve and so on, must be strictly seamless and properly designed. To test in advance such as water test and compressed air test without leaking or sucking.

The requirements for design include three items: The first is no gas dead area, second proper locations for the steam spraying points, and the third is explosion-proof steps.

12.1.5 *Watering Knocking and Explosion*

There was an explosion at intermediate reline of No. 3 BF of the Capital Steel. A lot of water sprayed to larger bell and throat to control the top temperature to be less than 500 °C when the stock line was descending. But the top temperature still rose to more than 600 °C at near 8 m of stock level and to 780 °C at 12.7 m, then blowing out to 0.2 kg/cm^2 of blast pressure and the explosion occurred at the moment of the large bell opening. The fire injected violently through tuyeres and snort valve. At last, 8 of tuyeres damaged and snort valve deformed. Fortunately, no one hurt.

Huge amount of water sprayed into the furnace and contact with hot coke to decompose H_2 and O_2 to form explosive water gas for explosion instead of air and gas mixing.

When stock line descended to a very low level and snort valve opened, the high-speed gas stream passed through the snort valve to make earth-shaking noise.

Mr. Guo and Mr. Jin of Ben Steel have continuously taken a series of measurement for hydrogen content in gas during stock line descending. It was considered that the content of hydrogen in gas is one of the main reasons for knocking. Hydrogen in the gas is produced mainly by the chemical reaction that occurs when water is pumped from the top of the furnace to the red-hot material surface for cooling in the process of stopping the furnace, in addition to the water-gas reaction that is carried out by the decomposition of hydrogen-containing compounds in the coke and the water in the drum. Figure 12.3 shows that as the height of stock line decreases, the temperature of stock surface increases, which is conducive to water-gas reaction, and the $H_{2\,\%}$ also shows an upward trend. Especially when the stock line drops to the belly, most of the burden is coke. It made water-gas reaction to produce a lot of hydrogen, which would lead to knocking if over the critical point. At this moment, the quantity of topwater spraying should be controlled properly to avoid too much hydrogen yielding.

In the process of blowing off in 1984, because of various reasons, the gas on the top of the furnace contained too much hydrogen. If the content of hydrogen increases over 4–5%, the knocking would happen. But the situation had some changes in 1987, the new spray method, atomizing spray instead of water stream injecting, played a role to control the water and reduce the content of Hydrogen to a lower level, especially at the later period of blowing off, almost no knocking happened [3].

The normal smoothly water spraying could avoid knocking because the decomposed H_2 and O_2 could ascend along with the gas stream. But the intensive water injecting made too much water entered into the furnace to yield water gas and is possible to knock. Actually, knock has the same meaning with explosion, may be less energy and lighter destroy.

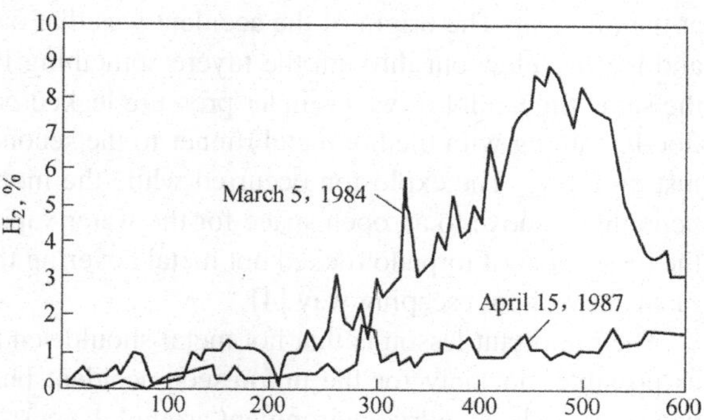

Fig. 12.3 Change of $H_{2\,\%}$ in gas along with the time of blowing off process

12.2 The Explosion from the Contact Between Hot Metal and Water

Another kind is the explosion that occurs when the hot metal contacts with water. If the water is put on the hot metal, an explosion occurs immediately because of the water evaporating, but if in contrast that the hot metal covers the water, the explosion must occur immediately. There are two explanations for the reason, one of which is the water decomposition under the hot metal and the energy recomposing; another one is supper high temperature made water convert into steam expending at a moment leading to explosion [4].

There was a sudden explosion after the western slag notch opening at No. 1 BF of the Capital Steel on 20:00 January 6, 1982. The casting house floor was vibrated and some slag notch cooler and holder damaged and renewed to stop the blast 9 h 55 min. Because of foundry pig iron yielding, graphite precipitated at hearth wall to shrink the volume of hearth, so hot metal ascends over the slag notch level before taping and pours out through the opening slag notch. Then a lot of water from the damaged slag coolers flood out to mix with hot metal leading to the explosion. Such an explosion happened once again at 11th 9:05 to stop blast for 6 h without tuyere cooler holder damaging.

There was a most miserable accident in the history at Victoria BF of British Steel that about 2 m^3 of water flooding into the torpedo with hot metal at 1400 °C, then the explosion happened when the torpedo car started to move. The result was that the flow nozzle of torpedo (1.25 t) blew up to 10 m above and flying 21 m away; the roof of casting house was broken (put through a big hole); 80 t hot metal and lining bricks flew and poured as dog and cat raining toward 23 workers, 11 of whom died at the moment. The origin of the accident was that one of the tuyere caps breaking and the fire blew out through the tuyere something like the volcanic explosion; at the same time, a lot of water under pressure leaked out from broken tuyere coolers flooding along with the hot metal runner to the second torpedo, in which hot metal just on filling. The explosion occurred while the mobile pulled the torpedo cars to be going to move to an open space for the water vaporizing. It was considered that the movement of torpedo rocked hot metal covering the water to yield a big amount of steam without escaping way [4].

The important lesson is that hot metal should send to steelmaking plant as soon as possible, not only for the unforeseen accident but also for heat retaining. It is necessary to keep a dry environment around the casting house to prevent hot metal from contacting water on the ground. All of these should be put in security regulation in production.

At the beginning of coal injection, the explosion used to happen in China. Now nitrogen and/or hot stove waste gas are used for coal transportation and injection to replace air. The explosion threatens had been eliminated by the improvement of temperature controlling and the pulverized coal accumulation avoiding. The lesson written in blood should not be ignored.

References

1. S. Jiu, Improve the Awareness of the Furnace, Ensure the Safety of the Blast Furnace—A Reflection on the 3.12 Accident of Jiu Steel No. 1 Blast Furnace, 20 Mar 2009 (internal information)
2. S. Shi, S. Wei, in *The Production of Blast Furnace*, ed. by C. An (Capital Steel, 1983), pp. 488–491
3. Y. Guo, Z. Jin, Changes law of gas composition during blast furnace shutdown. Iron-Making **5**, 45–48 (1995)
4. McMaster University, *Blast Furnace Iron-making Technology Seminar* (Metallurgical Industry Press, Beijing, 1980), pp. 260–261

The manufacturer's authorised representative in the EU is Springer Nature Customer Service Centre GmbH, Europaplatz 3, 69115 Heidelberg, Germany. If you have any concerns regarding our products, please contact ProductSafety@springernature.com

Printed and bound by CPI Group (UK) Ltd, Croydon, CR0 4YY
30/09/2026
02225558-0019